THEORETICAL AND APPLIED MATHEMATICS

A CLOSER LOOK AT BOUNDARY VALUE PROBLEMS

THEORETICAL AND APPLIED MATHEMATICS

Additional books and e-books in this series can be found on Nova's website under the Series tab.

THEORETICAL AND APPLIED MATHEMATICS

A CLOSER LOOK AT BOUNDARY VALUE PROBLEMS

MUSTAFA AVCI
EDITOR

nova
science publishers
New York

Library of Congress Cataloging-in-Publication Data

ISBN: 978-1-53617-857-9

Published by Nova Science Publishers, Inc. † New York

CONTENTS

PREFACE

Many problems arising in applied mathematics or mathematical physics can be formulated in differential equations with a set of additional constraints, called the boundary conditions.

Especially nonlinear partial differential equations (PDEs) are extremely attractive because they can be used to model many dynamical phenomena which arise from the studies in various disciplines.

Lack of general existence theorems for establishing solutions for nonlinear PDEs forces scientists to seek alternative pathways. In this regard, the present work presents different methods and approaches to obtain solutions to some class of differential equations given under different boundary conditions.

The aim of this book is to provide a platform where contemporary developments in the area of boundary value problems can be shared.

The present book is intended for the use of advanced undergraduates, graduate students, and researchers whose research interest aligns with the area of differential equations.

Mustafa Avci, PhD
Professor
Department of Science,
Grande Prairie Regional College,
Grande Prairie, AB, Canada
February, 2020

In: A Closer Look at Boundary Value Problems ISBN: 978-1-53617-857-9
Editor: Mustafa Avci

Chapter 1

BOUNDARY VALUE PROBLEM OF CO_2 PRODUCTION AND TRANSPORT IN FOREST SANDY SOIL

Yury V. Zaika* and Olga N. Bakhmet
Karelian Research Centre of the Russian Academy of Sciences,
Department of Multidisciplinary Sci Research, Petrozavodsk, Russia

Abstract

In the context of human-induced changes in global biospheric processes, research on carbon cycling at the 'atmosphere – plants – soil' scale and the corresponding mathematical models are of substantial interest. In addition to the global analysis of the balance of carbon fluxes, specialists in soil science argue that closer attention should be given to the mathematical description of the gaseous phase of soils at the 'local level', taking into account the diversity of soils and their 'life cycle' conditions. Thus, it is the local experimental data that are the basis for conclusions about an area being mainly a source or a sink of carbon, about the effects of soil reclamation, permafrost thawing, etc.

A mathematical model is suggested here for daily carbon dioxide production and transport in forest soils typical of Eastern Fennoscandia. The model performs a comparative quantitative analysis of the diffusion and convection components to estimate changes in soil respiration and transfer directions (to the atmosphere, deep horizons, groundwater). An attempt

*Corresponding Author's Email: zaika@krc.karelia.ru.

is made to minimize the number of boundary value problem parameters to be estimated using experimental data.

PACS: 51.20.+d, 89.60.-k

Keywords: boundary value problems in a porous medium, diffusion and convection, CO_2 production and transport in sandy soils

AMS Subject Classification: 35K57; 76R50; 35Q92

1. INTRODUCTION

Integrated planetary ‘atmosphere – plants – soil’ carbon cycling models, a thorough overview of the literature, tendencies in the development of mathematical modeling in this sphere are related in [1–4]. Closer scrutiny was applied to a physically substantiated mathematical description of the gaseous phase in soils in the books [4–7]. It is based on local experimental data on soil respiration that conclusions are drawn about whether an area is mainly a source or a sink of carbon, about the effect of soil reclamation, permafrost thawing, etc. This is a multidisciplinary issue, which needs to be approached from a variety of perspectives. Numerical modeling results are but one of the arguments in the discussion of debatable issues and problems arising, in particular, in relation to the analysis of carbon dioxide and methane flux balances in different soil and wetland ecosystems.

Understandably, there is no sense in attempting at a universal mathematical model of carbon dioxide transport down the soil profile. The model should ‘be fitted’ to a specific issue, describing only the effect of the limiting factors, considering the existing possibilities and the accuracy of measurements of certain fluxes, concentrations, characters (particle size distribution, physicochemical parameters, etc.). One can hardly expect to get a correct estimation of a dozen of *a priori* unknown parameters in detailed models from measurements of ‘external’ CO_2 fluxes from the soil, spatially discrete readings of temperature, moisture, etc. There is another consideration to be kept in mind. Coefficients (e. g. for diffusion) are not measured but calculated based on measurements within a given model (usually rather simplistic). Care should be taken when using these values in other (far more detailed) models, referring to the order of magnitude at best. One and the same (formally and nominally) coefficient will take different values when in different models (e. g. under different boundary conditions), even if the same 'experimental curve' is treated. Hence, when using data from the literature, not only the experimental conditions should be specified, but also the method by which one or another 'tabular' coefficient was calculated should be taken into account.

Proceeding from the general considerations above, we consider the following modeling problem. Imagine, for definiteness, an ordinary clear summer day in late July, in an ordinary pine forest in Eastern Fennoscandia. There is ordinary sandy soil with ordinary forest floor on top of it. Let us look into the quantitative comparison of the diffusion and convection components of CO_2 transport (it is traditionally postulated that molecular diffusion prevails). Scientific literature has offered analysis of other possible transfer mechanisms [6, Ch. V], even 'exotic' ones (Knudsen diffusion, solid-phase surface diffusion, thermal slip, etc.). These processes do contribute their share and this listing can be continued, but what matters is their commensurability (and how can the respective coefficients be obtained?). For instance, the input of thermal slip can hardly be noticeable unless we are dealing with super rarified gases [17, p. 357]. It can happen in micropores, but they are poorly interconnected and can scarcely sum up into a competitive transport channel. We will refrain from being too categorical however, since the model adopted below is based on integral, averaged characteristics, without detailing the versatile transfer mechanisms.

Proceeding from our task (which will be specified further as the model is explained) we shall generally keep to the following approach. Non-equilibrium thermodynamics (Onsager equations) stipulates a 'cross-estimation' of the gradients of concentration, temperature, pressure, moisture..., which in our case would unreasonably complicate the multiparameter model. We will model the dynamics of the gaseous phase (its characteristics are hard to measure at high precision). Temperature and moisture will be taken into account through their effect on 'effective gas parameters' of the transfer, relying on experimental data and their physically meaningful approximation (without writing partial differential equations for heat conductivity and moisture transfer).

The reasoning in general mathematical terms is the following. Consider the functional $J\big[t, z, c(t,\cdot)\big]$, whose value at the time moment t at the depth z is determined by the distribution of the concentration $c(t,\cdot)$ around the current point z. In the linear approximation, this distribution is locally characterized by the first two terms of Taylor's formula: the values of $c(t,z)$ and $\partial_z c(t,z)$. Classical molecular diffusion responds to the gradient, while the magnitude of convective movement of carbon dioxide (by gravity, in dissolved form in soil water, driven by water evaporation, etc.) depends on the medium velocity and concentration. The various processes of gas transport in soil are described in much detail in the book [6, Ch. V], while the handbook [10, p. 437] distinguishes ten possible major transfer mechanisms for isothermal conditions alone. The challenge of explicitly quantifying all the multiple processes being too boundless to handle, we shall classify these processes in the linear approximation by their contributions mainly to coefficients related to concentration or its gradient. The effective convection velocity and effective diffusion coefficient are 'cumulated'. To avoid cumbersome notation with the index $_{\text{eff}}$ we imply hereafter that the diffusion coefficient and convective transfer rate we speak of are integral, averaged, effective, cumulative in nature. The subject of the numerical analysis and comparison is 'types of transfer' (including their local directions inside the body of soil), not the numerous 'components', each with its own physical or biochemical meaning and mechanism of operation. In particular, we are interested in finding how much of the CO_2 produced in the soil (mainly through root respiration, chemical processes of organic matter decomposition, and activity of microorganisms) reaches the surface (where CO_2 emission can be recorded quite reliably), and how much is deposited by lower soil horizons (including removal by groundwater).

When the mathematical model is further represented in the form of a boundary value problem, the identified limiting processes will be analyzed one by one.

As regards references, the authors mainly referred to available Russian-language monographs, not attempting at an overview. All the necessary information is briefly provided in the text.

2. Mathematical Model

2.1. Temperature

Let us begin with the 'simplest' task — estimating the effect of daily temperature variations in the root layer. Inverted commas here signify that we apply the simplest approximation of the temperature $T(t, z)$ in time t and depth z, without the sophisticated mathematical toolkit of soil thermophysics [11]. The range of daily temperature fluctuations in East Fennoscandian pine woods in the summer season is quite narrow. Regarding the soil, the significant attenuating effect of the forest floor should be kept in mind. Compared to physicochemical processes of CO_2 production, which are usually exponentially dependent on temperature (according to the Arrhenius law), the coefficients of diffusion and convection are weakly (according to the power law, with sufficient precision) dependent on T. Yet, we will take this dependence into account too, for the sake of balance in the model and remembering that small oscillations are capable of initiating resonance phenomena. We are not aiming at ultimate simplifications which could potentially lead to analytical conclusions. Taking this refinement into account in numerical analysis is not a problem.

Let us build upon the qualitative characteristics of the soil temperature variation shown in Fig. 1 (see also [5, pp. 15,16]).

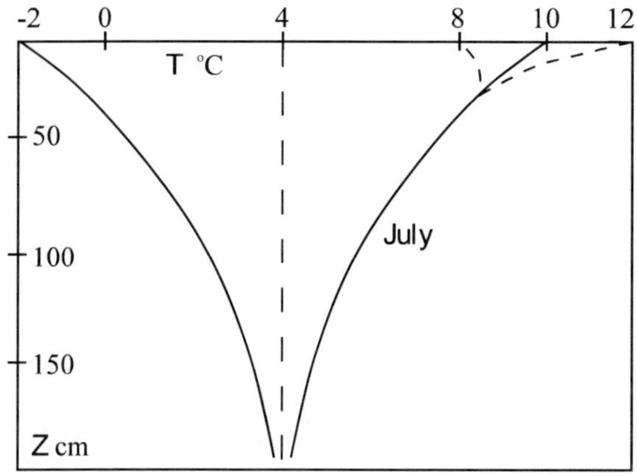

Figure 1. Seasonal temperature variations.

We use the solution of the classic 'earth's temperature problem' [16, §14], more specifically, the expressions for temperature waves [17, §55]:

$$T_y(t, z) = T_\infty + \Delta T_y \exp\{- \alpha_y z\} \cos\{\omega_y t - \alpha_y z\}, \quad \alpha_y \equiv \sqrt{\omega_y [2\chi_y]^{-1}}.$$

The function cos can be substituted with sin depending on the preferred time zero (to be specified later). We so far consider only mean annual fluctuations (index y denotes 'year'). The value of T_∞ corresponds to the temperature established at 3-4-meter depth (e. g., $T_\infty = 4$°C). The period of fluctuations $\tau_y = 2\pi/\omega_y$ is equal to one year, ω_y is the corresponding frequency. For numerical calculations based on the diffusion equation in a relatively small time interval it is convenient to take $[t] =$ s. As to the orders of magnitude [16, p. 101]: the period $\tau_y \approx 365 \cdot 24 \cdot 60 \cdot 60 \approx 3.15 \times 10^7$s, the thermal conductivity coefficient

$\chi_y \approx 2 \times 10^{-3}\text{cm}^2/\text{s}$. The temperature wave propagates with a phase delay, as shown in the formula. The uniform phase surface $\omega_y t - \alpha_y z = \text{const}$ moves inwards the soil at a rate of $v = dz/dt = \sqrt{2\omega_y \chi_y}$. At $z = 0$ we get the temperature fluctuation in the near-surface soil layer: $T_y^0(t) = T_\infty + \Delta T_y \cos\{\omega_y t\}$. If, for a demonstration, we fix the amplitude at $\Delta T_y = 6\,°\text{C}$, then $T_y^0(t) \in [-2, 10]$, and fluctuations almost die out at a depth of 3-4 meters. Formally, $\max T_y^0$ is attained when $t = 0$, i.e., the time zero here is conventionally set at the July-August interface. Being hereafter interested in daily temperature variations, we neglect the small $\omega_y t$ ($\omega_y \approx 2\times 10^{-7}$), and fix the 'basic distribution ($J = July$)'

$$T_J(z) = T_\infty + \Delta T_y \exp\{-\alpha_y z\} \cos\{-\alpha_y z\}.$$

Remark also that for the stated value of thermal conductivity χ_y (by the order of magnitude) $\alpha_y \approx 10^{-2}$, and the oscillatory behavior of the exponentially decaying function $T_J(z)$ will be inconspicuous.

The approximation in the daily range is taken similarly: $\alpha_d \equiv \sqrt{\omega_d [2\chi_d]^{-1}}$,

$$T_d(t, z) = T_J(z) - \Delta T_d \exp\{-\alpha_d z\} \cos\{\omega_d t - \alpha_d z\}. \tag{1}$$

Note that $\omega_d \gg \omega_y$ (approximately 365-fold) and $\alpha_d \gg \alpha_y$ (ca. 19-fold). The daily amplitude ΔT_d is quite small (e. g., $\Delta T_d = 2$) owing to the attenuating effect of the forest floor. For the given values we have $T_d^0(t) \in [8, 12]$°C. The minimum of near-surface ($z = 0$) temperature takes place at $t = 0$, so the time zero in the formula (1) is set at the onset of soil heating after nighttime cooling, i.e.,, conventionally, 6 a. m. For brevity, we speak of daylight onset (implying the light level at the 'cooling – heating' transition). Essentially, fluctuations affect only the roughly half-meter-thick near-surface root layer, where most of active soil respiration and CO_2 production through physical and chemical processes of organic matter decomposition and microorganisms activity take place.

Next, observe the following circumstances. Even on clear days, near-surface temperature fluctuations can be only very roughly described by the sine function. The mean annual amplitude ΔT_y is also rather conventional, whereas the measured ΔT_d would vary substantially even within several square meters under the forest floor. Thermal conductivities $\chi_{y,d}$ will have very different values, even for 'one and the same soil'. One can speak of some average warming or cooling of the soil over a season, but surface air temperature fluctuations over 24 hours can only alter the temperature of some soil mass 'around pore-space channels'. Not to mention that averaged (effective) values of $\chi_{y,d}$ do not portray the changes in the thermal conductivity of the soil as its structure and composition change with depth. Thus, we view the formula (1) as a physically meaningful variant of approximating (by the least squares method) experimental data in temperature by varying the parameters T_∞, $\Delta T_{y,d}$, $\chi_{y,d}$ within reasonable limits.

There is no need for high precision in the approximation, since the formula will only be used in power-law (low power) dependences of coefficients on temperature (in the context of an exponential outburst of CO_2 production in the root layer when temperature and moisture conditions are favorable).

On the other hand, analysis of the values of input variables allows a physically meaningful analysis of soils to be carried out even within a small area. A continuous function on an interval can be approximated, at any desired precision, by a polynomial in a uniform metric. It is however difficult to assign a conceptual meaning to coefficients in an applied problem.

2.2. Concentration of CO_2

We are interested in carbon dioxide concentrations in soil air and soil water (in dissolved state). Connectivity with atmospheric air is mainly via relatively large-diameter pore channels (micropores are largely isolated, and do not collectively constitute a competitive transport channel for the soil's gaseous phase). So, further reasoning assumes the equality of pressures of the atmospheric air and the gaseous phase of the soil in the layer with active CO_2 production. Minor fluctuations of atmospheric pressure over 24 h are disregarded.

The commonly accepted notations in the study of transfer processes in gases (such as diffusion processes in gas mixture) are the following [17, Ch. VII]: n is the total concentration (total number of gas molecules per unit gas volume, $[n] = 1/\mathrm{cm}^3$, or, alternatively, $\mathrm{mol}/\mathrm{cm}^3$, $\mathrm{mol}/\mathrm{m}^3$); if n_1 refers to a specific gas in a mixture then $c_1 = n_1/n$ is the relative concentration of the 'first gas' (in our case, CO_2). For distinction, denote mass concentration with the capital letter C, $[C] = \mathrm{mg}/\mathrm{cm}^3$, g/m^3. Note that in the studies of the soil gaseous phase [6] this implies the mass of the studied gas per unit volume of soil air. The gas dissolved in soil water and adsorbed on the solid phase free surface is estimated separately. Percent by volume composition of the atmosphere is 78 (nitrogen), 20.9 (oxygen), ..., 0.035 (carbon dioxide), ...(the percent by weight composition is 75.51, 23.14, ..., 0.05, ...). The situation in soil air is very different [5, p. 34]. At $10 - 20$ cm depth already, CO_2 concentration may increase in order of magnitude. The books [6, p. 29], [7, p. 298] indicate the 'limit' range of 0.035–20%.

Let us move on to the solubility of carbon dioxide in the liquid phase. Soil water has some specific features [12, p. 25]. Soil water can take the following forms (A. F. Lebedev's classification): 1) water vapor, 2) hygroscopic water, 3) pellicular water, 4) gravitational water, 5) solid-state water, 6) crystalline water, and 7) chemically bound water. The locally dissolved CO_2 does not have to 'diffuse down the water column'. The 'working range' of the gas concentration is within several percents. The dissolution process is fast on the scale of the characteristic time of the limiting molecular diffusion in the gaseous phase. Hence, in the linear approximation of the model we adopt a generally proportional relationship $C_w = \gamma(T)C_g$, where C_w is the mass concentration of CO_2 dissolved in a unit volume of liquid, $C_g \equiv C$. Here, $\gamma(T)$ should also be viewed as an effective coefficient, and the dissolution capacity 'from the table' can be taken for reference only as rough averages (for order of magnitude).

The monograph [6] poses the problem of estimating CO_2 adsorbed on the solid phase free surface. A possible transfer mechanism is surface diffusion. Quantitatively, the surface concentration (especially if the specific surface area is large) is substantial and registered in experiments with degassing of soil samples. Our interest however is transfer dynamics, not the total concentration. The amount of adsorbed gas is regarded as a low-mobility background (which must certainly not be neglected in the estimation of the total CO_2 content in the soil). By analogy, this is in agreement with a well-known opinion [20, p. 512] (although regarding a different gas): ...part of hydrogen is captured tightly in grid defects and must therefore be excluded from the estimation of the concentration whose gradient is included in Fick's equation, since part of hydrogen is a semi-inert impurity, Fick's law is only applicable to the remaining part.

Let us now introduce two requisite parameters: $\varepsilon(t, z)$ is the aeration porosity (proportion of gas in the bulk volume of soil), $\theta(t, z)$ is the proportion of moisture in the bulk volume. Then $1 - (\varepsilon + \theta) \equiv \zeta(t, z)$ is the proportion of the solid phase in the bulk volume of soil. Theoretically, $\varepsilon \in (0, 1)$, but near-limit values disagree with the concept of the soil, and we shall consider a realistic, narrower range. Starting with a relatively small depth of within a meter, ζ (and hence the pore space share $\varepsilon + \theta$) steadies down. So, the focus thereafter is on the major, mobile part of CO_2 with a total concentration of

$$C_m(t, z) = C_g(t, z)\varepsilon(t, z) + C_w(t, z)\theta(t, z) = C(t, z)\big[\varepsilon + \gamma(T)\theta\big] \quad (C \equiv C_g).$$

This mass amount of CO_2 in the gaseous phase and solution (the pore space in total) is expressed relative to the unit volume of soil.

2.3. Diffusion and Convection Transfer

Suppose atmospheric pressure is steady throughout a full day. Soil air pressure is in quasi-equilibrium with atmospheric pressure at the surface. Only the main channel of transfer via the system of interconnected macropores is taken into account. Micropores and microcapillaries, where thermal slip effects or Knudsen diffusion may take place, are regarded as a minor channel. Since N_2, O_2 are commensurate in mass and dimensions, while a CO_2 molecule is heavier (not by an order of magnitude) and carbon dioxide concentration is within several percents, one can make use of the theory for diffusion in binary gas mixtures.

First, for simplicity, picture the 'soil' as a set of vertical hollow cylinders (channels) whose diameters are much greater than the length of a free path of gas molecules. The aeration porosity will be included in the estimations after the preliminary consideration.

Denote by n_1 the concentration (density) of 'first gas' (CO_2) molecules, which is small relative to the 'second gas' ($N_2 + O_2$). Hence, n_2 is the number of $N_2 + O_2$ molecules per unit volume (for definiteness, cm^3), $n = n_1 + n_2$ is the total concentration (density) of soil air particles. Minor gaseous components can be neglected. Note that the soil under ordinary conditions will have the near-equality (vol.%) [7, p. 298]: $O_2 + CO_2 \approx 21\%$.

In the given approximation, the density of the particle diffusion flux will be written in the form [17, p. 344], [19, p. 117]:

$$J_1 = -Dn\frac{\partial c_1}{\partial z}, \quad c_1 = \frac{n_1}{n},$$

where c_1 is the relative concentration (CO_2). The value of n is not necessarily constant, it may vary spatially in response to some factors (e. g., field of force or temperature gradient, so thermodiffusion is indirectly taken into account). We still hold the ideal gas equation to be true: $p = nkT$. Here and below $[T] = K$. Transform the derivative of the fraction:

$$J_1 = -D\frac{\partial n_1}{\partial z} + Dn_1 n^{-1}\frac{\partial n}{\partial z} = -D\frac{\partial n_1}{\partial z} + \mathrm{v}_1 n_1, \quad \mathrm{v}_1 \equiv D[\ln n]'. \qquad (2)$$

The total concentration is steady under ordinary conditions, so the change of the function $\ln n$ can be neglected ($\mathrm{v}_1 \approx 0$) and the expression $J_1 \approx -D\partial_z n_1$ be used. Meanwhile [14, p. 357] (see also [19, p. 119]), the diffusion coefficient D increases along with the increase in temperature proportionally to $T^{3/2}$ (and is inversely proportional to pressure p). In the accepted units of measurement $[J_1] = 1/(\text{cm}^2\text{s})$, $[D] = \text{cm}^2/\text{s}$, $[\text{v}_1] = \text{cm}/\text{s}$, i.e., the term $\text{v}_1 n_1$ has the formal meaning of the convective flux with the velocity of the medium v_1. If the gas flow rate u is non-zero, then the expression for the flux is supplemented with a term (molar transfer) representing gas flow as a whole [19, p. 117]:

$$J_2 = -Dn\frac{\partial c_1}{\partial z} + nc_1 u = -D\frac{\partial n_1}{\partial z} + \text{v}_2 n_1, \quad \text{v}_2 \equiv \text{v}_1 + u. \tag{3}$$

Being cognisant of the difficulties involved in a detailed description of multiple transfer mechanisms, we will classify them formally, based on contribution to one or the other of the terms in the expression generalizing (3). Locally, the situation is defined not by the concentration in a point, but by the distribution in its surroundings. In the linear approximation, $n_1(t, \cdot)$ is defined by two numbers $n_1(t, z)$ and $\partial_z n_1(t, z)$. Equation (3) has a separate standard-form term for gradient-dependent molecular diffusion and a temperature-dependent convective term. The movement of water vapor, gravitational flow of the heavier gas, augmented transfer channel along roots, wind conditions at the surface, ... — all these will result in the accumulation of a generalized 'velocity of the medium' coefficient instead of v_2. Hence, we are from now on interested in the commensurability of the contributions of two transfer mechanisms, without detailing their component parts.

Multiplying the flux expression by CO_2 molecular weight, owing to linearity on n_1, we get the mass flux (the dimensionality for reference is $\text{mg}/(\text{cm}^2\text{s})$). Having isolated the 'basic' temperature dependence and the normal conditions $p_0 = 760\,\text{Torr}$, $T_0 = 273\,\text{K}$, we write down the following expression (so far excluding aeration porosity):

$$J^\circ = -D_0\frac{p_0}{\bar{p}} \cdot \left[\frac{T}{T_0}\right]^{3/2}\frac{\partial C}{\partial z} + \text{v}_0\frac{p_0}{\bar{p}} \cdot \left[\frac{T}{T_0}\right]^{1/2} C, \quad \bar{p} = p_{\text{air}},\ C = C_g,$$

$[C] = \text{mg}/\text{cm}^3$, $[T] = K$. Atmospheric pressure during the period of calculations is assumed to be constant ($p_{\text{air}} = \text{const}$). This pressure is deemed to be uniform for the gas mixture in the soil column in question, speaking of the major CO_2 transfer macrochannels.

The square-root dependence on temperature was identified with regard to the average rate of thermal motion of molecules. In the framework of the kinetic theory of ideal gases, the degree can be uprated ($1.5 \to 1.75$). The recommendation for real gases [9, p. 261] is to round off, so we accept

$$J^\circ = -D(\bar{p},T)\,\frac{\partial C}{\partial z} + \mathrm{v}(\bar{p},T)\,C, \tag{4}$$

$$D(\bar{p},T) \equiv D_0\,\frac{p_0}{\bar{p}}\cdot\left[\frac{T}{T_0}\right]^2, \quad \mathrm{v}(\bar{p},T) \equiv \mathrm{v}_0\,\frac{p_0}{\bar{p}}\cdot\frac{T}{T_0}.$$

The fluctuations of atmospheric pressure and soil temperature (dampened by the forest floor) are relatively small, and low-power relationships are usually neglected as incommensurate with exponential ones, characteristic of chemical reactions. Such a detailing does however made sense for maintaining the integrity of the 'concentration – temperature – moisture' model. D_0 and v_0 will now be deemed practically independent of $\bar{p}$, T.

As postulated in [9, p. 267], "in most cases, apart from hydrogen containing mixtures, thermal diffusion can be neglected". So, where the response of the flux to the temperature gradient needs to be portrayed, (4) should be supplemented with the following term:

$$J^\circ = -D(\bar{p},T)\,\frac{\partial C}{\partial z} - D\,\frac{C}{T}\cdot\frac{\partial T}{\partial z} + \mathrm{v}(\bar{p},T)\,C = \tag{5}$$

$$= -D\,\partial_z C + VC, \quad V(\bar{p},T,\partial_z T) \equiv \mathrm{v} - DT^{-1}\partial_z T.$$

Note that while neglecting the thermal diffusion mechanism per se in the additional term we 'retain' the temperature gradient. We concur to the following well-grounded opinion [9, p. 266]. "We believe that for non-isothermal diffusion... the division of the molecular flux of matter into concentration diffusion flux and thermal diffusion flux should be avoided, and the ordinary expression for diffusion should be used instead, except that the driving force of diffusion would be partial pressure rather than concentration. In this case, the thermal diffusion phenomenon is construed as relative molecular transfer in inhomogeneous gases (or gas mixtures) induced by a temperature gradient".

The second term emerging in (5) is defined by the ideal gas equation:

$$D_g\partial_z p_1 = D_g\partial_z(n_1 kT) = D_g kT\partial_z n_1 + D_g k n_1 \partial_z T \Rightarrow D = D_g kT.$$

In the model, the temperature $T = T_d(t,z)$ is represented by a composition of elementary functions (1), so that the gradient $\partial_z T(t,z)$ is defined by an explicit formula. We will look into each term in the formula (5) as a characteristic of an individual transfer mechanism: molecular (diffusional) and convective (in a generalized sense). Formally, the value V can be seen as the 'generalized rate' of movement (molar transfer). The heat transfer direction is followed also by water vapor diffusion, which can 'entrain' substantial amounts of CO_2 along. Hence, the change of sign of the gradient $\partial_z T(t, z)$ influences the overall transfer process significantly. This is particularly important in the analysis of daily temperature and moisture fluctuations. Furthermore, the heavier CO_2 molecules (among the gases in question) are diffused along the direction of heat flux. Thus, estimation of the dependence of V on the temperature gradient $\partial_z T$ appears to be a requisite element for the model.

2.4. Porosity of Aeration

In [5, p. 42] describes two options for choosing the diffusion coefficient: 'effective' D_B (for the total concentration gradient $C_m = C_g\varepsilon + C_w\theta$) and 'apparent' D_s (for the gradient of CO_2 concentration $C = C_g$ in the gaseous phase). We view the CO_2 adsorbed by the solid phase of the soil as the 'inert background' and disregard it as a transfer channel. Coming in contact with water it migrates to the solution and is quantified indirectly via the concentration C_w. Otherwise, a more sophisticated variant should be considered: $C_m = C_g\varepsilon + C_w\theta + C_s\zeta$.

The relationship $D_s(\varepsilon)$ has some distinctive features [5, p. 39]: slow growth (plateau) at the beginning is followed by an upsurge. As mentioned in [6,7], the power-law dependence is used as a rule. We will try to avoid formal approximations by polynomials and will use a physicochemical analogy.

On average, a plane section of unit area perpendicular to the vertical axis z has the same share of the 'gas gap' ε. If not, a division into thin-volume layers (and integral sum formation) would yield a volume percent different from ε. Therefore, no distinction will be made in the mathematical model we are constructing between porosity at the surface and in the bulk, and the same notation will be used for both: ε, θ.

In reality, it is not only porosity per se that matters, but also the so-called sinuosity of the pore space [5, p. 52]. Naturally, an increase in sinuosity (denoted here by δ) entails a reduction in gas flux. We therefore write the model flux density in the form $J = J^\circ\varepsilon/\delta$. The parameters ε and δ are not independent. As the average size of pores increases and a certain stage is reached they start merging in an avalanche manner, and sinuosity decreases quite rapidly. By analogy with the activational nature of chemical reactions, we consider an exponential (Arrhenius-law) relationship instead of a power-law relationship: $\exp\{-a/\varepsilon\}$, $a > 0$. Add the normalizing constant $\exp\{a\}$ to make the factor accounting for aeration porosity (including, in the generalized sense, the sinuosity of pore channels) $\varepsilon/\delta(\varepsilon) \equiv f_\varepsilon = \exp\{a\}\exp\{-a/\varepsilon\} = \exp\{-a/\sigma\}$ take the value of 1 in the formal substitution $\varepsilon = 1$. The parameter $\sigma \equiv \varepsilon/(1-\varepsilon) = \varepsilon/(\theta+\zeta)$ is the analog of temperature in the activation of chemical reactions. Needless to say, the asymptotics $\varepsilon \to 1$ corresponds to a degeneration of the concept of soil, so the approximation should be limited to the realistic range ε.

The parameter a acquires an 'activational' meaning (and can be used for comparing soils). The initial 'built-up period' and the upsurge with a physically-conditioned deceleration of the monotone S-shaped saturation curve (there is a point of inflection at $a < 1$) are secured. The characteristic form of the factor f_ε in relation to the parameter $a > 0$ is shown in Fig. 2.

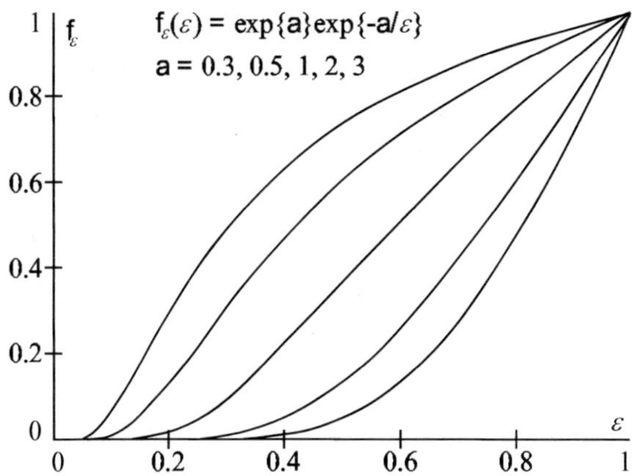

Figure 2. Dependence of the aeration porosity factor on the parameter.

If such a multiplier is used for adjusting the 'ideal' flux (5), then the parameters D_0 and v_0 acquire the meaning of boundary parameters — in the surface air space under normal conditions $(\{p_0, T_0\})$. We thus get the opportunity to take the order of magnitude of the coefficient of CO_2 diffusion in the atmosphere for reference. Finally, we adopt the following expression for the flux density:

$$J(t,z) = f_\varepsilon(\varepsilon)\big[-D(\bar{p},T)\,\partial_z C + V(\bar{p},T,\partial_z T)\,C(t,z)\big], \tag{6}$$

$$f_\varepsilon(\varepsilon) = \exp\Big\{-\frac{a}{\sigma}\Big\}, \quad \sigma(\varepsilon) \equiv \frac{\varepsilon}{1-\varepsilon} = \frac{\varepsilon}{\theta+\zeta}.$$

In this expression D_0 and the 'averaged convective velocity of the medium' v_0 are believed to be reasonably variable parameters. We assume at first approximation that $\mathrm{v}_0 = \bar{\mathrm{v}} = \mathrm{const}$ for daylight hours and night, while the sign of $\bar{\mathrm{v}}$ is not specified and is selected separately ($\bar{\mathrm{v}} = \bar{\mathrm{v}}_{d,n}$).

Remark. *Our assumption here is that there is no rain or abrupt weather changes during the considered summer days, and moisture distribution down the soil column is stable. Only a relatively slow change of the $\theta(t,z)$ profile is happening (specified below). The soil upper layers get somewhat drier during daylight hours, with plant roots contributing to this process. At night, the moisture profile partially recovers owing to an upward flux and condensation of water vapor. The minor CO_2 transfer by water vapor is formally taken into account by the integral convective term in* (6) *with regard to the transformation $C_w = \gamma(T)C_g$. There are no noticeable 'hydrodynamical' fluxes of soil moisture further down the column, so we can neglect this mechanism of dissolved CO_2 transfer in* (6) *in the explicit form.*

2.5. Production of CO_2 and Material Balance Equation

Let us specifically point out two components in the material balance equation — the transport term and the productivity of the upper root layer of soil:

$$\frac{\partial C_m}{\partial t} = \mathcal{F}_{\rm tr} + \mathcal{F}_{\rm pr}, \quad \mathcal{F}_{\rm tr} = -\frac{\partial J}{\partial z}, \quad C_m = C_g\varepsilon + C_w\theta, \ C \equiv C_g.$$

The characteristic form of the second term (productivity) $\mathrm{Pr} = \mathcal{F}_{\rm pr}$ is shown in Fig. 3 (where $[\mathrm{Pr}] = \mathrm{mg/m^2{\cdot}cm{\cdot}h}$ as in Fig. 3.9 [5]).

Modeling of organic matter decomposition, root system respiration, activity of microorganisms is extremely challenging (see e. g. the description of the ROMUL multiparameter model [3, 8]). On the other hand, the characteristic form of CO_2 productivity shown in Fig. 3 is transparent, and can be postulated across the soil profile. It is clear at what depth the maximum occurs, and the optimal temperature and moisture can be determined in the laboratory for any

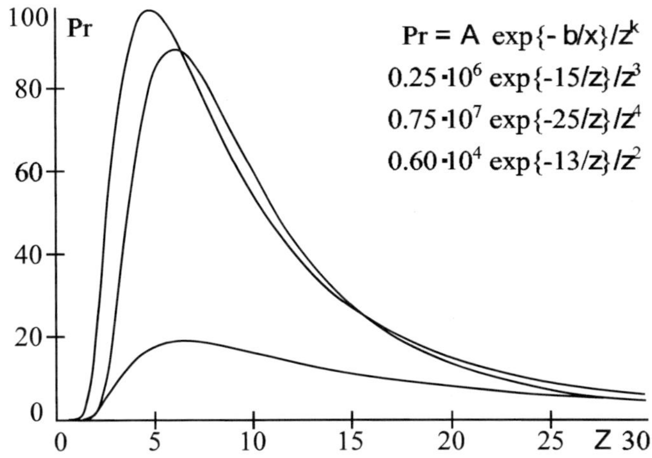

Figure 3. Functional approximation of CO_2 productivity.

specific type of soil. Take the relationship $\mathcal{F}_{\text{pr}} = A\exp\{-b/z\}/z^k$, $k > 0$, as the baseline. An Arrhenius-like upsurge in the subsurface layer (temperature and moisture are deemed to be as favorable as possible) is succeeded by a relatively slow decline at a power-law rate (the density of roots, bacteria, organic matter declines rapidly). The model parameters A, b, k need to be determined more accurately in the laboratory.

It remains to estimate the effect of temperature and moisture deviations from the optimum. We employ penalty functions (or factors) to this end. CO_2 production almost ceases outside of a certain interval of values of the volume percentage of moisture, and it therefore remains to consider $\theta \in [\theta^0_{\text{min}}, \theta^0_{\text{max}}]$. Choose, pursuant to [6], the moisture factor in the form

$$f_\theta(\theta) = \left[\frac{\theta - \theta^0_{\text{min}}}{\hat{\theta} - \theta^0_{\text{min}}}\right]^\alpha \cdot \left[\frac{\theta^0_{\text{max}} - \theta}{\theta^0_{\text{max}} - \hat{\theta}}\right]^\beta, \quad \alpha > 1, \ \beta > 0. \tag{7}$$

The optimal moisture value $\hat{\theta} = \left[\alpha\theta^0_{\text{max}} + \beta\theta^0_{\text{min}}\right]/[\alpha + \beta]$ (where temperature is favorable) for which $f_{\text{max}} = f_\theta(\hat{\theta}) = 1$, is determined from experimental data. The input variable is $\alpha > 1$, and the value $\beta > 0$ is calculated from the given $\hat{\theta}$, α. The condition $\alpha > 1$ provides for a 'smoother' rising branch, since $f_\theta(\theta^0_{\text{min}}) = f'_\theta(\theta^0_{\text{min}}) = 0$. The characteristic form of the function f_θ is shown in Fig. 4. For instance, $\theta^0_{\text{min}} = 0.1$, $\theta^0_{\text{max}} = 0.8$, $\hat{\theta} = 0.5, 0.6$.

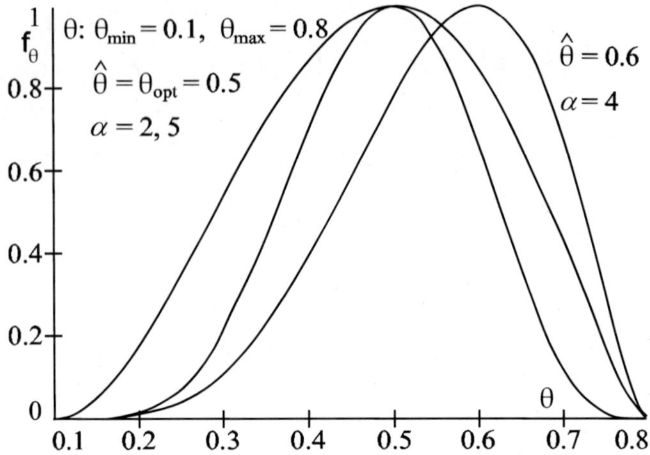

Figure 4. Dependence of the factor f_θ on the parameter.

The temperature factor $f_T(T)$ is determined similarly, and the productivity then takes the form of the relationship

$$\mathcal{F}_{\mathrm{pr}} = f_\theta(\theta) f_T(T)\, A \exp\{-bz^{-1}\}\, z^{-k}, \quad \theta = \theta(t,z),\ T = T(t,z).$$

Here we do not formally use the aeration porosity factor f_ε since it is no more than one of the many additional characteristics influencing the productivity of the 'inhabited' layer (the layer's 'identity' will be defined by A, b, k).

If the forest site has an open canopy and herbaceous vegetation is quite abundant, the daylight duration factor should additionally be taken into account:

$$f_L(t) = L_0 + (1 - L_0)\left[\frac{t}{t_L}\right]^p \cdot \left[\frac{t_* - t}{t_* - t_L}\right]^q, \quad p > 1,\ q > 0, \tag{8}$$

where $_L = {}_{Light}$; $t_L :\ f_{\max} = f_L(t_L) = 1$; t_* is the daily time period (until the onset of the next daylight period). During the white nights time, in July, we can assume $t_L = t_*/2$ (18 o'clock). At the same time, $q = q(p, t_L)$, similarly to (7). A positive value of the 'minimal level' $L_0 > 0$ corresponds to a noticeable CO_2 production around the clock, even at night (otherwise $L_0 = 0$).

2.6. Temporal Pattern of $\varepsilon(t, z)$ and $\theta(t, z)$

A characteristic variant of the depth dependence of aeration porosity and volumetric moisture content is shown in [5, p. 40]: porosity declines in a monotone manner, while moisture increases (reaching asymptotes). The values of ε, θ are related as $\varepsilon + \theta + \zeta = 1$, where ζ is the vol.% of dry matter of the soil. It is reasonable to first experimentally determine θ and ζ, and then calculate ε. If moisture is viewed as a nutrition resource, one can by analogy use the Monod (Michaelis–Menten) functions, frequently used in microbiology and biochemistry [13, p. 18], in the form $\theta = \theta_{\max} z/[z + h]$. It is a saturation curve, with the parameter h in the meaning of depth, where the function reaches the 'asymptotic' half-maximum. In fact, the same function is used to describe the temperature dependence of the average length of free path in gas dynamics: $\lambda = \lambda_\infty T/[T + \text{const}]$ (Sutherland's law [18, p. 377]). The denominator can be supplemented with an 'inhibiting' term in the form rz^2 [13, p. 30], permitting to model a maximum at a certain depth (caused by yesterday's rainfall), or use a 'smoother' quadratic relationship (see Fig. 5, 6).

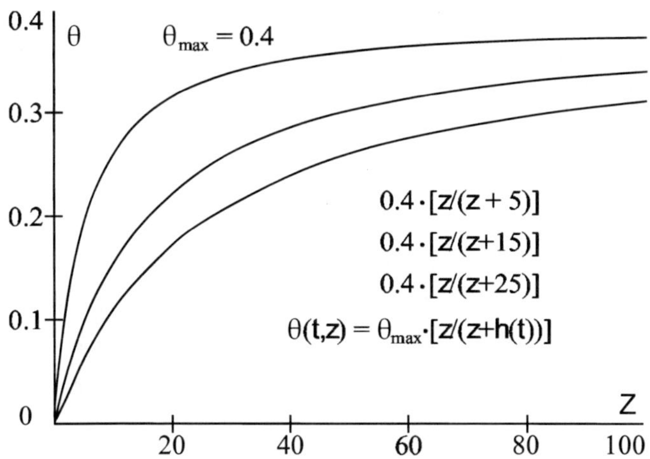

Figure 5. The temporal pattern of θ dependence on the parameter h.

The time dependence of volumetric moisture content is best portrayed by $h = h(t)$, since the processes of evaporation, microorganism- and root activity are quite intensive (while the gravitational flow of moisture down the profile has almost ceased). To wit, during daytime $\theta(t, z) = \theta_{\max} z/[z + h(t)]$ with a

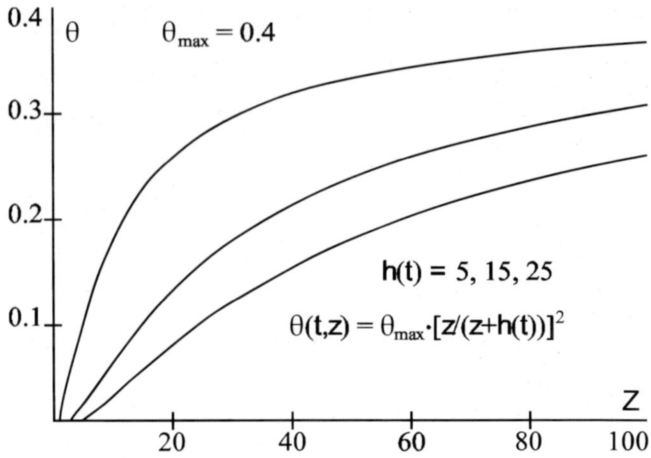

Figure 6. The temporal pattern of θ quadratic dependence.

monotonously growing function $h(t)$, e.g., $h(t) = h_0 + v_h t,\ v_h = v_{h,d} > 0$. Attainment of the half-maximum will then be delayed in time deeper into the soil, which corresponds to the process of moisture level 'subsidence'. During nighttime $v_h = v_{h,n} < 0$ and the volumetric moisture content is partially recovered.

This is the simplest variant of modeling the temporal pattern of $\theta(t, z)$. Application of an advanced theory of soil moisture movement [12] would make the model too complicated. The parameter representing the rate of moisture level subsidence v_h depends on the intensity of evaporation and absorption by the root system. Detailing of these processes is a challenging task per se. Moisture and temperature measurements (depthwise and timewise) for the considered class of sandy soils are far more doable than for the 'evasive' gaseous phase. This helps avoid excessive model complexity.

2.7. Initial and Boundary Conditions

Time zero corresponds to the onset of daylight (beginning of soil warming, roughly 6 a. m.). The initial conditions are generated using the same saturation functions (similarly to the relationships in Fig. 5, 6) from experimental data. The concentration profile is 'unloaded' overnight, and the concentration diagram becomes more monotone depthwise.

First-type boundary conditions are set for the boundary value problem: CO_2 concentrations at 1-2 meter depth (a stable value, which can be determined experimentally) and at the surface (atmospheric carbon dioxide level) are prescribed. Temperature, volumetric moisture content and CO_2 flux from soil surface to the atmosphere are measured. The model parameters are to be tuned (identified) based on these data. The task for the modeling is to compare the diffusive and the convective components of the flux and estimate which part of CO_2 production is released into the atmosphere, and how much of it reaches the groundwater and can be deposited for a long term.

Aeration porosity at a 'reference depth' (plane horizontal section is partitioned according to the volumetric indices of ε, θ and ζ) can be used to estimate the gaseous component of the CO_2 flux downward the soil, and lysimeter readings will provide data on the flux of CO_2 dissolved in soil water after heavy rainfall. This information should be taken into account when solving the inverse problem of model parameter identification, and when calculating concentrations and mass balance of fluxes.

3. Concise Representation of the Model

3.1. Main Assumptions

We are considering a 24 h time interval in the summer season, at around late July. The weather conditions are steady, cloudiness is minor. There was no rainfall in the preceding several days, and the 'hydrodynamical' downward flow of soil water has become negligible. The soil has a relatively shallow (ca. 5 cm) forest floor layer on the top, followed by the root layer (roughly 20 cm) changing downwards into a nearly homogeneous sandy 'monolith' until the studied depth (1 meter). The time zero $t_0 = 0$ is the onset of daylight (6 a. m.).

3.2. Temperature

The forest floor has a high 'damping' capacity, so that temperature fluctuations recorded in the root layer are within several degrees. The complex heat transfer model appears unnecessary. Yet, the temperature dependence needs to be identified explicitly for the sake of the model's integrity and because minor oscillations may lead to noticeable (although not decisive) resonance phenomena. Besides, the model aims to analyze the 'evasive' gaseous phase, and temperature dependences can be used after direct measurements.

For minor oscillations and small depths it suffices to use a model of thermal waves with an exponentially decaying amplitude and a phase shift (see Fig. 1):

$$T_d(t,z) = T_J(z) - \Delta T_d \exp\{-\alpha_d z\}\cos\{\omega_d t - \alpha_d z\},$$

$$T_J(z) = T_\infty + \Delta T_y \exp\{-\alpha_y z\}\cos\{-\alpha_y z\}, \quad \alpha_y \equiv \sqrt{\omega_{y,d}[2\chi_{y,d}]^{-1}}.$$

Here, ¡¡$d = day,\ J = July,\ y = year$¿¿, T_∞ is constant temperature at 1-2-meter depth, $\Delta T_{d,y}$ are the amplitudes of daily and annual fluctuations. As a matter of fact, the thermophysical parameters $\alpha_{d,y}$ differ (the daily one largely has to do with the space around pores at small depths, whereas the annual one — with the mineral matrix of the soil). We view them as variables manipulated for approximation in the treatment of measurements. This approximation is, however, not abstract, but has a clear physical meaning.

3.3. Aeration Porosity and Volumetric Moisture

Denote by $\varepsilon(t,z)$ the aeration porosity (proportion of gas in the bulk volume of soil), and by $\theta(t,z)$ the proportion of moisture in the bulk volume of soil. Then $1-(\varepsilon+\theta) \equiv \zeta(t,z)$ is the proportion of solid material in the bulk volume of soil. Starting with a relatively small depth of within one meter the values of ζ (and hence the proportion of the pore space $\varepsilon+\theta$) steady down. We are interested in the main labile part of carbon dioxide with the total concentration of

$$C_m(t,z) = C_g(t,z)\varepsilon(t,z) + C_w(t,z)\theta(t,z) = C(t,z)\big[\varepsilon + \gamma(T)\theta\big] \ \big(C \equiv C_g\big).$$

This mass amount of CO_2 in the gaseous phase and dissolved in soil water (in the pore space in total) is normalized to unit volume of soil. Here, C_g is the mass concentration of CO_2 in the gaseous phase, $C_w = \gamma(T)C_g$ — in the solution, $T \equiv T_d(t,z)$. The carbon dioxide adsorbed by the solid phase is regarded as an almost immobile 'background'. Coming in contact with moisture it is dissolved and indirectly estimated within the concentration C_w.

If moisture is generally perceived as a nutritional resource for microorganisms and roots, the slow evaporation-condensation process and lack of a tangible gravitionational flow are taken into account, then the simplest relationship $\theta = \theta_{\max} z/[z + h]$ will suffice. This is a saturation curve, with the parameter h meaning depth, on which the function reaches the 'asymptotic' half-maximum. During daytime $\theta(t, z) = \theta_{\max} z/[z + h(t)]$ with a monotonously growing function $h(t)$, e.g., $h(t) = h_0 + v_h t$, $v_h = v_{h,d} > 0$. Attainment of the half-maximum will then be delayed in time deeper into the soil, which corresponds to the process of moisture level 'subsidence'. During nighttime $v_h = v_{h,n} < 0$ and the volumetric moisture concent is partially recovered.

3.4. CO_2 Flux Density

Let us adopt the following generalized expression for the flux, taking the numerous processes into account integrally, based on their characteristics as transfer mechanisms ('response' to concentration and temperature gradients, convection):

$$J(t, z) = f_\varepsilon(\varepsilon)\big[-D(\bar{p}, T)\,\partial_z C + V(\bar{p}, T, \partial_z T)\,C(t, z)\big],$$

$$f_\varepsilon(\varepsilon) \equiv \exp\Big\{-\frac{a}{\sigma}\Big\}, \quad \sigma(\varepsilon) \equiv \frac{\varepsilon}{1-\varepsilon} = \frac{\varepsilon}{\theta + \zeta},$$

$$D(\bar{p}, T) \equiv D_0\,\frac{p_0}{\bar{p}} \cdot \left[\frac{T}{T_0}\right]^2, \quad \mathrm{v}(\bar{p}, T) \equiv \mathrm{v}_0\,\frac{p_0}{\bar{p}} \cdot \frac{T}{T_0},$$

$$V(\bar{p}, T, \partial_z T) \equiv \mathrm{v}(\bar{p}, T) - DT^{-1}\partial_z T, \quad T \equiv T_d(t, z).$$

The aeration porosity factor f_ε (including the effect of the sinuosity of pore channels) has an Arrhenius form with the 'activation energy' a and the temperature analog $\sigma(\varepsilon)$. The index $_0$ corresponds to normal conditions (pressure-temperature), $[T] = K$, $\bar{p} = p_{\mathrm{air}}$. D_0 and the 'averaged convective velocity of the medium' v_0 are regarded as parameters varying within physically reasonable limits. We assume at first approximation that $\mathrm{v}_0 = \bar{\mathrm{v}} = \mathrm{const}$ for daylight hours and night, and the sign of $\bar{\mathrm{v}}$ is not specified ($\bar{\mathrm{v}} = \bar{\mathrm{v}}_{d,n}$).

3.5. CO_2 Production and Material Balance Equation

Let us single out two components in the material balance equation — the transport term and the root layer productivity:

$$\frac{\partial C_m}{\partial t} = \mathcal{F}_{\mathrm{tr}} + \mathcal{F}_{\mathrm{pr}}, \quad \mathcal{F}_{\mathrm{tr}} = -\frac{\partial J}{\partial z}, \quad C_m \equiv C_g\varepsilon + C_w\theta \quad (C_g \equiv C),$$

$$f_\theta(\theta) = \left[\frac{\theta - \theta^0_{\min}}{\hat{\theta} - \theta^0_{\min}}\right]^\alpha \cdot \left[\frac{\theta^0_{\max} - \theta}{\theta^0_{\max} - \hat{\theta}}\right]^\beta \quad (\alpha > 1,\ \beta > 0),\ f_{\max} = f_\theta(\hat{\theta}) = 1,$$

$$f_T(T) = \ldots, \quad \big(f_L(t) = \ldots\big), \quad \mathcal{F}_{\mathrm{pr}} = f_\theta(\theta) f_T(T)\, \frac{A}{z^k} \exp\left\{-\frac{b}{z}\right\}.$$

The factors f_θ, f_T are penalty functions for deviation from the optimal conditions of CO_2 production: $\theta = \hat{\theta}$, $T = \hat{T}$. The indexes of power are not independent: namely, β is defined by setting α and $\hat{\theta}$. Where necessary, one should additionally take into account the daylight factor $f_L(t)$, which is determined by the formula (8). The characteristics of the root layer are determined by the values of the parameters A, b, k from experimental data.

3.6. Initial and Boundary Conditions

Time zero corresponds to the onset of daylight. The initial conditions $C_0(z) = C(0, z)$ are generated using the same saturation functions from experimental data. The concentration profile is 'unloaded' during the night. First-type boundary conditions are adopted for the boundary value problem: CO_2 concentrations at 1 meter depth (a stable value) and at the surface (atmospheric carbon dioxide level) are prescribed.

Temperature, volumetric moisture content and CO_2 flux from soil surface to the atmosphere are measured. The model parameters are tuned based on these data. The task for the modeling is to compare the diffusive and the convective components of the flux and estimate which part of CO_2 production is released into the atmosphere, and how much of it reaches the groundwater and can be deposited. Aeration porosity at a 'reference depth' (plane horizontal section is partitioned according to the volumetric indices of ε, θ and ζ) can be used to estimate the gaseous component of the CO_2 flux downward the soil, and lysimeter readings will provide data on the flux of CO_2 dissolved in soil water after heavy rainfall.

ACKNOWLEDGMENTS

The study was carried out under state order to the Karelian Research Centre of the Russian Academy of Sciences.

REFERENCES

[1] Tarko, A. 2005. *Anthropogenic Changes in Global Biospheric Processes. Mathematical Modeling*. Moscow: Fizmatlit.

[2] Poluektov, R., et al. 2006. *Models of the Crop Production Process*. St. Petersburg: St. Petersburg State University Press.

[3] Chertov, O., et al. 2007. *Dynamical Modeling of Soil Organic Matter Transformation Processes. ROMUL Model*. St. Petersburg: St. Petersburg State University Press.

[4] Shein, E., and Ryzhova, I. 2016. *Mathematical Modeling in Soil Science*. Moscow: Markushev.

[5] Richter, J. 1987. *The Soil as a Reactor: Modelling Processes in the Soil*. Hannover: Catena Verlag.

[6] Smagin, A. 2005. *The Gaseous Phase of Soils*. Moscow: Moscow State University Press.

[7] Shein, E. 2005. *A Course in Soil Physics*. Moscow: Moscow State University Press.

[8] Komarov, A., et al. 2017. "Romul Hum Model of Soil Organic Matter Formation Coupled with Soil Biota Activity". *Ecological Modelling* 345: 113–124.

[9] Lykov, A. 1954. *Transfer Phenomena in Capillary-Porous Bodies*. Moscow: GITTL.

[10] Lykov, A. 1978. *Heat and Mass Exchange*. Moscow: Energiia.

[11] Chudnovsky, A. 1976. *Thermophysics of Soils*. Moscow: Nauka.

[12] Polubarinova-Kochina, P. 1977. *Theory of Groundwater Movement*. Moscow: Nauka.

[13] Riznichenko, G. 2003. *Mathematical Models in Biophysics and Ecology*. Moscow-Izhevsk: ICR.

[14] Landau, L., Achiezer, A., and Lifschitz, E. 1969. *A Course in General Physics. Mechanics and Molecular Physics*. Moscow: Nauka.

[15] Lifschitz, E., and Pitaevskii, L. 1979. *Physical Kinetics // Landau, L., and Lifschitz, E. Theoretical Physics. Vol. X*. Moscow: Nauka.

[16] Sommerfeld, A. 1950. *Partial Differential Equations in Physics*. Moscow: IL.

[17] Sivukhin, D. 2005. *A General Physics Course. Vol. II. Thermodynamics and Molecular Physics*. Moscow: Fizmatlit.

[18] Savelyev, I. 1970. *A Course in General Physics. Vol. I.* Moscow: Nauka.

[19] Kirichenko, P. 2005. *Thermodynamics, Statistical and Molecular Physics*. Moscow: Fizmatkniga.

[20] Darken, L., and Gurry, R. 1960. *Physical Chemistry of Metals*. Moscow: NTI.

In: A Closer Look at Boundary Value Problems ISBN: 978-1-53617-857-9
Editor: Mustafa Avci

Chapter 2

BOUNDARY VALUE PROBLEM OF HYDROGEN THERMAL DESORPTION: REDUCTION TO FRACTIONAL DIFFERENTIAL EQUATION

Yury V. Zaika* and Olga V. Fomkina
Karelian Research Centre of the Russian Academy of Science,
Institute of Applied Math Research, Petrozavodsk, Russia
Petrozavodsk State University, Petrozavodsk, Russia

Abstract

One of the technological challenges for hydrogen materials science (including the ITER project) is the currently active search for structural materials with various potential applications that will have predetermined limits of hydrogen permeability. One of the experimental methods is thermal desorption spectrometry (TDS). A hydrogen-saturated sample is degassed under vacuum and monotone heating. The desorption flux is measured by mass spectrometer to determine the character of interactions of hydrogen isotopes with the solid. We are interested in such transfer parameters as the coefficients of diffusion, dissolution, desorption. The paper presents a distributed boundary-value problem of thermal desorption with nonlinear dynamical boundary conditions and corresponding fractional differential equation (or Riccati functional differential equation of neutral type with integrable weak singularity). A new numerical method for TDS spectrum computer simulation is proposed.

*Corresponding Author's Email: zaika@krc.karelia.ru.

Keywords: hydrogen thermal desorption, dynamical nonlinear boundary value problems, fractional differential equations

PACS: 67.63.-r, 66.30.-h, 68.43.Nr

AMS Subject Classification: 34A08, 35K60, 35Q92

1. INTRODUCTION

Interaction of hydrogen isotopes with solids has many applications [1–6]. It is sufficient to mention power engineering, protection of metals from hydrogen corrosion, design of chemical reactors, rocket engineering. Hydrides help to retain substantial amounts of hydrogen: environmental-friendly energy source. Some special topics are considered in [7–9]. Transfer parameters depend on the process characteristics of producing the material batch, and one needs effective algorithms for processing measured curve instead of focusing on "tabular data". We focus on the thermal desorption method and take into account only the limiting factors and the information capacities of the TDS experiment.

2. MATHEMATICAL MODEL OF TDS EXPERIMENT

We consider hydrogen transfer through a test metal sample (plate of thickness ℓ). For brevity we speak of a metal plate, although it may be a multialloy or an intermetallic compound. Suppose heating is relatively slow and uniform so that the diffusion flow can be assumed to be proportional to the concentration gradient. Concentration of diffusible hydrogen is relatively low. The material is sufficiently homogeneous to neglect interaction of H with traps (in particular with micro-defects of the crystal structure). Let us assume a standard model for diffusion:

$$c_t(t,x) = D(T)c_{xx}(t,x), \quad (t,x) \in Q_{t_*}, \tag{1}$$

where t is time, $Q_{t_*} = (0,t_*) \times (0,\ell)$; $c(t,x)$ is concentration of dissolved atomic hydrogen. Diffusion coefficient D is an Arrhenius function of temperature $T(t)$: $D = D_0 \exp\{-E_D/[RT(t)]\}$, D_0 and E_D are the preexponential factor and activation energy (R is the gas constant).

More detailed transfer models are known. For example, different diffusion channels can be considered: transcrystalline, the grain boundary diffusion, along the defects, with interchange between them. However, a significant increase of number of unknown a priori coefficients leads to the complex problem of parameter identification. TDS-experiment provides limited amount of information; therefore we use integral parameters (effective coefficients). The main difficulty is due to dynamic boundary conditions. Suppose the plate contacts with gas H_2 and surface is potential barrier (see [2, pp. 177–206]). Taking into account (de)sorption processes, we obtain the following boundary conditions:

$$c(0,x) = \bar{c}(x), \quad x \in [0,\ell],\ t \in [0,t_*], \tag{2}$$

$$c_0(t) = g(T)q_0(t), \quad c_\ell(t) = g(T)q_\ell(t), \tag{3}$$

$$\dot{q}_0(t) = 2\mu s(T)p_0(t) - b(T)q_0^2(t) + D(T)c_x(t,0), \tag{4}$$

$$\dot{q}_\ell(t) = 2\mu s(T)p_\ell(t) - b(T)q_\ell^2(t) - D(T)c_x(t,\ell), \tag{5}$$

$b(T) = b_0 \exp\{-E_b/[RT]\}$, $s(T) = s_0 \exp\{-E_s/[RT]\}$. Here $c_0(t) \equiv c(t,0)$, $c_\ell(t) \equiv c(t,\ell)$ are boundary bulk concentrations of dissolved diffusing atomic hydrogen; $q_0(t), q_\ell(t)$ are concentrations of H on the surfaces ($x = 0, \ell$); $g(T)$ is the coefficient of local equilibrium between concentrations on the surface and near-surface bulk; μ is the gas kinetic coefficient; $p_0(t), p_\ell(t)$ are pressures of molecular (H_2) gas; $b(T)$ is the desorption coefficient; $s(T)$ is the parameter indicating that only a small part of incident hydrogen find its way on the surface (we keep in mind that this coefficient describes the result of concurrent molecular (H_2) physisorption, dissociation, and chemisorption of H on the surface). One can write s (as a parameter of the model) instead of $2s$, but it is more convenient to interpret the dimensionless probability factor s as the fraction of adsorbed H atoms within the $2s$ notation.

Although hydrogen can be both in atomic and molecular forms, we use atoms as the unit of amount: $[c] = 1/\mathrm{cm}^3$, $[q] = 1/\mathrm{cm}^2$, $[Dc_x] = [J] = 1/\mathrm{cm}^2\mathrm{s}$ $\left(J = bq^2\right)$. According to the kinetic gas theory, μp is the number of H_2 particles colliding with unit area per unit time; then $2\mu sp$ is the flux density of atoms precipitating on the surface. This quantity is a general parameter, applicable without distinguishing the processes. The quick dissolution model (3) is obtained from the more universal formulas

$$k^-(T)q_0(t) - k^+(T)c_0(t) = -Dc_x(t,0), \quad k^-(T)q_\ell(t) - k^+(T)c_\ell(t) = Dc_x(t,\ell).$$

Coefficients k^-, k^+ describe the intensity of the dissolution in the bulk and transfer to surface. If these processes are significantly faster than diffusion (for considered temperature range) then it is assumed that $Dc_x \approx 0$. Hence we obtain (3), where $g = k^-/k^+$. If the surface is isotropic (in terms of $E_{k^-} \approx E_{k^+}$) then the coefficient $g(T)$ has weak temperature dependence. In addition, we can take into account degree of the surface covering and near-surface bulk saturation:

$$k^-(T)\big[1 - c_{0,\ell}(t)c_{\max}^{-1}\big]q_{0,\ell}(t) - k^+(T)\big[1 - q_{0,\ell}(t)q_{\max}^{-1}\big]c_{0,\ell}(t) = \mp D(T)c_x\big|_{0,\ell}.$$

The value $\theta(t) = q(t)/q_{\max}$ means the degree of the surface covering. Flux density of atom H adsorption (dissociative hydrogen chemisorption on the surface) can be simulated by formula $2\mu s(T)p(t)[1-\theta(t)]^2$ in the flux balance equations (4), (5). But we have $\theta \ll 1$ for low level of hydrogen. It's in good agreement with quadratic desorption, linear diffusion equation, $D \neq D(c)$. In the context of TDS experiment it is convenient to take $[p] = \mathrm{Torr}$ and so $\mu(T) = (2\pi mkT)^{-1/2} \approx 2.484 \cdot 10^{22}/\sqrt{T}$. Here we have $[\mu] = 1_{H_2}/(\mathrm{Torr\,cm^2 s})$, $[T] = \mathrm{K}$, k is the Boltzmann constant, m is mass of a hydrogen molecule. Variation of the kinetic constant with temperature ($\mu \propto 1/\sqrt{T}$) is usually neglected against exponential relation for $s(T)$. Saturation means that pressure $\bar{p}$ of H_2 on both sides of the plate is constant at $T = \mathrm{const}$. Then (after a time) saturation concentration ($\bar{c}$) of dissolved atomic diffusivity hydrogen is established. Assuming zero derivatives in (3)–(5), we obtain $\bar{c} = \Gamma\sqrt{\bar{p}}$, $\Gamma \equiv g\sqrt{2\mu s/b}$ ($\bar{c} = g\bar{q}$). The model is according to the Sieverts law adequacy range ($\bar{c} \propto \sqrt{\bar{p}}$).

Let us clarify the TDS-experiment. Plate of the material is placed into a chamber. Hydrogen gas is pumped into the chamber under relatively high pressure. The sample is heated to hasten sorption until the sample absorbs enough hydrogen (equilibrium gas saturation). Then heating current is shut off and the material is cooled. In such a case, the physico-chemical processes slow down sharply and significant amount of hydrogen remains inside the sample. The sample is monotonically and rather slowly reheated in vacuum. The heating rate is low and constant. Molecular hydrogen pressure in the chamber is measured by a mass-spectrometer. The desorption flux from the surface is defined by pressure. The flux density is denoted by $J(t) = b(t)q^2(t)$ (usually $J(t) \propto p(t)$). From now on the contracted notation will be used for simplification:

$$b(t) \equiv b(T(t)), \quad D(t) \equiv D(T(t)), \quad s(t) \equiv s(T(t)).$$

Parameter g of rapid dissolution (local surface-bulk equilibrium) is taken to be constant ($E_{k^-} = E_{k^+}$) near the TDS-peak. Symmetry conditions are fulfilled for the TDS method: $\bar{c}(x) = \bar{c} = \mathrm{const}$,

$$p(t) = p_{0,\ell}(t),\ q(t) = q_{0,\ell}(t),\ c_0(t) = c_\ell(t),\ D(t)c_x(t,0) = -D(t)c_x(t,\ell).$$

The experiment end time t_* is defined by the degassing: $p \approx 0,\ c(t_*, x) \approx 0$.

3. Functional Differential TDS Equation

Suppose the TDS-degassing model: $r(t) \equiv 2\mu s(t)p(t),\ J(t) \equiv b(t)q^2(t)$,

$$c_t = D(t)c_{xx}\,,\ c(0,x) = \bar{c},\ c_{0,\ell}(t) = gq(t),\ \dot{q}(t) = r(t) - b(t)q^2(t) + D(t)c_x(t,0),$$

$g = \mathrm{const}$. Vacuum system is assumed to be powerful enough. Hence resorption is neglected ($r(t) = 0$) for degassing. Values of $J(t)$ are input data for the inverse problem of parametric identification. In the paper we only present numerical simulation of TDS spectrum $J = J(T)$. Heating is usually linear: $T(t) = T_0 + vt$. Heating rate v is less than 2–3 K/s. Heating is stopped (although degassing is continued) if upper temperature limit $T(t) = T_{\max}$ is reached.

Let us formulate the numerical modeling problem. Difference schemes for the boundary-value problem (with different types of defects) solution was presented in [8]. But a curve of TDS spectrum is determined by surface concentration ($J = bq^2$). Hence, we try to avoid the iterative solution of the boundary-value problem for the current approximations of the model parameters D_0, E_D, b_0, E_b, s_0, E_s, g. For this aim we perform the problem transformation. Finally, the modeling requires only integration of a nonlinear ODE.

We replace time by $t' = \int_0^t D ds$ (keep previous notation t):

$$c_t(t,x) = c_{xx}(t,x), \quad c(0,x) = \bar{c}, \quad c_{0,\ell} = gq(t), \tag{6}$$

$$c_x|_0 = -c_x|_\ell = \dot{q}(t) + [J(t) - r(t)]D^{-1}(t). \tag{7}$$

Let $q(t)$ be functional parameter, formula (7) is additional relation to the linear problem (6). Let us change variables so that boundary conditions in (6) become homogeneous: $\hat{c} = c(t,x) - gq(t)$,

$$\hat{c}_t(t,x) = \hat{c}_{xx}(t,x) + f(t), \quad f(t) = -g\dot{q}(t), \quad \hat{c}(0,x) = \hat{\varphi}(x) = 0,\ \hat{c}|_{0,\ell} = 0.$$

Write the solution using function of instantaneous source (Green's function):

$$\hat{c}(t,x) = \int_0^t \int_0^\ell G_1(x,\xi,t-\tau) f(\tau)\, d\xi\, d\tau,$$

$$G_1(x,\xi,t) = \frac{2}{\ell} \sum_{n=1}^{\infty} \exp\Big\{ -\frac{n^2\pi^2}{\ell^2} t \Big\} \sin\frac{n\pi x}{\ell} \sin\frac{n\pi\xi}{\ell}.$$

Dynamical boundary conditions contain $\hat{c}_x(t,0)$:

$$\hat{c}_x|_0 = -\frac{4g}{\ell} \int_0^t \dot{q}(\tau) {\sum}' \exp\Big\{ \frac{n^2\pi^2}{\ell^2}(\tau - t) \Big\}\, d\tau, \quad {\sum}' \equiv \sum_{n=1,3,5...}. \tag{8}$$

The series is divergent for $\tau = t$ so that term-by-term integration is meant. For original time t we obtain $c_x(t,0) = \hat{c}_x(t,0) = c_x(t,\ell) = \hat{c}_x(t,\ell)$,

$$c_x(t,0) = -\frac{4g}{\ell} {\sum}' \int_0^t \dot{q}(\tau) \exp\Big\{ -\frac{n^2\pi^2}{\ell^2} \int_\tau^t D(s)\, ds \Big\}\, d\tau \ \Rightarrow$$

$$\frac{dq}{dt} = r(t) - b(t) q^2(t) - \frac{4gD}{\ell} {\sum}' \int_0^t \dot{q}(\tau) \exp\Big\{ -\frac{n^2\pi^2}{\ell^2} \int_\tau^t D(s)\, ds \Big\}\, d\tau. \tag{9}$$

The resultant equation (9) with quadratic nonlinearity will be classified as a functional differential Riccati equation of the neutral type. The equation is equivalent to the original boundary value problem in that the solution $q(t)$ uniquely determines the solution $c(t,x)$. The analogy with the functional differential equation of the form $\dot{\mathrm{x}}(t) = \mathcal{F}\big[t, \mathrm{x}(t), \dot{\mathrm{x}}(t), \mathrm{x}_{[0,t]}, \dot{\mathrm{x}}_{[0,t]}\big]$ of the neutral type [10] is that it is impossible to eliminate the derivative $\dot{q}$ on the right side of the equation through integration by parts lest a divergent series arises. We are concerned with the time interval $[t_1, t_2] \subset (0, t_*)$, which corresponds to the TDS peak. Measurement for $t \approx 0, t_*$ yields little information. There is a voluminous body of literature on Riccati-type equations (including matrix equations for the optimal control theory).

Dimensionless parameters are more convenient for numerical modeling. We use replaces: $t' = \int_0^t D(s)\, ds/\ell^2$, $x' = x/\ell$, $v = q/\bar{q}$ ($\bar{c} = g\bar{q}$). But notation of t is not changed. We obtain $\big(r(t) = 0\big)$:

$$\dot{v}(t) = -\tilde{b}(t) v^2(t) - 4g\ell {\sum}' \int_0^t \dot{v}(\tau) \exp\big\{ -n^2\pi^2[t-\tau] \big\} d\tau. \tag{10}$$

Here $v(0) = 1$ is the initial condition, $\dot{v} \equiv dv/dt$, $\tilde{b}(t) \equiv b(t)\ell^2\bar{q}/D(t)$. We only use first k terms of the series. Denote

$$z_i(t) = \int_0^t \dot{v}(\tau) \exp\{-(2i-1)^2\pi^2[t-\tau]\}\, d\tau.$$

Differentiate $z_i(t)$ and use $\dot{v}(t)$ from formula (10) ($\max n = 2k-1$):

$$\dot{z}_i = -\tilde{b}v^2 - \big[(2i-1)^2\pi^2 + 4g\ell\big]z_i - 4g\ell(z_1 + ... + z_{i-1} + z_{i+1} + ... + z_k).$$

The result is the following system of ordinary differential equations:

$$\frac{d}{dt}\begin{pmatrix} v \\ z_1 \\ \cdots \\ z_k \end{pmatrix} = -\tilde{b}(t)v^2(t)\begin{pmatrix} 1 \\ 1 \\ \cdots \\ 1 \end{pmatrix} - 4g\ell\begin{pmatrix} 1 & 1 & \ldots & 1 \\ \alpha_1 & 1 & \ldots & 1 \\ \cdots & \cdots & \cdots & \cdots \\ 1 & 1 & \ldots & \alpha_k \end{pmatrix}\begin{pmatrix} z_1 \\ z_2 \\ \cdots \\ z_k \end{pmatrix},$$

$\alpha_i \equiv 1 + (2i-1)^2\pi^2/(4g\ell)$, $v(0) = 1$, $z_i(0) = 0$ (matrix size is $(k+1) \times k$). The system is integrated using classic Runge-Kutta fourth-order method. We make calculations using Scilab 6. But for the required accuracy of calculations, 15–20 equations are necessary because of the slow convergence of the series.

Initial saturation is conducted at high temperature $T = \overline{T} = \text{const}$ and pressure $\bar{p} = \text{const}$ (to stimulate sorption). After establishing of equilibrium concentration we obtain: $2\mu(\overline{T})s(\overline{T})\bar{p} = b(\overline{T})\bar{q}^2 \Rightarrow \bar{q} \Rightarrow \bar{c} = g\bar{q}$ ($g = \text{const}$).

In the experiment, the initial data are estimated by the results of sample degassing. If during initial saturation the surface concentration for the given model parameters is $\bar{q} > 10^{15}$, the degree of surface coverage must be taken into account in numerical modeling. Then, to calculate the initial value of $\bar{q}$ we take the ratio $2\mu s\bar{p}\,[1 - \bar{q}q_{\max}^{-1}]^2 = b\bar{q}^2$ instead of $2\mu s\bar{p} = b\bar{q}^2$.

4. Fractional Differential TDS Equation

For extracting singularity (the series in (8) is divergent for $\tau = t$) let us clarify why the quite accurate approximation of the thermal desorption functional differential equation (10) still requires a large number of terms in $\sum'$. The function

$$\Theta(s) = 4\sum{}' \exp\{-n^2\pi^2 s\}, \quad \sum{}' \equiv \sum_{n=1,3,5\ldots},$$

has finite values for $s > 0$. The series is rapidly converging for large s. Formally, substituting $s = 0$ (the integration variable reaches the upper limit of t) we obtain a divergent series. This can be "fixed" by term-by-term integration. We have $\max |z_i(t)| = O(n^{-2})$ ($n = 2i - 1$, $i \geq 1$), wherefore convergence is slow. Let us run some transformations using the theory of Jacobi theta functions. More specifically, we will focus on

$$\theta_3(t, x) = 1 + 2\sum_{n=1}^{\infty} \exp\{-n^2\pi^2 t\} \cos(2n\pi x), \; t > 0.$$

We have an alternative presentation [11] $\theta_3(t, 0) = \sum_{-\infty}^{+\infty} \exp\{-n^2/t\}/\sqrt{\pi t}$ for $x = 0$. The series on the left is rapidly converging for large t. But the series on the right is rapidly converging for small t. Here we are only concerned with small t. Defining $\theta(t) = \sum \exp\{-\pi n^2 t\}$ ($n \in \mathbb{Z}$) we obtain $\theta(1/t) = \sqrt{t}\theta(t)$ or

$$\sqrt{t} \sum \exp\{-\pi n^2 t\} = \sum \exp\{-\pi n^2 t^{-1}\}, \quad (n \in \mathbb{Z}, \, t > 0).$$

This relation is known as functional equation for theta function [11].

Let us run transformations. The series is split into odd and even terms:

$$[\pi t]^{-1/2} \sum_{-\infty}^{+\infty} \exp\{-n^2 t^{-1}\} = 1 + 2\sum_{n=1}^{\infty} \exp\{-n^2\pi^2 t\}$$

$$= 1 + 2\sum{}' + 2\sum_{k=1}^{\infty} \exp\{-k^2\pi^2 4t\} = 2\sum{}' + [4\pi t]^{-1/2} \sum_{-\infty}^{+\infty} \exp\{-n^2[4t]^{-1}\}.$$

With this in mind, the last series is subtracted from the first series and the result is doubled. The following expansion is obtained for $s > 0$:

$$\Theta(s) \equiv 4\sum{}' \exp\{-n^2\pi^2 s\} = \frac{1 - Q}{\sqrt{\pi s}}, \quad Q(s) \equiv -2\sum_{n=1}^{\infty} q^{n^2}, \, q \equiv -\exp\{-(4s)^{-1}\}.$$

The series Q is rapidly converging for small s. As $s \to +0$ we have $Q \to 0$ and the integrable singularity $\Theta \approx 1/\sqrt{\pi s}$. Since the series is alternating, we have the inequalities $Q > 0$ (more exactly $Q \in (0, 1)$) and $Q \leq 2|q|$. The graph for function $Q(s)$ is an S-shaped saturation curve and $Q(1) \approx 0.999$. Using the above-described presentation of $\Theta(s)$ we obtain the TDS functional differential equation of neytral type (terminology from [10]):

$$\frac{dv}{dt} = -\tilde{b}(t)v^2(t) - \varkappa \int_0^t \frac{1 - Q(t-\tau)}{\sqrt{\pi[t-\tau]}}\,\dot{v}(\tau)\,d\tau \quad \big(t \geq 0,\ v(0) = 1,\ \varkappa \equiv g\ell\big). \quad (11)$$

The fraction (weak singularity under the integral) rapidly decreases from infinity ($\tau = t$) to near zero ($\tau = t-1$). For $t > 1$, the lower limit of the integral can be replaced with $t-1$. Thus, the neglectable background can be easily identified in the original physical time using the relation $1 = t_2' - t_1' = \int_{t_1}^{t_2} D(T(\tau))d\tau/\ell^2$. The compact functional differential TDS equation (11) with initial data $v(0) = 1$ replaces the original nonlinear boundary value problem with dynamic boundary conditions in the sense that formally only the dynamics of the surface concentration (desorption) is required for TDS spectrum construction. Note, that equation (11) contains a Caputo fractional derivative:

$$\frac{d^{1/2}v(t)}{dt^{1/2}} = \int_0^t \dot{v}(\tau)[\pi(t-\tau)]^{-1/2}\,d\tau \ \Big[t \equiv t'\big(D(\cdot), T(\cdot), \ell, t\big) = \ell^{-2}\int_0^t D\big[T(s)\big]ds\Big],$$

$$\frac{dv}{dt} + \varkappa\frac{d^{1/2}v}{dt^{1/2}} = -\tilde{b}(t)v^2(t) + \varkappa\int_0^t \frac{Q(t-\tau)}{\sqrt{\pi[t-\tau]}}\,\dot{v}(\tau)\,d\tau, \quad t \geq 0,\ v(0) = 1. \quad (12)$$

To simplify notation, we use dimensionless time $t \equiv t'$. The relationship to the physical time for conversions to the desorption flux density $J(t) = b(T(t))q^2(t)$ and TDS spectrum $J(T)$ $\big(t \leftrightarrow T = T_0 + vt\big)$ that we are interested in is given in brackets $[\ldots]$ (implying dimensionless time on the left, and stating physical time in the integral).

The nonlinear equation (12) (integro-differential, functional differential, neutral type, with fractional derivatives, with integrable weak singularity, Riccati type, ...) is physically meaningful and has important applications in hydrogen materials science. At the same time, it is an interesting mathematical object worthy of a theoretical investigation: whether a solution exists and is unique, extendability (forward in time), qualitative analysis.

5. Numerical Method and Computer Simulation

Note that due to the singularity, the "main contribution" under the integral in the equation (11) (as $\tau \to t$) will come from the values of $\dot{v}(\tau) \approx \dot{v}(t)$. We therefore take the following equation integrated in quadratures as the first approximation:

$$\dot{v}_0(t) = -\tilde{b}(t)v_0^2(t) - \varkappa \dot{v}_0(t)\psi(t), \quad v_0(0) = 1,$$

$$\psi(t) \equiv \int_0^t \frac{1 - Q(t-\tau)}{\sqrt{\pi[t-\tau]}}\, d\tau > 0, \;\; t \geq 0.$$

Formally, the equation $\dot{v}(t) = -\tilde{b}(t)v^2(t)$ corresponds to hydrogen desorption from the surface in the absence of H atoms diffusing from the bulk to the surface. Diffusion would slow down the decline of the surface concentration, as demonstrated by the equation $\dot{v}_0(t) = -\hat{b}(t)v_0^2(t)$ obtained above, which has a lower coefficient of "dimensionless desorption" (see (10)) $\hat{b} = \tilde{b}/[1 + \varkappa\psi]$. Hence, the first approximation is physically meaningful.

Next, the following system of successive approximations can be used:

$$\dot{v}_{k+1}(t) = -\tilde{b}(t)v_k(t)v_{k+1}(t) - \varkappa \int_0^t \frac{1 - Q(t-\tau)}{\sqrt{\pi[t-\tau]}}\, \dot{v}_{k+1}(\tau)\, d\tau \quad \big(v_{k+1}(0) = 1\big).$$

These equations can be viewed as linear equations with fractional derivatives relative to the functions $v_{k+1}(t)$, or as linear Volterra integral equations (with integrable weak singularity) relative to $w_{k+1}(t) = \dot{v}_{k+1}(t)$ (replacing v_{k+1} on the right with the integral of w_{k+1}). In a short enough time interval $[0, t_*]$ the applicability conditions of the contraction mapping theorem will be fulfilled. Extendability forward in time follows from the constraint $0 < \tilde{b}^- \leq \tilde{b}(t) \leq \tilde{b}^+$.

This system is convenient for proving the theorem of existence and uniqueness of solu-tions, but is not efficient as a numerical method due to slow convergence. We shall therefore pursue another approach, proceeding from the properties of the function $Q(s)$. In particular, since $Q(1) \approx 0.999$, the background of the length > 1 no longer has any significant effect on how the process is running. For reference: where $D \sim 10^{-5}\,\mathrm{cm}^2/\mathrm{s}$, $\ell \sim 0.1\,\mathrm{cm}$ of the time $t = t' = 1$ corresponds to 10^3 s of the physical time t (sufficiently long duration of the experiment). Hence, the lower limit of 0 at $t = t' > 1$ can be replaced in the integral with $t - 1$, thus reducing the memory array to be manipulated.

Choose a step $h < 1$, keeping in mind that too small a step is inexpedient as it enlarges the amount of computations. Proceeding from the monotonicity interval of the function $(1-Q(s))/\sqrt{s}$ we fix the range of $h \approx 0.2-0.3$ (to enable further efficient application of the integral formula of the mean). Consider the segment $t \in [kh, (k+1)h)$ $(k \geq 0)$ and isolate the integrable singularity at $\tau \approx t$:

$$\int_0^t \frac{1-Q(t-\tau)}{\sqrt{\pi[t-\tau]}}\, \dot{v}(\tau)\, d\tau = \int_0^{kh} \ldots + \int_{kh}^t \ldots,$$

$$\int_{kh}^t \frac{1-Q(t-\tau)}{\sqrt{\pi[t-\tau]}}\, \dot{v}(\tau)\, d\tau \approx \dot{v}(t)\varphi_k(t), \quad t \geq kh,$$

$$\varphi_k(t) \equiv \int_{kh}^t \frac{1-Q(t-\tau)}{\sqrt{\pi[t-\tau]}}\, d\tau = \int_0^{t-kh} \frac{1-Q(\tau)}{\sqrt{\pi\tau}}\, d\tau \equiv \varphi(t-kh).$$

Since $\dot{v}(\tau) < 0$ and $|\dot{v}(\tau)|$ declines monotonically (in contrast to the flux $J(t) = b(T(t))q^2(t)$, where the Arrhenius-like desorption coefficient has a monotonically rising S-shaped saturation curve based on temperature T), let us slightly increase the fraction in $\varphi_k(t)$ "to compensate". To do so, it is enough to use 3–5 members of the rapidly converging series for the function $Q(s)$. Fraction integration yields the following analytical expressions:

$$\varphi_k(t) = \varphi(t-kh) \approx \frac{2}{\sqrt{\pi}}\sqrt{t-kh}$$

$$-\frac{2}{\sqrt{\pi}}\sum_{n=1}^{5}(-1)^{n^2+1} n \left[\frac{2\exp\left\{-\frac{n^2}{4(t-kh)}\right\}}{n} - \sqrt{\pi} + \sqrt{\pi}\,\mathrm{erf}\left(\frac{n}{2\sqrt{t-kh}}\right)\right].$$

It remains to approximate the integral on the interval $[0, kh]$ using the theorem of the mean:

$$\int_0^{kh} \frac{1-Q(t-\tau)}{\sqrt{\pi[t-\tau]}}\, \dot{v}(\tau)\, d\tau = \sum_{j=0}^{k-1} \int_{jh}^{(j+1)h} \frac{1-Q(t-\tau)}{\sqrt{\pi[t-\tau]}}\, \dot{v}(\tau)\, d\tau$$

$$= \sum_{j=0}^{k-1} \frac{1 - Q(t - jh - \xi_j)}{\sqrt{\pi[t - jh - \xi_j]}} \cdot \int_{jh}^{(j+1)h} \dot{v}(\tau)\, d\tau$$

$$\approx \sum_{j=0}^{k-1} \frac{1 - Q(t - jh - \xi)}{\sqrt{\pi[t - jh - \xi]}} \cdot \big[v((j+1)h) - v(jh)\big].$$

Working further with the computational algorithm we assume the a priori unknown values ξ_j to be equal to the method's universal parameter $\xi \in (0, h)$, which is subject to adjustment. For time intervals $t = t' \in [kh, (k+1)h)$ we apply a standard Runge-Kutta 4th order numerical method with a step size of $\Delta t \in [0.01h, 0.1h]$ (the authors used freeware Scilab 6).

Written consecutively in compact form, the equations of the suggested method appear as follows:

$$\frac{dv}{dt} = -\hat{b}(t)v^2(t), \quad v(0) = 1, \quad t \in [0, h)\ (k = 0),$$

$$\frac{dv}{dt} = -\hat{b}(t)v^2(t) - \hat{\varkappa}(t)\sum_{j=0}^{k-1} \frac{1 - Q(t - jh - \xi)}{\sqrt{\pi[t - jh - \xi]}} \cdot \big[v((j+1)h) - v(jh)\big],$$

$$t \in [kh, (k+1)h),\ k \ge 1, \quad \hat{b}(t) \equiv \frac{\tilde{b}(t)}{1 + \varkappa\varphi_k(t)}, \quad \hat{\varkappa}(t) \equiv \frac{\varkappa}{1 + \varkappa\varphi_k(t)}.$$

Data Simulation

We use experimental data for nickel, steel (12Cr18Ni10Ti) and beryllium [4]. The assumed values of model parameters are the following: (steel) $\ell = 0.1$ cm, $T_0 = 300$ K, $b_0 = 1.3 \times 10^{-9}$ cm^2/s, $E_b = 80 \times 10^3$ J/mol, $D_0 = 3.1 \times 10^{-4}$ cm^2/s, $E_D = 31 \times 10^3$, $g = 50$ cm^{-1}, $s_0 = 0.6$, $E_s = 110 \times 10^3$, $\bar{c} = 5 \times 10^{17}$ at.H/cm^3, $\overline{T} = 1170$ K, $\dot{T} = 1$ K/s; (nickel) $\ell = 0.1$, $T_0 = 300$, $b_0 = 1.53 \times 10^{-14}$, $E_b = 43.2 \times 10^3$, $D_0 = 7.5 \cdot \times 10^{-3}$, $E_D = 40 \times 10^3$, $g = 100$, $s_0 = 1.8 \times 10^{-2}$, $E_s = 61.4 \times 10^3$, $\bar{c} = 9.9 \times 10^{17}$, $\bar{p} = 37.4$ Torr, $\overline{T} = 770$, $\dot{T} = 0.5$; (beryllium) $\ell = 0.1$, $T_0 = 300$, $b_0 = 3.1 \times 10^{-9}$, $E_b = 57.43 \times 10^3$, $D_0 = 3 \times 10^{-3}$, $E_D = 28 \times 10^3$, $g = 200$, $s_0 = 1.44 \times 10^{-4}$, $E_s = 1.82 \times 10^3$, $\bar{c} = 5.2 \times 10^{17}$, $\bar{p} = 760$ Torr, $\overline{T} = 1150$, $\dot{T} = 5$.

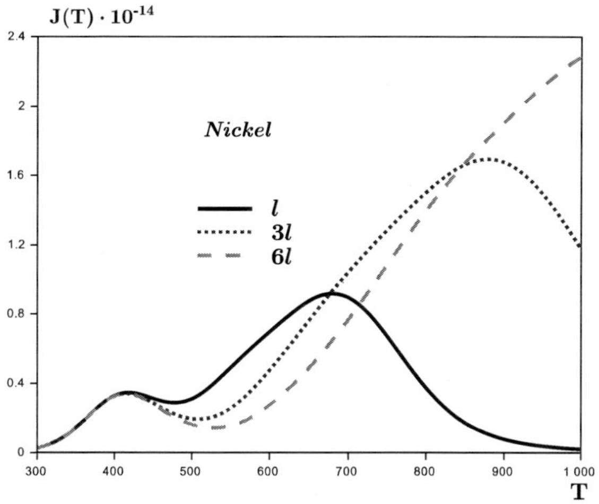

Figure 1. Nickel, $\ell \to 3, 6\ell$.

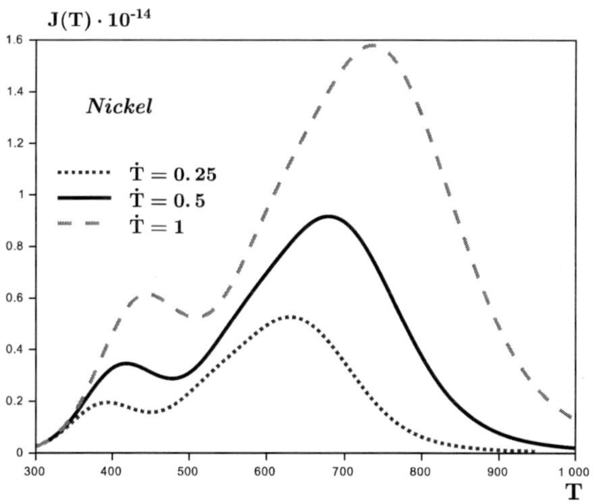

Figure 2. Nickel, heating rates.

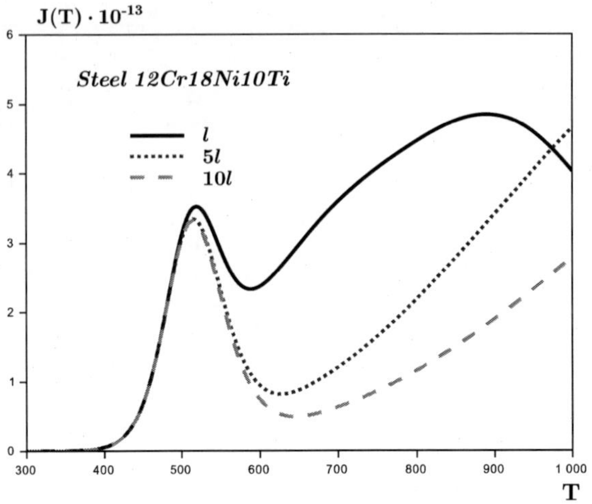

Figure 3. Steel, $\ell \to 5, 10\ell$.

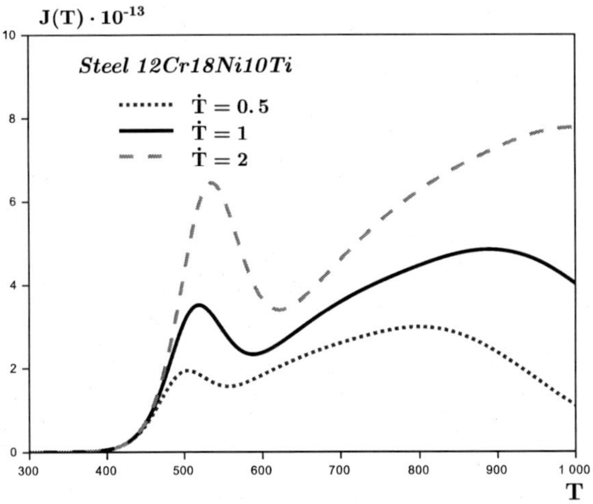

Figure 4. Steel, heating rates.

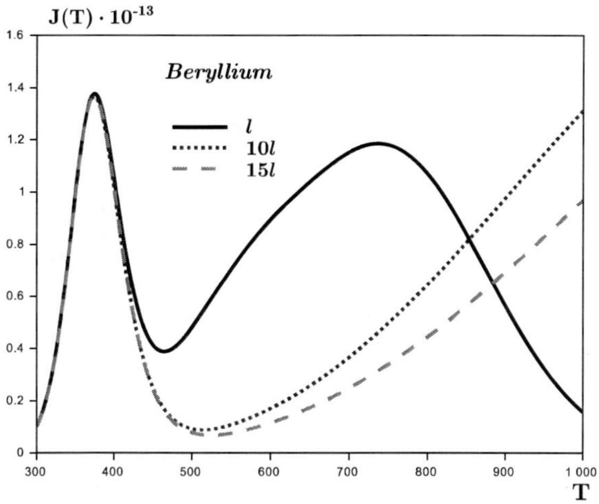

Figure 5. Beryllium, $\ell \to 10, 15\ell$.

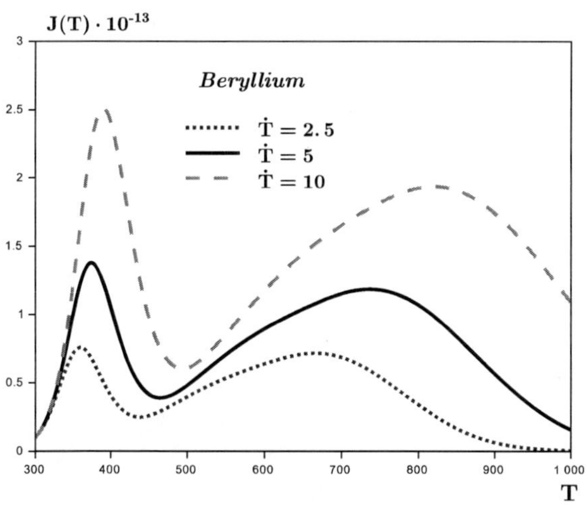

Figure 6. Beryllium, heating rates.

Simulation Results

To illustrate the modeling process let us play with parameter values to mimick changes in TDS spectrum. A phenomenon that may show up is a two-peak structure. As a rule, this phenomenon is explained by the release of hydrogen from traps with different binding energies. Other possible causes are not considered since zero boundary conditions are taken for the concentration in the bulk (and the effect of surface processes dynamics is thus artificially excluded). It is possible to simulate the required number of TDS-peaks using different binding energies: numerical algorithms based on difference schemes and modeling results were presented in [7, 8].

In this chapter we use only the diffusion equation (1) ($D = D_{\text{eff}}$). For thin membranes of quickly permeable material used in experiments this approximation is usually sufficient. Additional experimental studies are needed to discriminate the two-peak structure caused by differences in the binding energies of the trapped diffusant from the dynamic interplay between the surface and the bulk. Both effects are certainly present, but the question is their comparability.

As an example, we ran the simulation with the membrane thickness enlarged by an order of magnitude to increase the capacity of the bulk. This would not have affected the commensurability of the peaks (i.e. the ratio of the areas under the respective parts of the graph) had they been generated by different energies of capture in the bulk. We see that the first peak keeps losing magnitude as ℓ increases ($S\ell\bar{c} \gg 2S\bar{q}$, $V = S\ell$). Hence, capture in the bulk is not the primary factor (as we already know from the model (1)). The stability of the first peak (S persists as ℓ grows) suggests it is mainly generated by surface processes. This is one of the possible ways to differentiate between the above mechanisms of peak generation.

CONCLUSION

The paper presents a thermal desorption functional differential equation of neutral type with integrable weak singularity (a generalized Riccati equation with fractional derivative) for modeling the degassing TDS-spectrum of hydrogen-presaturated structural materials. The proposed numerical method allows using a standard procedure for ODE system solving as an alternative to solving the original non-linear problem with dynamical boundary conditions. One can thus quickly model TDS-spectra for a wide range of parameters, reveal the limiting factors and explore the material also under extreme conditions.

Solving of the direct problem is need: 1) for qualitative comparison with experimental data; 2) as an iterative component of the parametric identification algorithm; 3) for material pretesting under different conditions. The variation of parameters allows estimating the TDS spectrum's sensitivity to changes in the material composition. The main physical sense of the results is the following. One cannot a priori rule out fluctuating thermal desorption, not even for defect-free materials with minor trapping of the diffusing hydrogen (the diffusion model in this paper is free of additives associated with traps). The problem for future research is to differentiate the peaks arising from "surface-volume" interaction dynamics from those induced by hydrogen release from traps.

ACKNOWLEDGMENTS

Partially the study (computer simulation) was supported by the RFBR (Project 18-29-19149 mk).

REFERENCES

[1] Alefeld, G., and Völkl, J., eds. 1978. *Hydrogen in Metals.* Berlin: Springer-Verlag.

[2] Zakharov, A., ed. 1987. *Interactions of Hydrogen with Metals.* Moscow: Nauka.

[3] Pisarev, A., *et al.* 2008. *Hydrogen Permeability Through Metals.* Moscow: MEPhI.

[4] Yukhimchuk, A., ed. 2009. *Hydrogen Isotopes. Fundamental and Applied Studies.* Sarov: Russian Federal Nuclear Center.

[5] Ball, M., and Wietschel, M., eds. 2009. *The Hydrogen Economy: Opportunities and Challenges.* New York: Cambridge University Press.

[6] Hirscher, M., ed. 2010. *Handbook of Hydrogen Storage: New Materials for Future Energy Storage.* Weinheim: Wiley-VCH.

[7] Zaika, Yu., and Rodchenkova, N. 2013. "Hydrogen-Solid Boundary-Value Problems with Dynamical Conditions on Surface." In *Mathematical modelling*, edited by Brennan, C. R., 269–302. New York: Nova Science Publishers.

[8] Zaika, Yu., and Kostikova, E. 2014. "Modelling of Hydrogen Thermodesorption." *Adv. Mater. Sci. Appl.* 3:120–9.

[9] Zaika, Yu., Sidorov, N., and Fomkina, O. 2019. " Identification of Hydrogen Permeability Parameters of Membrane Materials in an Aggregated Experiment." *Int. J. Hydrogen Energy*. doi:10.1016/j.ijhydene.2019.04.098.

[10] Hale, J. 1977. *Theory of Functional Differential Equations*. New York: Springer-Verlag.

[11] Lang, S. 1973. *Elliptic Functions*. London: Addison-Wesley publishing.

In: A Closer Look at Boundary Value Problems ISBN: 978-1-53617-857-9
Editor: Mustafa Avci

Chapter 3

EXACT ABSORBING CONDITIONS FOR INITIAL BOUNDARY VALUE PROBLEMS OF COMPUTATIONAL ELECTRODYNAMICS: A REVIEW

***Yuriy K. Sirenko*[1,2,*], *Vadym L. Pazynin*[1], *Kostyantyn Yu. Sirenko*[1] and *Nataliya P. Yashina*[1]**

[1]Department of Diffraction Theory and Diffraction Electronics, O.Ya. Usikov Institute for Radiophysics and Electronics, Kharkiv, Ukraine

[2]Department of Applied Mathematics, V.N. Karazin Kharkiv National University, Kharkiv, Ukraine

ABSTRACT

One of the most difficult theoretical problems of computational electrodynamics is the correct limitation of computational space of initial boundary value problems which have an unbounded domain of interest. In this chapter, we detail the approach to this problem based on the construction and practical application of the so-called exact absorbing conditions. The proper enforcement of these conditions

---

* Corresponding Author's Email: yks2002sky@gmail.com.

on virtual boundaries of new, smaller and bounded computation domains does not lead neither to any distortions in the physics of processes being modeled nor to introduction of any additional errors into numerical solutions.

Keywords: initial boundary value problems, waveguide units, nodes and junctions, compact inhomogeneities in the free space, outgoing waves, exact absorbing conditions

1. INTRODUCTION

Practically all initial boundary value problems of electrodynamics of non-sinusoidal waves are open, i.e., have unbounded domains of interest: the sought-for perturbation $U(g,t)$ ($\mathbf{U}(p,t)$), which is generated by compact sources, can propagate infinitely far in the surrounding locally inhomogeneous space as the time t marches. In such a situation, the direct application of universal and powerful grid methods that are most suitable for the numerical analysis of such problems is impossible as it is necessary to take care of the effective and correct limitation of their computational space in advance. Here, $U(g,t)$, $g \in \mathrm{Q} \subset \mathrm{R}^2$ ($\mathbf{U}(p,t)$, $p \in \mathrm{Q} \subset \mathrm{R}^3$) is a scalar (vector) function defining components of electromagnetic field; Q is an analysis domain of an open problem; t is a time variable. All physical quantities in the chapter have dimensions of the SI system, and only the time is measured in meters as t is the product of the true time and the speed of light in vacuum.

The known heuristic and approximate solutions to the problem of analysis domain truncation are mainly based on the use of so-called Absorbing Boundary Conditions (ABCs) [1 – 5] and Perfectly Matched Layers (PMLs) [6–8]. The principal disadvantage of these solutions is the unpredictable behavior of computational errors for large values of the observation time and, as a result, no guarantee of correctness for situations involving resonant wave scattering.

In this chapter, we consider an approach that allows us to estimate and minimize errors arising when replacing open initial boundary problems with closed ones. This approach is based on the construction of exact absorbing conditions (EACs) and inclusion them in standard computational schemes of finite difference [9] or finite element [10] methods. The incorporation of EACs into the original initial boundary value problem does not affect its solution.

The velocity of the disturbance $U(g,t)$ ($\mathbf{U}(p,t)$) is finite, thus for any finite T it is always possible to construct such compact virtual boundary $\mathrm{L}_T \subset \mathrm{Q}$ that is simply out of reach for the disturbance at times $t \in [0,T]$; i.e.,

$$U(g,t)\big|_{g\in \mathrm{L}_T,\, t\in[0,T]} = 0 \quad \left(\mathbf{U}(p,t)\big|_{p\in \mathrm{L}_T,\, t\in[0,T]} = 0\right). \tag{1}$$

If the boundary L_T together with pieces of physical boundaries of the domain Q (some metal surfaces, for example) covers the domain $\mathrm{Q}_{\text{int}} \subset \mathrm{Q}$ containing all sources and scattering inhomogeneities of the medium of wave propagation (these may be any compact objects with sufficiently arbitrary geometric and electrical parameters), then application of (1) as an exact boundary condition imposed on the waves freely propagating into the domain $\mathrm{Q}_{\text{ext}} = \mathrm{Q} \setminus (\mathrm{Q}_{\text{int}} \cup \mathrm{L}_T)$, turns the original open initial boundary value problem into a closed one. Obviously, these two problems are equivalent for all $t \in [0,T]$. It is also obvious that such a simple way of limitation of the computational space of open problems in practice only works when a very short history of space-time transformations $U(g,t)$ ($\mathbf{U}(p,t)$) is of interest in the immediate vicinity of the generating sources. But such a short history is not of a big interest for a researcher, it cannot be used even for a transform to standard amplitude-frequency characteristics represented by integrals like

$$\tilde{U}(g,k) = \int_0^\infty U(g,t)\mathrm{e}^{ikt}dt \quad \left(\tilde{\mathbf{U}}(p,k) = \int_0^\infty \mathbf{U}(p,t)\mathrm{e}^{ikt}dt\right) \tag{2}$$

($k = 2\pi/\lambda$ is a wave number, λ is a wavelength in a free space).

Apparently, the problem can be radically solved by "moving" condition (1) to a stationary compact virtual boundary L that does not depend on the length of the interval $[0,T]$, covering (perhaps, together with physical boundaries of the domain Q) fairly densely the part Q_{int} of the open problem analysis domain containing all sources of electromagnetic waves and all inhomogeneities determining the nature of their further free propagation. And such a solution was found: over the past thirty years, for almost all types of initial boundary value problems of computational electrodynamics, various versions of exact radiation conditions have been constructed for waves propagating through the boundary L. The exact absorbing conditions (EACs) follow from these radiation conditions connecting magnitudes of functions $U(g,t)$, $g \in Q_{ext}$ ($\mathbf{U}(p,t)$, $p \in Q_{ext}$) with their values on the virtual boundary L

$$D\left[U(g,t)\right]\Big|_{g\in L} = 0 \quad \left(D\left[\mathbf{U}(p,t)\right]\Big|_{p\in L} = 0\right); \quad t \geq 0. \tag{3}$$

Here, $D[...]$ is a certain differential (in the case of local conditions) or integro-differential (in the case of non-local conditions) operator. We call these conditions *exact* because they follow directly from the Maxwell equations formulated for electromagnetic waves propagating into the region Q_{ext}. We call them *absorbing* because waves passing across the boundary L are completely "absorbed" by this boundary or external domain located behind it Q_{ext}, there is no reflection back into the domain Q_{int}, the virtual boundary L with the condition (3) formulated on it does not violate the natural course of the disturbance propagation. The correctness classes of the original (open) and modified (closed) problem coincide, their solutions at all points of the domain Q and at all moments of time $t > 0$ are completely the same.

The analytical form of the operator $D[...]$ is defined by a geometry of the domain Q_{ext}, by the dimension of the problem and the coordinate system where it is considered. Usually, its construction is based on the following sequence of simple operations widely used in the theory of boundary value problems for hyperbolic equations:

- allocation of a regular (admitting incomplete separation of variables) domain Q_{ext} where the perturbation $U(g,t)$ ($\mathbf{U}(p,t)$), formed by sources and scattering objects located in the domain Q_{int}, propagates freely;
- partial separation of variables (separation of the coordinate that is longitudinal relatively the direction of perturbation propagation and time) in the original initial boundary problem for the domain Q_{ext};
- transition to initial boundary problems for one-dimensional Klein-Gordon equations for the space-time perturbation amplitudes $U(g,t)$ ($\mathbf{U}(p,t)$);
- integral transformations of one-dimensional Klein-Gordon equations and the solution of auxiliary boundary value problems for ordinary differential equations with respect to the transformed space-time amplitudes;
- inverse integral transforms.

These actions result in non-local (in space and in time) EACs for the coordinate virtual boundary L, which separates the domains Q_{int} and Q_{ext}. Very often, these non-local EACs can be transformed into local ones by replacing several integral forms with corresponding differential forms. This usually requires a solution of an additional initial boundary problem with respect to some auxiliary function of time and transverse coordinates.

The chapter has the following structure. In the second section, we consider non-local and local EACs for virtual boundaries in the cross-section of hollow metal waveguides. In the third and fourth sections, non-local and local EACs for coordinate virtual boundaries in free space are formulated. The focus is on the technique of EACs construction and a brief overview of the key results: who, when and how have them obtained. Problems of improvement of computational efficiency when using EACs in standard finite difference methods are discussed in the fifth section. The sixth and seventh sections are devoted to results confirming the efficiency and reliability of EACs in solving complicated electrodynamic problems, the possibility of a significant reduction in computer's resources required for calculations due to the so-called history thinning and by storing its significant, rarely updated volumes on computer's hard disk.

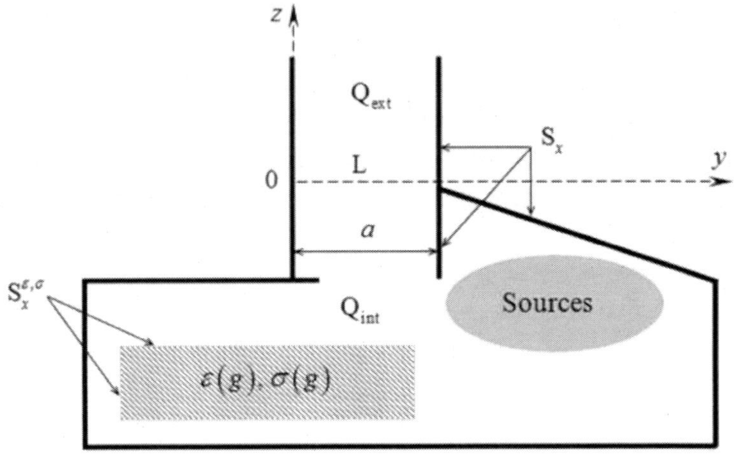

Figure 1. Geometry of the problem (4).

To describe the key points in the derivation of EACs, we consider simple scalar problems. Usually, this is enough for dealing with the specifics of vector problems. All necessary additional information for understanding these issues can be obtained in the original works cited where needed. We also point out that EACs for waveguides are used with almost no changes in problems of the electrodynamic theory of gratings (the role of guiding structures here is assigned to the Floquet channels), and vice versa [11–15]. Therefore, the problematics of the grating theory is not considered in this chapter.

The authors hope that the material presented in the chapter will be interesting for those who are engaged in the construction and implementation of robust algorithms for the numerical solution of initial boundary value problems of computational physics, for those who acquire new knowledge using modern advanced methodology based on mathematical modeling and computational experiments.

2. Waveguide Units. Conditions for Virtual Boundaries in the Cross-Section of Hollow Metal Waveguides

2.1. Non-Local Exact Absorbing Conditions

In 1986, A. Maikov, A. Sveshnikov, and S. Yakunin published the work [16], where they formulated for the first time exact non-local condition for a virtual boundary in the cross-section of a regular hollow waveguide, which is a channel where signals generated by some waveguide junction propagate. In that work, the derivation of condition is not explained, but it cannot differ considerably from the one described in [14] and given below.

E - and H -polarized pulsed fields at nodes in a waveguide with constant in any plane $x = const$ cross-section (Figure 1) are determined by the solutions of a two-dimensional scalar initial boundary value problem

$$\begin{cases} \left[-\varepsilon(g)\partial_t^2 - \sigma(g)\eta_0\partial_t + \partial_y^2 + \partial_z^2\right]U(g,t) = F(g,t); \\ \quad t > 0, \quad g = \{y,z\} \in \mathrm{Q} \\ U(g,0) = \varphi(g), \quad \partial_t U(g,0) = \psi(g); \quad g \in \bar{\mathrm{Q}} \\ \mathbf{E}_{\mathrm{tg}}(p,t)\big|_{p=\{x,y,z\}\in \mathrm{S}} = 0 \quad \text{and} \quad \mathbf{E}_{\mathrm{tg}}(p,t), \quad \mathbf{H}_{\mathrm{tg}}(p,t) \\ \quad \text{are continuous when crossing} \quad \mathrm{S}^{\varepsilon,\sigma}; \quad t \geq 0. \end{cases} \tag{4}$$

Here, $U(g,t) = E_x$ in the case of E -polarized field ($\partial_x \equiv 0$, $H_x = E_y = E_z = 0$) and $U(g,t) = H_x$ in the case of H -polarization ($\partial_x \equiv 0$, $E_x = H_y = H_z = 0$); $\mathbf{E} = \{E_x, E_y, E_z\}$ and $\mathbf{H} = \{H_x, H_y, H_z\}$ are the electric and magnetic field strength vectors; the waves propagate in a non-dispersive and non-magnetic medium with relative dielectric constant and specific conductivity given by the piecewise-continuous (in E -case) or piecewise-constant (in H -case) functions $\varepsilon(g)$ and $\sigma(g)$; $\eta_0 = \sqrt{\mu_0/\varepsilon_0}$ is the free space impedance.

The analysis domain $\mathrm{Q}=\mathrm{Q}_{\mathrm{int}}\cup\mathrm{L}\cup\mathrm{Q}_{\mathrm{ext}}$ of the problem (4) is a part of the plane $y0z$, limited by the contours S_x. $\mathrm{S}=\mathrm{S}_x\times[|x|\leq\infty]$, is a surface of perfect conductors; on the surface $\mathrm{S}^{\varepsilon,\sigma}=\mathrm{S}_x^{\varepsilon,\sigma}\times[|x|\leq\infty]$, the functions $\varepsilon(g)$ and $\sigma(g)$ have discontinuities. It is supposed, that the functions $F(g,t)$, $\varphi(g)$, $\psi(g)$, $\sigma(g)$ and $\varepsilon(g)-1$ are finite in the closure of the domain Q and satisfy the theorem on the unique solvability of the problem (4) in the Sobolev space $\mathrm{W}_2^1(\mathrm{Q}^T)$, $\mathrm{Q}^T=\mathrm{Q}\times(0,T)$; $T<\infty$ is observation time [15, 17]. These functions' supports belong to the domain $\mathrm{Q}_{\mathrm{int}}$. In a regular waveguide (in the domain $\mathrm{Q}_{\mathrm{ext}}$), where the field formed by the structure propagates freely and infinitely far, $\varepsilon(g)\equiv 1$, and $\sigma(g)=\varphi(g)=\psi(g)=F(g,t)\equiv 0$ as there are no sources and effective scatterers. The virtual boundary L (in Figure 1 it is indicated by the dashed line) coincides with the cross-section $z=0$ of this waveguide.

Assume that the perturbation $U(g,t)$ has not yet reached the boundary $z=0$ by the time $t=0$. Using the separation of variables, we obtain the following representation for the solution $U(g,t)$ to the problem (4):

$$U(g,t)=\sum_{n\in\{n\}}u_n(z,t)\mu_n(y);\quad g=\{y,z\}\in\mathrm{Q}_{\mathrm{ext}},\quad t\geq 0. \tag{5}$$

Complete (in the space $\mathrm{L}_2(0,a)$ of square integrable functions $f(y):f(0)=f(a)=0$ or $d_yf(0)=d_yf(a)=0$), orthonormal system of transverse functions $\{\mu_n(y)\}$ and their own eigenvalues system $\{\lambda_n\}$ are determined by nontrivial solutions of homogeneous (spectral) problem

$$\begin{cases}\left[d_y^2+\lambda_n^2\right]\mu_n(y)=0; & 0<y<a\\ \mu_n(0)=\mu_n(a)=0\quad (E-\text{case})\quad\text{or} & ,\\ \quad d_y\mu_n(0)=d_y\mu_n(a)=0\quad (H-\text{case}) & \end{cases} \tag{6}$$

and the space-time amplitudes $\{u_n(z,t)\}$ (evolutionary basis) of the signal $U(g,t)$ are given by solutions of initial boundary value problem

$$\begin{cases} \left[-\partial_t^2+\partial_z^2-\lambda_n^2\right]u_n(z,t)=0; \quad t>0 \\ u_n(z,0)=0, \quad \partial_t u_n(z,t)\big|_{t=0}=0 \end{cases}; \quad z\geq 0, \quad n\in\{n\}. \tag{7}$$

Here, $\lambda_n = n\pi/a$; in the case of E-polarized field $\{n\}=1,2,...$, $\mu_n(y)=\sqrt{2/a}\sin(n\pi y/a)$ and in the case of H-polarization $\{n\}=0,1,2,...$, $\mu_n(y)=\sqrt{(2-\delta_0^n)/a}\cos(n\pi y/a)$; δ_0^n is the Kronecker symbol.

Applying the Fourier cosine transform

$$\tilde{f}(\omega)=\sqrt{\frac{2}{\pi}}\int\limits_0^\infty f(z)\cos(\omega z)dz \leftrightarrow f(z)=\sqrt{\frac{2}{\pi}}\int\limits_0^\infty \tilde{f}(\omega)\cos(\omega z)d\omega \tag{8}$$

to (7) over the variable z on the semi axis $z\geq 0$, we obtain the following Cauchy problem for the images $\tilde{u}_n(\omega,t)\leftrightarrow u_n(z,t)$:

$$\begin{cases} \left[\partial_t^2+\left(\lambda_n^2+\omega^2\right)\right]\tilde{u}(\omega,t)=-\sqrt{2/\pi}u_n\partial_z(0,t); \quad \omega>0, \quad t>0 \\ \tilde{u}(\omega,0)=0, \quad \partial_t\tilde{u}(\omega,0)=0; \quad \omega\geq 0. \end{cases} \tag{9}$$

Here, $\partial_z u_n(b,t)=\partial_z u_n(z,t)/\partial z\big|_{z=b}=\int_0^a \partial_z U(g,b)\mu_n(y)dy$, and in the derivation of (9) it was taken into account that $-\omega^2\tilde{f}(\omega)-\sqrt{2/\pi}\,d_z f(z)\big|_{z=0} \leftrightarrow d_z^2 f(z)$ [18], and the wave $U(g,t)$ in the considered part of the domain Q_{ext} does not contain components propagating in the direction of decreasing z. Components propagating towards $z=\infty$ at any finite moment of time $t=T$ are equal to zero for sufficiently large values of z.

Continuing the functions $\tilde{u}_n(\omega,t)$ and $\partial_z u_n(0,t)$ with zero on the semi axis $t<0$, we proceed to the generalized formulation of Cauchy problem (9) [19]:

$$\begin{gathered}\left[\partial_t^2+\left(\lambda_n^2+\omega^2\right)\right]\tilde{u}_n(\omega,t)=-\sqrt{2/\pi}\,\partial_z u_n(0,t)+\delta^{(1)}(t)\tilde{u}_n(\omega,0)+\\+\delta(t)\partial_t\tilde{u}_n(\omega,0)=-\sqrt{2/\pi}\,\partial_z u_n(0,t);\quad \omega>0,\quad -\infty<t<\infty.\end{gathered}\tag{10}$$

Here, $\delta(...)$ and $\delta^{(m)}(...)$ are the Dirac function and its generalized derivative of order m. Convolution of the fundamental solution $G(\lambda,t)=\chi(t)\lambda^{-1}\sin\lambda t$ ($\chi(...)$ is the Heaviside step function) of the operator $D(\lambda)\equiv\left[\partial_t^2+\lambda^2\right]$ [12, 14, 19] with the right side of equation (10) allows to write $\tilde{u}_n(\omega,t)$ in the form

$$\tilde{u}_n(\omega,t)=-\sqrt{\frac{2}{\pi}}\int_0^t \sin\left[(t-\tau)\sqrt{\lambda_n^2+\omega^2}\right]\frac{\partial_z u_n(0,\tau)}{\sqrt{\lambda_n^2+\omega^2}}d\tau;\quad \omega\geq 0,\quad t\geq 0.\tag{11}$$

Returning back to the originals in (11) (applying the inverse Fourier cosine transform), we obtain exact radiation condition

$$u_n(z,t)=-\int_0^{t-z} J_0\left[\lambda_n\sqrt{(t-\tau)^2-z^2}\right]\partial_z u_n(0,\tau)d\tau;\quad z\geq 0,\quad t\geq 0\tag{12}$$

for the space-time amplitudes $u_n(z,t)$ of the wave $U(g,t)$, $g\in \mathrm{Q}_{\mathrm{ext}}$ ($J_n(...)$ is the cylindrical Bessel functions). Obviously, the result (12) can also be represented in the form

$$u_n(z,t)=-\int_0^{t-(z-z_0)} J_0\left[\lambda_n\sqrt{(t-\tau)^2-(z-z_0)^2}\right]\partial_z u_n(z_0,\tau)d\tau;\quad z\geq z_0\geq 0,\quad t\geq 0.\tag{13}$$

The relation (13) defines the so-called transport operator [11, 12], which acts in the space of the evolutionary basis $\{u_n(z,t)\}$ of signals $U(g,t)$ propagating in a segment of a regular waveguide. The transport operator allows recalculation of the pulse field from one cross-section of this waveguide ($z=z_0$) to another ($z\geq z_0$).

Putting in (12) the observation point on the imaginary boundary L ($z=0$), we obtain

$$u_n(0,t) = -\int_0^t J_0\left[\lambda_n(t-\tau)\right]\partial_z u_n(0,\tau)\,d\tau; \quad t \ge 0. \tag{14}$$

It is historically first exact non-local condition (1.5) from [16], or, taking into account (5):

$$U(y,0,t) = -\sum_{n\in\{n\}}\left\{\int_0^t J_0\left[\lambda_n(t-\tau)\right]\left[\int_0^a \partial_z U(\tilde{y},0,\tau)\mu_n(\tilde{y})\,d\tilde{y}\right]d\tau\right\}\mu_n(y); \tag{15}$$
$$0 \le y \le a, \quad t \ge 0.$$

Non-local EAC (14) on the virtual boundary L connects values of the space-time amplitudes of all partial components (modes) of waves channeling by the regular waveguide Q_{ext} to the direction $z \to \infty$, and EAC (15) connects values of the fields of such waves as a whole.

Differentiating (14) by t, we obtain EAC

$$\partial_t u_n(0,t) + \partial_z u_n(0,t) = \lambda_n \int_0^t J_1\left[\lambda_n(t-\tau)\right]\partial_z u_n(0,\tau)\,d\tau; \quad t \ge 0. \tag{16}$$

Applying the Laplace transform

$$\tilde{f}(s) = \int_0^\infty f(t)\exp(-st)\,dt \leftrightarrow f(t) = \frac{1}{2\pi i}\int_{\alpha-i\infty}^{\alpha+i\infty} \tilde{f}(s)\exp(st)\,ds$$

to (16) and taking into account the known relationships $\tilde{f}_1(s)\tilde{f}_2(s) \leftrightarrow \int_0^t f_1(t-\tau)f_2(\tau)\,d\tau$, $s\tilde{f}(s) - f(0) \leftrightarrow d_t f(t)$, $\lambda^2\left[\sqrt{s^2+\lambda^2}\left(\sqrt{s^2+\lambda^2}+s\right)\right]^{-1} \leftrightarrow \lambda J_1(\lambda t)$ [20], move on to the next representation in the space of images $\tilde{u}_n(z,s)$:

$$\partial_z \tilde{u}_n(0,s) = -\left[s + \lambda_n^2 \left(s + \sqrt{s^2 + \lambda_n^2} \right)^{-1} \right] \tilde{u}_n(0,s). \tag{17}$$

Returning to the originals in (17) and using the relation $\left(s+\sqrt{s^2+\lambda^2}\right)^{-1} \leftrightarrow (\lambda t)^{-1} J_1(\lambda t)$ from [21], we derive EAC

$$\partial_t u_n(0,t) + \partial_z u_n(0,t) = -\lambda_n \int_0^t J_1\left[\lambda_n (t-\tau)\right](t-\tau)^{-1} u_n(0,\tau)\,d\tau; \quad t \geq 0. \tag{18}$$

Non-local EAC (16) first appeared in [22] and EAC (18) in [23]. Later, in [24], EAC (18) was derived within the approach proposed and developed by I. Sofronov [25].

2.2. Local Exact Absorbing Conditions

Now we use the representation $J_0(x) = (2/\pi)\int_0^{\pi/2} \cos(x \sin\varphi)\,d\varphi$ [18] and transform (14) into

$$u_n(0,t) = -\frac{2}{\pi} \int_0^{\pi/2} \left\{ \int_0^t \cos\left[\lambda_n (t-\tau)\sin\varphi\right] \partial_z u_n(0,\tau)\,d\tau \right\} d\varphi; \quad t \geq 0. \tag{19}$$

Denote

$$w_n(t,\varphi) = -\int_0^t \frac{\sin\left[\lambda_n (t-\tau)\sin\varphi\right] \partial_z u_n(0,\tau)}{\lambda_n \sin\varphi}\,d\tau; \quad t \geq 0, \quad 0 \leq \varphi \leq \pi/2. \tag{20}$$

Then

$$\partial_t w_n(t,\varphi) = -\int_0^t \cos\left[\lambda_n (t-\tau)\sin\varphi\right] \partial_z u_n(0,\tau)\,d\tau,$$

and from (19) we obtain

$$u_n(0,t) = \frac{2}{\pi} \int_0^{\pi/2} \partial_t w_n(t,\varphi) d\varphi; \quad t \geq 0. \tag{21}$$

The integral form (20) is equivalent to the differential form

$$\begin{cases} \left[\partial_t^2 + \lambda_n^2 \sin^2 \varphi\right] w_n(t,\varphi) = -\partial_z u_n(0,t); \quad t > 0 \\ w_n(0,\varphi) = \partial_t w_n(0,\varphi) = 0 \end{cases}. \tag{22}$$

Indeed, if in (22) we proceed to the generalized formulation of the corresponding Cauchy problem and use the fundamental solution $G(\lambda,t) = \chi(t)\lambda^{-1} \sin \lambda t$ of the operator $D(\lambda) \equiv \left[d_t^2 + \lambda^2\right]$, then it is easy to verify that the representations (20) and (22) define the same function $w_n(t,\varphi)$.

Multiply now (21) and (22) by $\mu_n(y)$ and sum over all $n \in \{n\}$. As a result, taking into consideration that $\sum_n \lambda_n^2 w_n(t,\varphi)\mu_n(y) = -\partial_y^2 W(y,t,\varphi)$ when $W(y,t,\varphi) = \sum_n w_n(t,\varphi)\mu_n(y)$ holds (see the problem (6)), we obtain the relations

$$U(y,0,t) = \frac{2}{\pi} \int_0^{\pi/2} \partial_t W(y,t,\varphi) d\varphi; \quad t \geq 0, \quad 0 \leq y \leq a, \tag{23a}$$

$$\begin{cases} \left[\partial_t^2 - \sin^2(\varphi)\partial_y^2\right] W(y,t,\varphi) = -\partial_z U(y,0,t); \quad 0 < y < a, \quad t > 0 \\ W(y,0,\varphi) = \partial_t W(y,0,\varphi) = 0; \quad 0 \leq y \leq a \\ W(0,t,\varphi) = W(a,t,\varphi) = 0 \quad (E-\text{case}) \quad \text{or} \\ \quad \partial_y W(0,t,\varphi) = \partial_y W(a,t,\varphi) = 0 \quad (H-\text{case}); \quad t \geq 0 \end{cases} \tag{23b}$$

that are local (both for spatial and time variables) EAC. Here and below, $W(y,t,\varphi)$ is an auxiliary function, determined by the solution of the initial boundary value problem (23b); $0 \leq \varphi \leq \pi/2$ is a numeric parameter.

Transforming EACs (16) and (18) in the same way, we obtain the following local EACs connecting the values of derivatives of the function $U(g,t)$ on the boundary L ($z=0$):

$$[\partial_t+\partial_z]U(y,0,t)=\frac{2}{\pi}\int_0^{\pi/2}W(y,t,\varphi)\cos^2\varphi d\varphi;\quad t\geq 0,\quad 0\leq y\leq a, \tag{24a}$$

$$\begin{cases}\left[\partial_t^2-\cos^2(\varphi)\partial_y^2\right]W(y,t,\varphi)=-\partial_y^2\left[\partial_z U(y,0,t)\right]; & 0<y<a,\quad t>0\\ W(y,0,\varphi)=\partial_t W(y,0,\varphi)=0; & 0\leq y\leq a\\ W(0,t,\varphi)=W(a,t,\varphi)=0 \quad (E-\text{case}) \quad \text{or} & \\ \quad \partial_y W(0,t,\varphi)=\partial_y W(a,t,\varphi)=0 \quad (H-\text{case}); & t\geq 0,\end{cases} \tag{24b}$$

$$[\partial_t+\partial_z]U(y,0,t)=\frac{2}{\pi}\int_0^{\pi/2}\partial_t W(y,t,\varphi)\sin^2\varphi d\varphi;\quad t\geq 0,\quad 0\leq y\leq a, \tag{25a}$$

$$\begin{cases}\left[\partial_t^2-\cos^2(\varphi)\partial_y^2\right]W(y,t,\varphi)=\partial_y^2 U(y,0,t); & 0<y<a,\quad t>0\\ W(y,0,\varphi)=\partial_t W(y,0,\varphi)=0; & 0\leq y\leq a\\ W(0,t,\varphi)=W(a,t,\varphi)=0 \quad (E-\text{case}) \quad \text{or} & \\ \quad \partial_y W(0,t,\varphi)=\partial_y W(a,t,\varphi)=0 \quad (H-\text{case}); & t\geq 0.\end{cases} \tag{25b}$$

The above results are presented in the same way as in the book [14]. But for the first time they appeared in [23]. The materials presented in this section reveals far from all possible approaches: minor changes in technical details or in the sequence of basic operations can lead to results that are substantially complemental to already presented (see, for example, [14, 26, 27]). Updating EACs is a continuous process. Permanently growing needs of computing practice will always essentially stimulate this process.

The following important differences between non-local and local EACs should be noted. Non-local EACs require complete information about the systems of eigenfunctions and eigenvalues of the Sturm-Liouville operator in the region coinciding with the cross-section of a waveguide, and with the

Dirichlet and Neumann conditions on its external boundary. The solution of the corresponding spectral problems can be too cumbersome and will require a significant consumption of computer resources. Local EACs are free from this drawback, they should be preferred when analyzing structures with energy output channels, the transverse functions for which cannot be determined analytically (see next subsection). Nevertheless, even in the case when transverse functions of regular waveguide channels are known, the utilization of local EACs provides much faster calculations with much smaller amounts of computer memory involved. The analysis confirming this statement was first carried out in the works [28, 29].

2.3. Exact Absorbing Conditions for Vector Waveguide Problems

Scalar two-dimensional problems are suitable for a quick view on possible situations, for obtaining qualitatively correct ideas about the physics of real wave processes, for selecting promising design solutions, etc. Such research work with three-dimensional vector problems is barely within the power of modern publicly available computer technology. Vector problems are most often formulated in connection with the calculation of a specific electrodynamic object with a small number of variable parameters, but with fairly strict requirements for the accuracy of numerical results. One of the resources for reducing the computational error consists in the application of accurate absorbing conditions for finite virtual boundaries in unbounded areas of analysis of open initial boundary value problems. For vector problems of the theory of open waveguide resonators such conditions were first strictly formulated in [23]. Here, we present them, following the version presented in [14]. The main goal is to demonstrate the possibility of effective modification and generalization of approaches developed initially for solving two-dimensional scalar problems.

Considering an arbitrary waveguide node, let us forget for a while about that limited part of the space in which all inhomogeneities and sources are concentrated and where, when solving the corresponding vector problems, it is only necessary to follow the standard well-developed schemes of grid methods. We focus on the channels (closed regular waveguides) which guide the signals formed by the node. For analysis, let us consider one of these channels, see Figure 2.

The field $\mathbf{E}(p,t)$ is present in the domain $Q_{ext} = \{p = \{x, y, z\} : \{x, y\} \in \text{int}\, S_z,\ z \geq 0\}$ only at the moments of time $t > 0$, this field is a wave going in the direction $z \to \infty$ and satisfies the following vector initial boundary value problem:

$$\begin{cases} \left(-\partial_t^2 + \Delta\right)\mathbf{E}(p,t) = 0; & t > 0, \quad p \in Q_{ext} \\ \mathbf{E}(p,t)\big|_{t=0} = \partial_t \mathbf{E}(p,t)\big|_{t=0} = 0; & p \in \overline{Q_{ext}} \\ E_z(p,t)\big|_{p\in S} = \left(\boldsymbol{\tau} \cdot \mathbf{E}_\perp(p,t)\right)\big|_{p\in S} = 0; & t \geq 0 \end{cases}. \tag{26}$$

Here, (see Figure 2) $\mathbf{E} = E_z\mathbf{z} + \mathbf{E}_\perp$, $\mathbf{E}_\perp = E_x\mathbf{x} + E_y\mathbf{y}$; S_z is the boundary contour of the cross-section of a regular waveguide; $\mathbf{x}$, $\mathbf{y}$, $\mathbf{z}$, $\boldsymbol{\tau}$ and $\mathbf{n}$ are unit vectors of the coordinate system, the tangent and the normal to the contour S_z; $\text{int}\, S_z$ is the domain in the pane $z = const$, limited by the contour S_z; $S = S_z \times (0 < z < \infty)$.

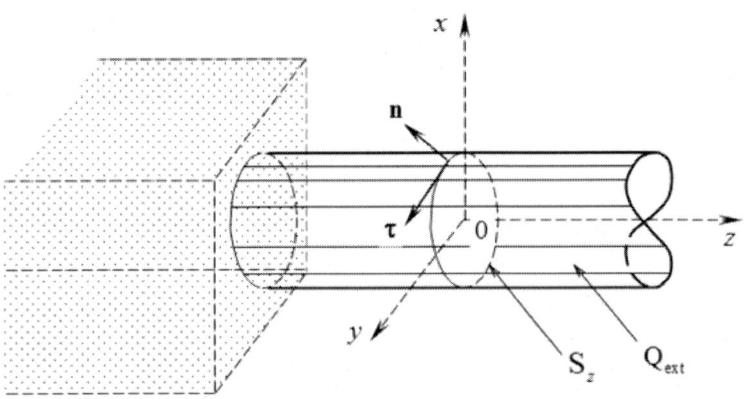

Figure 2. Geometry of a vector waveguide problem.

Let us use the Borgnis functions $U^E(p,t)$ and $U^H(p,t)$ (see [12, 14]) such that

$$\begin{cases} \left(-\partial_t^2+\Delta\right)\partial_t U^{E,H}(p,t)=0; & t>0, \quad p\in \mathrm{Q}_{\mathrm{ext}} \\ U^E(p,t)\big|_{p\in \mathrm{S}}=\partial_{\mathbf{n}} U^H(p,t)\big|_{p\in \mathrm{S}}=0; & t\geq 0 \end{cases}, \tag{27}$$

and present the general solution to the vector differential equation of the problem (26) in the form

$$E_x=\partial_{xz}^2 U^E-\partial_{yt}^2 U^H, \quad E_y=\partial_{yz}^2 U^E+\partial_{xt}^2 U^H, \quad E_z=\partial_z^2 U^E-\partial_t^2 U^E. \tag{28}$$

It is easy to verify that with such a representation, the field $\mathbf{E}(p,t)$ also satisfies the boundary conditions of the problem (26).

Now separate the transverse variables x and y in the problem (27) and write its solutions in the form

$$U^{E,H}(x,y,z,t)=\sum_{n=1,0}^{\infty} u_n^{E,H}(z,t)\mu_n^{E,H}(x,y). \tag{29}$$

Here, $\left\{\mu_n^E(x,y)\right\}_{n=1}^{\infty}$ and $\left\{\mu_n^H(x,y)\right\}_{n=0}^{\infty}$ are complete orthonormal systems of solutions to the Sturm-Liouville problem with equation $\left(\partial_x^2+\partial_y^2+\lambda^2\right)\mu=0$ in the domain $\operatorname{int}\mathrm{S}_z$ with the Dirichlet ($\mu^E(x,y)\big|_{\{x,y\}\in \mathrm{S}_z}=0$) and Neumann ($\partial_{\mathbf{n}}\mu^H(x,y)\big|_{\{x,y\}\in \mathrm{S}_z}=0$) conditions on its boundary S_z. The eigenvalues λ_n^E and λ_n^H are connected with these solutions.

Substitution of (29) into (28) allows writing the field $\mathbf{E}(p,t)$ in the form

$$E_z=\sum_{n=1}^{\infty} v_{n,z}(z,t)\xi_{n,z}(x,y), \quad \mathbf{E}_\perp=\sum_{n=-\infty}^{\infty} v_{n,\perp}(z,t)\boldsymbol{\xi}_{n,\perp}(x,y); \quad t\geq 0, \quad z\geq 0, \tag{30}$$

and to reduce the problem of determining scalar functions $v_{n,z}(z,t)$ and $v_{n,\perp}(z,t)$ to solving the following initial boundary value problems:

$$\begin{cases} \left[-\partial_t^2+\partial_z^2-\alpha_{n,z}^2\right]v_{n,z}(z,t)=0; & t>0,\quad z>0 \\ v_{n,z}(z,0)=\partial_t v_{n,z}(z,t)\big|_{t=0}=0; & z\geq 0,\quad n=1,2,... \end{cases}, \tag{31}$$

$$\begin{cases} \left[-\partial_t^2+\partial_z^2-\alpha_{n,\perp}^2\right]v_{n,\perp}(z,t)=0; & t>0,\quad z>0 \\ v_{n,\perp}(z,0)=\partial_t v_{n,\perp}(z,t)\big|_{t=0}=0; & z\geq 0,\quad n=0,\pm 1,\pm 2,... \end{cases}. \tag{32}$$

Here, $\alpha_{n,z}=\lambda_n^E$; $\alpha_{n,\perp}=\lambda_n^E$ for $n=1,2,...$ and $\alpha_{n,\perp}=\lambda_{-n}^H$ for $n=0,-1,-2,...$; $\xi_{n,z}=\mu_n^E$; $\boldsymbol{\xi}_{n,\perp}=\left(\partial_x\mu_n^E\right)\mathbf{x}+\left(\partial_y\mu_n^E\right)\mathbf{y}$ for $n=1,2,...$ and $\boldsymbol{\xi}_{n,\perp}=-\left(\partial_y\mu_{-n}^H\right)\mathbf{x}+\left(\partial_x\mu_{-n}^H\right)\mathbf{y}$ for $n=0,-1,-2,...$

Exploring properties of the functions $\left\{\mu_n^{E,H}(x,y)\right\}$, we obtain the inversion formulas for (30):

$$\begin{aligned} v_{n,z}(z,t)&=\int\limits_{\operatorname{int}S_z} E_z(x,y,z,t)\xi_{n,z}(x,y)\,dxdy, \\ v_{n,\perp}(z,t)&=\left(\alpha_{n,\perp}\right)^{-2}\int\limits_{\operatorname{int}S_z}\left(\mathbf{E}_\perp(x,y,z,t)\cdot\boldsymbol{\xi}_{n,\perp}(x,y)\right)dxdy. \end{aligned} \tag{33}$$

The first of the formulas (33) is an obvious consequence of the fact that the system $\left\{\mu_n^E(x,y)\right\}$ consists of orthogonal functions whose norm is equal to one. Considering integrals

$$\begin{aligned} &\int\limits_{\operatorname{int}S_z}\left(\left(\left(\partial_x\mu_n^E\right)\mathbf{x}+\left(\partial_y\mu_n^E\right)\mathbf{y}\right)\cdot\left(\left(\partial_x\mu_m^E\right)\mathbf{x}+\left(\partial_y\mu_m^E\right)\mathbf{y}\right)\right)dx\,dy= \\ &=\int\limits_{\operatorname{int}S_z}\left(\operatorname{grad}\mu_n^E\cdot\operatorname{grad}\mu_m^E\right)dx\,dy= \\ &=-\int\limits_{\operatorname{int}S_z}\mu_m^E\Delta\mu_n^E\,dx\,dy+\int\limits_{S_z}\mu_m^E\left(\partial_{\mathbf{n}}\mu_n^E\right)ds=\begin{cases} 0, & m\neq n \\ \left(\lambda_n^E\right)^2, & m=n \end{cases}, \end{aligned}$$

$$\int\limits_{\mathrm{int\,S}_z}\left(\left(-\left(\partial_y\mu_n^H\right)\mathbf{x}+\left(\partial_x\mu_n^H\right)\mathbf{y}\right)\cdot\left(-\left(\partial_y\mu_m^H\right)\mathbf{x}+\left(\partial_x\mu_m^H\right)\mathbf{y}\right)\right)dx\,dy=$$

$$=\int\limits_{\mathrm{int\,S}_z}\left(\left[\mathrm{grad}\,\mu_n^H\times\mathbf{z}\right]\cdot\left[\mathrm{grad}\,\mu_m^H\times\mathbf{z}\right]\right)dx\,dy=$$

$$=\int\limits_{\mathrm{int\,S}_z}\left(\mathrm{grad}\,\mu_n^H\cdot\mathrm{grad}\,\mu_m^H\right)dx\,dy=\begin{cases}0, & m\neq n\\ \left(\lambda_n^H\right)^2, & m=n\end{cases},$$

$$\int\limits_{\mathrm{int\,S}_z}\left(\left(\left(\partial_x\mu_n^E\right)\mathbf{x}+\left(\partial_y\mu_n^E\right)\mathbf{y}\right)\cdot\left(-\left(\partial_y\mu_m^H\right)\mathbf{x}+\left(\partial_x\mu_m^H\right)\mathbf{y}\right)\right)dx\,dy=$$

$$=\int\limits_{\mathrm{P}}\left(d\mathbf{P}\cdot\mathrm{rot}\,\boldsymbol{\mu}_m^H\right)=\int\limits_{\mathrm{S}_z}\left(d\mathbf{S}_z\cdot\boldsymbol{\mu}_m^H\right)=0,$$

we get convinced in the validity of the second one. Here, $\boldsymbol{\mu}_m^H=\mu_m^H\mathbf{z}$; P is a surface of function $z=\mu_n^E(x,y)$, which is stretched on the contour S_z; $d\mathbf{P}=\left(-\left(\partial_x\mu_n^E\right)\mathbf{x}-\left(\partial_y\mu_n^E\right)\mathbf{y}+\mathbf{z}\right)dx\,dy$ is a vector of the surface piece P; $d\mathbf{S}_z$ is a vector contour element S_z.

The problems (31), (32) are identical to the problem (7), which was considered in detail in Subsection 2.1. Therefore, omitting the details, we proceed immediately to the analysis of their solutions:

$$v_n(0,t)=-\int\limits_0^t J_0\left[\alpha_n(t-\tau)\right]\partial_z v_n(0,\tau)d\tau;\quad t\geq 0, \tag{34}$$

$$\left[\partial_z+\partial_t\right]v_n(z,t)\Big|_{z=0}=\alpha_n\int\limits_0^t J_1\left[\alpha_n(t-\tau)\right]\partial_z v_n(0,\tau)d\tau;\quad t\geq 0, \tag{35}$$

$$\left[\partial_z+\partial_t\right]v_n(z,t)\Big|_{z=0}=-\alpha_n\int\limits_0^t J_1\left[\alpha_n(t-\tau)\right](t-\tau)^{-1}v_n(0,\tau)dt;\quad t\geq 0. \tag{36}$$

Putting in (34)–(36) $\alpha_n=\alpha_{n,z}$, $n=1,2,...$, we obtain EACs for the elements $v_{n,z}(z,t)$ of the evolutionary basis of the component $E_z(p,t)$ of the outgoing wave $\mathbf{E}(p,t)$. As $\alpha_n = \alpha_{n,\perp}$, $n = 0, \pm 1, \pm 2, \ldots$ we get exactly the same representations for the elements $v_{n,\perp}(z,t)$ that are the amplitudes of the partial components of the transverse electric field $\mathbf{E}_\perp(p,t)$ of the wave $\mathbf{E}(p,t)$. Further on identifiers z and $\perp$ in indices of the eigenvalues α_n and functions v_n, ξ_n, E, $\mathbf{E}$ are omitted; the algorithm for reading any result that follows from (34), (35) and (36), both in terms of the longitudinal field $E_z(p,t)$ and in terms of the transverse field $\mathbf{E}_\perp(p,t)$, is absolutely transparent.

EACs (34)–(36) are non-local in time. Their incorporation into finite difference method for determining the field $\mathbf{E}(p,t)$ at all points $p \in \mathrm{Q}_{\mathrm{int}}$ of the waveguide node also generates spatial nonlocality: at each time step, it is necessary to move from the series (30) to the amplitudes (33) and vice versa. Using the technique tested in Subsection 2.2, we proceed from (35), (36) to EACs which are local in space and time:

$$\left[\partial_t + \partial_z\right] E(p,t)\Big|_{z=0} = \frac{1}{\pi}\int_0^\pi W(x,y,t,\varphi)\,d\varphi; \quad \{x,y\} \in \overline{\mathrm{int}\,\mathrm{S}_z}, \quad t \geq 0, \qquad (37\mathrm{a})$$

$$\begin{cases} \left[\partial_t^2 - \sin^2\varphi\left(\partial_x^2 + \partial_y^2\right)\right] W(x,y,t,\varphi) = -\sin^2\varphi\left(\partial_x^2 + \partial_y^2\right)\left[\partial_z E(p,t)\big|_{z=0}\right]; \\ \quad \{x,y\} \in \mathrm{int}\,\mathrm{S}_z, \quad t > 0 \\ W(x,y,0,\varphi) = \partial_t W(x,y,t,\varphi)\big|_{t=0} = 0; \quad \{x,y\} \in \overline{\mathrm{int}\,\mathrm{S}_z}, \end{cases} \qquad (37\mathrm{b})$$

$$\left[\partial_t + \partial_z\right] E(p,t)\Big|_{z=0} = \frac{2}{\pi}\int_0^{\pi/2} \partial_t W(x,y,t,\varphi)\sin^2\varphi\,d\varphi; \quad \{x,y\} \in \overline{\mathrm{int}\,\mathrm{S}_z}, \quad t \geq 0, \qquad (38\mathrm{a})$$

$$\begin{cases} \left[\partial_t^2 - \cos^2\varphi\left(\partial_x^2 + \partial_y^2\right)\right] W(x,y,t,\varphi) = \left(\partial_x^2 + \partial_y^2\right) E(x,y,L,t); \\ \quad \{x,y\} \in \mathrm{int}\,\mathrm{S}_z, \quad t > 0 \\ W(x,y,0,\varphi) = \partial_t W(x,y,t,\varphi)\big|_{t=0} = 0; \quad \{x,y\} \in \overline{\mathrm{int}\,\mathrm{S}_z}. \end{cases} \qquad (38\mathrm{b})$$

Auxiliary scalar or vector functions $W(x,y,t,\varphi)$ are represented in the same transverse basis functions $\xi_n(x,y)$ as the fields $E(p,t)$:

$$W(x,y,t,\varphi)=\sum_n w_n(t,\varphi)\xi_n(x,y). \tag{39}$$

Formally, it is necessary to supplement the internal problems in (37b), (38b) also with boundary (on the contour S_z) conditions for functions $W(x,y,t,\varphi)$ and $\mathbf{W}(x,y,t,\varphi)$, $t\geq 0$: $W(x,y,t,\varphi)\big|_{\{x,y\}\in S_z}=0$ in the case when function $E(p,t)$ in (37), (38) corresponds to the component $E_z(p,t)$ of the field $\mathbf{E}(g,t)$ ($E(p,t)\to E_z(p,t)$), and $\left(\boldsymbol{\tau}\cdot\mathbf{W}(x,y,t,\varphi)\right)\big|_{\{x,y\}\in S_z}=0$ as $E(p,t)\to\mathbf{E}_\perp(p,t)$. These conditions are an obvious consequence of the representations (39).

As for the technique to obtain the above EACs, in the case of (38), it repeats the technique used to derive EAC (25), and during the transition from (35) to (37), the intermediate representations used earlier in obtaining EAC (24) had been changed: here $J_1(x)=(\pi)^{-1}\int_0^\pi \sin(x\sin\varphi)\sin\varphi\, d\varphi$ and

$$w_n(t,\varphi)=\alpha_n\sin\varphi\int_0^t \sin\left[\alpha_n(t-\tau)\sin\varphi\right]\partial_z v_n(0,\tau)d\tau;\quad t\geq 0,\quad 0\leq\varphi\leq\pi.$$

The final result has changed accordingly (compare (24) and (37)).

3. Spherical Boundaries in Free Space

3.1. Historically First Non-Local Exact Absorbing Condition

In 1989, A. Maykov, A. Poezd, A. Sveshnikov, and S. Yakunin published work [30], and again, they were first to formulate an exact non-local condition for electromagnetic waves propagating into free space through a spherical virtual boundary. Details of this condition's derivation are not revealed. Below we present the approach [15, 31] which leads not only to the condition from [30], but also to other conditions (see Subsection 3.2), that are preferable for application in a number of important cases.

Figure 3 shows a compact axially symmetric structure's cut by the half-plane $\phi = const$. Here, $\{\rho,\phi,z\}$ are cylindrical and $\{r,\vartheta,\phi\}$ are spherical coordinates. The pulsed axial symmetrical TE_0- ($\partial_\phi \equiv 0$, $E_\rho = E_z = H_\phi \equiv 0$) and TM_0-waves ($\partial_\phi \equiv 0$, $H_\rho = H_z = E_\phi \equiv 0$) in open structures of this type are determined by solutions of the two-dimensional scalar initial boundary value problem

$$\begin{cases} \left[-\varepsilon(g)\partial_t^2 - \sigma(g)\eta_0\partial_t + \partial_z^2 + \partial_\rho\left(\rho^{-1}\partial_\rho\rho\right)\right]U(g,t) = F(g,t); \quad t>0, \quad g \in \mathrm{Q} \\ U(g,0) = \varphi(g), \quad \partial_t U(g,0) = \psi(g); \quad g = \{\rho,z\} \in \overline{\mathrm{Q}} \\ \mathbf{E}_{\mathrm{tg}}(p,t) \text{ and } \mathbf{H}_{\mathrm{tg}}(p,t) \text{ are continuous when crossing } \mathrm{S}^{\varepsilon,\sigma} \\ \quad \text{and } \mathbf{E}_{\mathrm{tg}}(p,t)\Big|_{p=\{\rho,\phi,z\}\in\mathrm{S}} = 0, \quad U(0,z,t) = 0 \quad |z| < \infty; \quad t \geq 0. \end{cases} \tag{40}$$

Here, $U(g,t) = E_\phi(g,t)$ in the case of TE_0-waves and $U(g,t) = H_\phi(g,t)$ in the case of TM_0-waves. The analysis domain Q is a part of the half plane $\phi = const$, bounded by the contours S_ϕ. The domains $\mathrm{Q}_{\mathrm{int}} = \{g = \{r,\vartheta\} \in \mathrm{Q} : r < L\}$ and $\mathrm{Q}_{\mathrm{ext}} : \mathrm{Q} = \mathrm{Q}_{\mathrm{int}} \cup \mathrm{Q}_{\mathrm{ext}} \cup \mathrm{L}$ (free space) are separated by the virtual boundary $\mathrm{L} = \{g = \{r,\vartheta\} \in \mathrm{Q} : r = L\}$.

It is assumed that the functions $F(g,t)$, $\varphi(g)$, $\psi(g)$, $\sigma(g)$, and $\varepsilon(g)-1$ with supports compact in the closure of domain Q, satisfy the theorem on the unique solvability of the problem (40) in the Sobolev space $\mathrm{W}_2^1\left(\mathrm{Q}^T\right)$, $\mathrm{Q}^T = \mathrm{Q}\times(0,T)$, $T < \infty$ [15, 17]. Current and instantaneous sources $F(g,t)$ and $\varphi(g)$, $\psi(g)$, as well as all scattering objects defined by the piecewise constant functions $\varepsilon(g)$, $\sigma(g)$ and piecewise smooth contours S_ϕ and $\mathrm{S}_\phi^{\varepsilon,\sigma}$, are concentrated in the domain $\mathrm{Q}_{\mathrm{int}}$. $\mathrm{S} = \mathrm{S}_\phi \times [0 \leq \phi \leq 2\pi]$ is a surface of perfect conductors, and $S^{\varepsilon,\sigma} = S_{\phi}^{\varepsilon,\sigma} \times [0 \le \phi \le 2\pi]$ is a surface where the functions $\varepsilon(g)$ and $\sigma(g)$ are discontinuous.

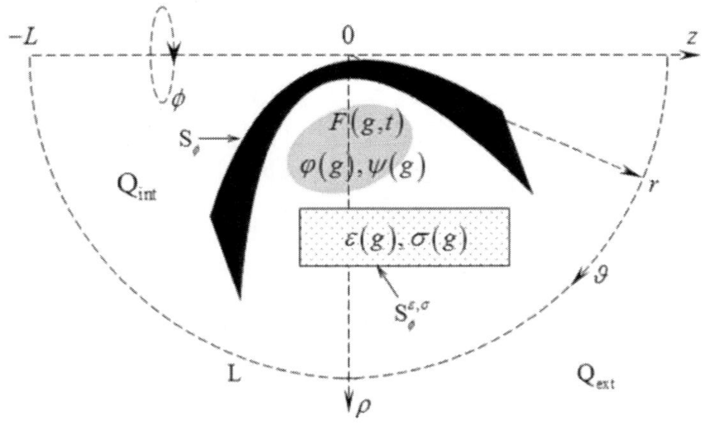

Figure 3. Geometry of the problem (40).

In the domain Q_{ext}, where wave $U(g,t)$ propagates freely and infinitely far when $t \to \infty$, the problem (40) in spherical coordinates can be written as [12, 14]:

$$\begin{cases} \left[-\partial_t^2 + r^{-1}\partial_r^2 r + r^{-2}\partial_\vartheta \left((\sin\vartheta)^{-1} \partial_\vartheta \sin\vartheta \right) \right] U(g,t) = 0; \quad t > 0, \quad g \in Q_{ext} \\ U(g,0) = 0, \quad \partial_t U(g,0) = 0; \quad g = \{r,\vartheta\} \in \bar{Q}_{ext} \\ U(r,0,t) = U(r,\pi,t) = 0; \quad r \ge L, \quad t \ge 0. \end{cases} \tag{41}$$

Incomplete separation of variables in the problem (41) allows us to represent its solution in the form

$$\begin{aligned} U(r,\vartheta,t) &= \sum_{n=1}^{\infty} u_n(r,t)\mu_n(\cos\vartheta); \quad r \ge L, \\ u_n(r,t) &= \int_0^{\pi} U(r,\vartheta,t)\mu_n(\cos\vartheta)\sin\vartheta d\vartheta, \end{aligned} \tag{42}$$

where the complete (in the space $L_2(0<\vartheta<\pi)$) orthonormal (with the weight $\sin\vartheta$) system of transverse functions $\mu_n(\cos\vartheta)$ and the system of eigenvalues λ_n corresponding to them are determined by nontrivial solutions of the homogeneous problem

$$\begin{cases} \left[d_\vartheta^2+\operatorname{ctg}\vartheta d_\vartheta-(\sin\vartheta)^{-2}+\lambda^2\right]\mu(\cos\vartheta)=0; \quad 0<\vartheta<\pi \\ \mu(\cos\vartheta)\big|_{\vartheta=0,\pi}=0, \end{cases} \tag{43}$$

and the space-time amplitudes $u_n(r,t)$ satisfy the initial boundary value problem

$$\begin{cases} \left[-\partial_t^2+\partial_r^2-r^{-2}\lambda_n^2\right]r\,u_n(r,t)=0; \quad r\ge L, \quad t>0 \\ u_n(r,0)=\partial_t u_n(r,0)=0; \quad r\ge L. \end{cases} \tag{44}$$

From (43), using the standard set of information about linearly independent solutions of the corresponding differential equation [32, 33], we obtain $\lambda=\lambda_n=\sqrt{n(n+1)}$, $n=1,2,3,...$ and $\mu_n(\cos\vartheta)=\sqrt{(2n+1)/(2n(n+1))}\,P_n^1(\cos\vartheta)$. Here, $P_n^1(...)$ are associated Legendre functions of the first kind of order n.

We now construct the exact radiation condition for the amplitudes $u_n(r,t)$ of the outgoing wave (42). Introducing notation $w_n(r,t)=ru_n(r,t)$ and taking into consideration the relation $\lambda_n^2=n(n+1)$, rewrite the equation (44) in the form

$$\left[-\partial_t^2+\partial_r^2-r^{-2}n(n+1)\right]w_n(r,t)=0; \quad r\ge L, \quad t>0. \tag{45}$$

Refer now to the integral transformation

$$\tilde{f}(\omega)=\int_L^\infty f(r)\,Z_\gamma(\omega,r)\,dr; \quad \omega\ge 0, \tag{46}$$

with the kernel $Z_\gamma(\omega,r)=r^a\left[\alpha(\omega)J_\gamma(\omega r)+\beta(\omega)N_\gamma(\omega r)\right]$ satisfying the equation $\left[\partial_r^2+(1-2a)r^{-1}\partial_r+\omega^2+\left(a^2-\gamma^2\right)r^{-2}\right]Z_\gamma(\omega,r)=0$ [18]. Here, $J_\gamma(...)$ and $N_\gamma(...)$ are the Bessel and Neumann cylindrical functions, $\alpha(\omega)$ and $\beta(\omega)$ are arbitrary functions, and a, $\operatorname{Im}a=0$ is a fixed constant.

Applying (46) to (45) with the parameters $a=1/2$ and $\gamma=n+1/2$, and taking into account that the "signal" $w_n(r,t)$ propagates in the space $r>L$ with a finite velocity, we obtain

$$\int_L^\infty\left[-\partial_t^2-\omega^2\right]w_n(r,t)Z_\gamma(\omega,r)dr-Z_\gamma(\omega,L)\partial_r w_n(L,t)+w_n(L,t)\partial_r Z_\gamma(\omega,L)=0 \quad (47)$$

($\partial_r f(L)=\partial_r f(r)\big|_{r=L}$). Simple differential equations

$$\left[\partial_t^2+\omega^2\right]\tilde{w}_n(\omega,t)=w_n(L,t)\partial_r Z_\gamma(\omega,L)-Z_\gamma(\omega,L)\partial_r w_n(L,t);\quad \omega\geq 0,\quad t>0 \quad (48)$$

with respect to the images $\tilde{w}_n(\omega,t)$ of functions $w_n(r,t)$ follow from (47).

In these equations, the values $\alpha(\omega)$ and $\beta(\omega)$ are functional parameters in the kernel $Z_\gamma(\omega,r)$ of the transform (46) and are not defined yet. Assuming $\alpha(\omega)=\sqrt{\omega}$ and $\beta(\omega)=0$, we obtain $Z_\gamma(\omega,r)=\sqrt{r\omega}J_\gamma(\omega r)$ and

$$\begin{gathered}\tilde{f}(\omega)=\int_0^\infty\left[f(r)\chi(r-L)\right]\sqrt{r\omega}J_\gamma(\omega r)dr \quad\leftrightarrow\\ \leftrightarrow\quad f(r)\chi(r-L)=\int_0^\infty\tilde{f}(\omega)\sqrt{r\omega}J_\gamma(\omega r)d\omega.\end{gathered} \quad (49)$$

The transform (46) turns into the Hankel transform [18], it is inverse to itself. The equations (48) can now be written in the form

$$\left[\partial_t^2+\omega^2\right]\tilde{w}_n(\omega,t)=g(\omega,t);$$
$$g_n(\omega,t)=\left\{w_n(L,t)\left[\left(2\sqrt{L}\right)^{-1}J_\gamma(\omega L)+\omega\sqrt{L}J_\gamma{}'(\omega L)\right]-\right.$$
$$\left.-\sqrt{L}J_\gamma(\omega L)\partial_r w_n(L,t)\right\}\sqrt{\omega},$$

where the prime on the superscript position in $J_\gamma{}'(\omega L)$ means derivative of the function $J_\gamma(\omega L)$ with respect to the full argument ωL. And the result of the integral transformation of the problem (44) is written in the form of differential equation [19]

$$\begin{aligned}\left[\partial_t^2+\omega^2\right]\tilde{w}_n(\omega,t)&=g_n(\omega,t)+\delta^{(1)}(t)\tilde{w}_n(\omega,0)+\delta(t)\partial_t\tilde{w}_n(\omega,0)=\\&=g_n(\omega,t);\quad -\infty<t<\infty\end{aligned}\tag{50}$$

(functions $\tilde{w}_n(\omega,t)$ and $g_n(\omega,t)$ are extended by zero on the half-line $t\leq 0$). Solutions $\tilde{w}_n(\omega,t)$ to the equation (50) are determined by the convolution $\tilde{w}_n(\omega,t)=[G*g_n]$, $G(\omega,t)=\chi(t)\omega^{-1}\sin\omega t$ is a fundamental solution of the operator $\left[\partial_t^2+\omega^2\right]$:

$$\tilde{w}_n(\omega,t)=\int_0^\infty G(\omega,t-\tau)g_n(\omega,\tau)d\tau=\frac{1}{2\sqrt{\omega L}}J_\gamma(\omega L)\int_0^t w_n(L,\tau)\sin\left[\omega(t-\tau)\right]d\tau+$$
$$+\sqrt{\omega L}J_\gamma{}'(\omega L)\int_0^t w_n(L,\tau)\sin\left[\omega(t-\tau)\right]d\tau-$$
$$-\frac{\sqrt{L}}{\sqrt{\omega}}J_\gamma(\omega L)\int_0^t\partial_r w_n(L,\tau)\sin\left[\omega(t-\tau)\right]d\tau.$$

Applying the inverse transformation (49) to this result, we obtain

$$
\begin{aligned}
& w_n(r,t)\chi(r-L)=\int_0^\infty \tilde{w}_n(\omega,t)\sqrt{\omega r}J_\gamma(\omega r)d\omega= \\
& =\sqrt{r}\int_0^t\int_0^\infty J_\gamma(\omega r)J_\gamma(\omega L)\sin\left[\omega(t-\tau)\right]d\omega\left[\frac{1}{2\sqrt{L}}w_n(L,\tau)-\sqrt{L}\partial_r w_n(L,\tau)\right]d\tau+ \\
& +\sqrt{rL}\int_0^t\int_0^\infty \omega J_\gamma(\omega r)J_\gamma'(\omega L)\sin\left[\omega(t-\tau)\right]d\omega\, w_n(L,\tau)d\tau.
\end{aligned}
\tag{51}
$$

Let us introduce the notation $I_1(L)=\int_0^\infty J_\gamma(\omega r)J_\gamma(\omega L)\sin\left[\omega(t-\tau)\right]d\omega$ and $I_2(L)=\int_0^\infty \omega J_\gamma(\omega r)J_\gamma'(\omega L)\sin\left[\omega(t-\tau)\right]d\omega$. In [33] it was noted that for $r>L>0$

$$
I_1(L)=\begin{cases}
0; \quad 0<t-\tau<r-L \\
\left(2\sqrt{rL}\right)^{-1}P_{\gamma-1/2}\left(\left(r^2+L^2-(t-\tau)^2\right)(2rL)^{-1}\right); \quad r-L<t-\tau<r+L \\
-\cos\gamma\pi\left(\pi\sqrt{rL}\right)^{-1}Q_{\gamma-1/2}\left(-\left(r^2+L^2-(t-\tau)^2\right)(2rL)^{-1}\right); \quad t-\tau>r+L.
\end{cases}
$$

Here, $P_\gamma(...)$ and $Q_\gamma(...)$ are the Legendre functions of the first and second kind. For values $\gamma=n+1/2$, this formula takes the form

$$
I_1(L)=\left(2\sqrt{rL}\right)^{-1}P_n(q)\chi\left[(t-\tau)-(r-L)\right]\chi\left[(r+L)-(t-\tau)\right]; \quad t-\tau>0, \tag{52}
$$

where $q=\left(r^2+L^2-(t-\tau)^2\right)(2rL)^{-1}$ and $P_n(...)$ are the Legendre polynomials. Considering that $I_2(L)=\partial_L I_1(L)$ and $d_x\chi(x)=\delta(x)$ [33], we can write also

$$
\begin{aligned}
& I_2(L)=\frac{1}{2\sqrt{rL}}\chi\left[(t-\tau)-(r-L)\right]\chi\left[(r+L)-(t-\tau)\right]\left[-\frac{P_n(q)}{2L}+\right. \\
& \left.+\partial_q P_n(q)\left(\frac{1}{r}-\frac{q}{L}\right)\right]+\frac{1}{2\sqrt{rL}}P_n(q)\left[\delta(t-\tau-r+L)+\delta(r+L-t+\tau)\right].
\end{aligned}
\tag{53}
$$

In summary, taking into account the particular values $P_n(1)=1$, $P_n(-1)=(-1)^n$ and the relation $w_n(r,t)=ru_n(r,t)$, we obtain from (51)–(53)

$$u_n(r,t)=\frac{L}{2r}\left\{\int\limits_{t-(r+L)}^{t-(r-L)}\left[\left(\frac{L-rq}{rL}\partial_q P_n(q)-\frac{1}{L}P_n(q)\right)u_n(L,\tau)-P_n(q)\partial_r u_n(L,\tau)\right]d\tau+\right.$$
$$\left.+u_n(L,t-(r-L))+(-1)^n u_n(L,t-(r+L))\right\},\quad r>L. \tag{54}$$

This is exact radiation condition (1.7) from [30]. Non-local EAC follows from (54), (42):

$$U(L,\vartheta,t)=\sum_{n=1}^{\infty}\left\{\int\limits_{t-2L}^{t}\left[\left(\frac{(t-\tau)^2}{2L^3}\partial_q P_n\left(1-\frac{(t-\tau)^2}{2L^2}\right)-\frac{1}{L}P_n\left(1-\frac{(t-\tau)^2}{2L^2}\right)\right)\times\right.\right.$$
$$\times\int\limits_0^{\pi}U(L,\vartheta_1,\tau)\mu_n(\cos\tilde{\vartheta})\sin\tilde{\vartheta}d\tilde{\vartheta}-P_n\left(1-\frac{(t-\tau)^2}{2L^2}\right)\int\limits_0^{\pi}\partial_r U(L,\tilde{\vartheta},\tau)\mu_n(\cos\tilde{\vartheta})\sin\tilde{\vartheta}d\tilde{\vartheta}\Bigg]d\tau+ \tag{55}$$
$$\left.+(-1)^n\int\limits_0^{\pi}U(L,\vartheta_1,t-(r+L))\,\mu_n(\cos\tilde{\vartheta})\sin\tilde{\vartheta}d\tilde{\vartheta}\right\}\mu_n(\cos\vartheta);\quad 0\le\vartheta\le\pi,\quad t>0.$$

3.2. Radiation Conditions Not Containing Derivatives of the Unknown Function of the Longitudinal Coordinate

Let's go back to that place in Subsection 3.1, where we defined values of the functions $\alpha(\omega)$ and $\beta(\omega)$ (between (48) and (49)), and set now $\alpha(\omega)=-N_\gamma(\omega L)$ and $\beta(\omega)=J_\gamma(\omega L)$. The function $Z_\gamma(\omega,L)=$ $=\sqrt{L}\left[\alpha(\omega)J_\gamma(\omega L)+\beta(\omega)N_\gamma(\omega L)\right]$ in (48) vanishes, (46) turns into the Weber-Orr transform [32] with the known inverse

$$\tilde{f}(\omega)=\int\limits_L^{\infty}\left[J_\gamma(\omega L)N_\gamma(\omega r)-N_\gamma(\omega L)J_\gamma(\omega r)\right]f(r)\sqrt{r}\,dr$$
$$\leftrightarrow\quad f(r)=\sqrt{r}\int\limits_0^{\infty}\frac{J_\gamma(\omega L)N_\gamma(\omega r)-N_\gamma(\omega L)J_\gamma(\omega r)}{J_\gamma^2(\omega L)+N_\gamma^2(\omega L)}\tilde{f}(\omega)\omega\,d\omega, \tag{56}$$

and the factor $\partial_r Z_\gamma(\omega,L)$ at $w_n(L,t)$ in the first term of the right-hand side of (48) takes the following form:

$$\begin{aligned}\partial_r Z_\gamma(\omega,L) &= \left(2\sqrt{L}\right)^{-1}\left[J_\gamma(\omega L)N_\gamma(\omega L) - N_\gamma(\omega L)J_\gamma(\omega L)\right] + \\ &+\omega\sqrt{L}\left[J_\gamma(\omega L)N_\gamma'(\omega L) - N_\gamma(\omega L)J_\gamma'(\omega L)\right] = \\ &= \omega\sqrt{L}\,W\left\{J_\gamma(\omega L), N_\gamma(\omega L)\right\} = 2\left(\pi\sqrt{L}\right)^{-1}.\end{aligned} \tag{57}$$

Here, $W\left\{J_\gamma(\omega L), N_\gamma(\omega L)\right\} = 2/(\pi\omega L)$ is the Wronskian [18]; $N_\gamma'(\omega L)$ and $J_\gamma'(\omega L)$ are derivatives of the functions $N_\gamma(\omega L)$ and $J_\gamma(\omega L)$ over complete argument ωL.

Applying the transform (56) to the problem (44), we turn to the differential equation (50), in which now $g_n(\omega,t) = 2w_n(L,t)/\left(\pi\sqrt{L}\right)$. The solutions

$$\begin{aligned}\tilde{w}_n(\omega,t) &= \int_0^\infty G(\omega,t-\tau)g_n(\omega,\tau)d\tau = \\ &= \frac{2}{\pi\omega^2\sqrt{L}}\left[w_n(L,t) - \int_0^t \cos\left[\omega(t-\tau)\right]\partial_\tau w_n(L,\tau)d\tau\right]; \quad \omega \ge 0, \quad t \ge 0\end{aligned} \tag{58}$$

of this equation are used to reconstruct the originals $w_n(r,t)$, $r \ge L$. Taking into account that [34]

$$\int_0^\infty \frac{J_\gamma(\omega L)N_\gamma(\omega r) - N_\gamma(\omega L)J_\gamma(\omega r)}{\omega\left[J_\gamma^2(\omega L) + N_\gamma^2(\omega L)\right]}d\omega = \frac{\pi}{2}\left(\frac{L}{r}\right)^\gamma; \quad r > L,$$

we obtain:

$$w_n(r,t) = \left(\frac{L}{r}\right)^{\gamma-1/2} w_n(L,t) - \frac{2}{\pi}\sqrt{\frac{r}{L}}\int_0^t F_\gamma(r,L,t-\tau)\partial_\tau w_n(L,\tau)d\tau; \quad r > L, \tag{59}$$

$$F_{\gamma}(r,L,t-\tau)=\int_{0}^{\infty}\frac{\cos\left[\omega(t-\tau)\right]\left[J_{\gamma}(\omega L)N_{\gamma}(\omega r)-N_{\gamma}(\omega L)J_{\gamma}(\omega r)\right]}{\omega\left[J_{\gamma}^{2}(\omega L)+N_{\gamma}^{2}(\omega L)\right]}d\omega.$$

For efficient implementation of the representation (59) it is necessary to find a simple method to calculate the function

$$\tilde{F}_{\gamma}(r,L,t-\tau)=\int_{-\infty}^{\infty}\frac{H_{\gamma}^{(1)}(zr)}{H_{\gamma}^{(1)}(zL)}\frac{\cos\left[z(t-\tau)\right]}{z}dz=2iF_{\gamma}(r,L,t-\tau)-\pi i\left(\frac{L}{r}\right)^{\gamma}. \quad (60)$$

The solution to this problem is given by the formula [14, 15]

$$\tilde{F}_{\gamma}(r,L,t-\tau)=\begin{cases}-\pi i\left[\left(\frac{L}{r}\right)^{\gamma}+\sum_{s}\frac{H_{\gamma}^{(1)}(z_{s}r)}{H_{\gamma-1}^{(1)}(z_{s}L)}\frac{\exp\left[-iz_{s}(t-\tau)\right]}{z_{s}L}\right]= \\ \quad =-\pi i\left[\left(\frac{L}{r}\right)^{\gamma}+S_{\gamma}(r,L,t-\tau)\right]; \quad \tau<t-(r-L) \\ 0; \quad \tau>t-(r-L),\end{cases} \quad (61)$$

where $H_{\gamma}^{(1)}(...)$ are the Hankel functions of the first kind, and summation is carried out over a finite number of simple poles $z=z_{s}:\operatorname{Im}z_{s}<0; s=1,2,...,n$ of the function $H_{\gamma}^{(1)}(zr)/H_{\gamma}^{(1)}(zL)$, coinciding with zeros of the function $H_{\gamma}^{(1)}(zL)$.

From (59), (60), (61), accounting the relation $w_{n}(r,t)=ru_{n}(r,t)$, we derive

$$\begin{aligned}u_{n}(r,t)&=\left(\frac{L}{r}\right)^{n+1}u_{n}(L,t-(r-L))+\sqrt{\frac{L}{r}}\int_{0}^{t-(r-L)}S_{n+1/2}(r,L,t-\tau)\,\partial_{\tau}u_{n}(L,\tau)d\tau= \\ &=\left[\left(\frac{L}{r}\right)^{n+1}+\sqrt{\frac{L}{r}}S_{n+1/2}(r,L,r-L)\right]u_{n}(L,t-(r-L))- \\ &-\sqrt{\frac{L}{r}}\int_{0}^{t-(r-L)}u_{n}(L,\tau)\partial_{\tau}S_{n+1/2}(r,L,t-\tau)d\tau; \quad r>L, \quad t\geq(r-L), \quad n=1,2,3,...\end{aligned} \quad (62)$$

that is exact radiation condition for the space-time amplitudes of a pulsed wave going into free space through the virtual spherical boundary $r = L$. From (42) and (62), exact radiation condition for waves $U(g,t)$ follows:

$$U(g,t) = \sum_{n=1}^{\infty} \left\{ \left(\frac{L}{r} \right)^{n+1} \int_0^{\pi/2} U(L,\vartheta,t-(r-L)) \mu_n(\cos\tilde{\vartheta}) \sin\tilde{\vartheta} d\tilde{\vartheta} + \right.$$
$$\left. + \sqrt{\frac{L}{r}} \int_0^{t-(r-L)} S_{n+1/2}(r,L,t-\tau) \left[\int_0^{\pi/2} \partial_\tau U(L,\tilde{\vartheta},\tau) \mu_n(\cos\tilde{\vartheta}) \sin\tilde{\vartheta} d\tilde{\vartheta} \right] d\tau \right\} \mu_n(\cos\vartheta); \quad (63)$$
$$g = \{r,\vartheta\} \in \mathrm{Q}_{\mathrm{ext}}, \quad t \geq (r-L).$$

The last two conditions were first published in [35], but their analogues for hemispherical boundaries appeared earlier in [36, 37].

When $r = L$, (62) and (63) become identities. Possible benefits of this fact when (62) and (63) are used as EACs on the virtual boundary L were considered in [11, 14]. Note that non-local conditions (62) and (63) do not contain derivative of the functions $u_n(r,t)$ and $U(g,t)$ with respect to r, and therefore can be implemented on coordinate grid $g = \{\rho, z\}$ with minimal errors.

3.3. Other Results

An original approach to the derivation of EAC (the author calls it transparent boundary condition) on a spherical boundary for solutions of the three-dimensional wave equation has been proposed by I. Sofronov in [25] and developed in [38]. The author himself describes it as an approach based on high-precision approximation of Green's functions by Fourier sums over spatial variables and sums of exponents over time [39].

In the review [40], an approach in which the operator $D[...]$ of EAC (3) is constructed for the solution $\tilde{U}(g,k)$ to the Dirichlet problem in the domain $\mathrm{Q}_{\mathrm{ext}}$ using the direct and inverse integral Laplace transforms in the so-called Dirichlet-to-Neumann mapping $\partial_r \tilde{U}(g,k) = -B[\tilde{U}(g,k)]$, $g \in \mathrm{L}$ is considered as the most frequently used. The operator $B[...]$ is determined by the Green function of this problem with a homogeneous boundary condition on L.

In [41], EACs are constructed on the basis of exact relations of the theorem on the representation of a radiation field in free space through equivalent currents on a surface, covering primary and secondary (scattering objects) sources. Unlike the conditions presented in Subsections 3.1 and 3.2, the boundary for non-local EACs proposed in [41] can be non-coordinate and the computational domain Q_{int} can be made more compact.

Special attention should be paid to [42], which presents a quasi-exact local absorbing condition that does not contain higher-order derivatives and, therefore, is easily included in standard computational schemes of grid methods. The condition is exact for solutions containing a finite number of combinations of spherical harmonics. In the general case, its error is estimated by $O\left(L^{-2N-1}\right)$, N is a number of harmonics taken into account. The approach implemented in [42] is based on the representation of solutions to the problem (44) in the form of finite series of inverse powers of r and the construction of simple recurrence formulas for coefficients of the corresponding expansions.

The result from [30] is summarized in the work of A. Maikov, A. Sveshnikov and S. Yakunin [43], where EAC is constructed for a sphere placed in medium, which is an inhomogeneous dielectric at some finite distance, with the permeability $\varepsilon(...)$ given by a function depending only on the coordinate r. Integral kernels of EAC are not defined explicitly, to define them it is necessary to solve numerically auxiliary discrete problems and Cauchy problems for differential equations of the same type. We also note that in [43], the question of the history truncation was also mentioned (information on time dependence of the perturbation $U(g,\tau)$ ($\mathbf{U}(p,\tau)$), $0 \leq \tau \leq t$ in non-local conditions). For the first time this was considered as a real way to reduce the computational resources needed for an accurate solution.

In the same paper, for the first time, the problem of the equivalence of initial open problems and problems with limited domains of analysis Q_{int} and EACs on their boundaries (modified, closed problems) was seriously discussed. The essence of the problem is as follows: (a) the formulation of the original problem assumes its unique solvability, and narrowing the corresponding solution to Q_{int} is the solution of the modified problem (by construction); (b) for the equivalence of open and closed problems, it suffices to prove the uniqueness of the solution of the latter. In [30], one sentence was given to such a proof, which states that any solution of the modified problem continued into the domain Q_{ext} with the help of exact radiation condition coincides with the solution of the original problem, and this solution is unique. A correct and complete proof of the equivalence of open and closed problems related to waveguide nodes and compact inhomogeneities in free space is given, respectively, in [15] and [44].

The generalization of the results, presented in Sections 2 and 3 of this chapter, allows to expand significantly the range of accurately solved fundamental and applied radiophysical problems. In this regard, it should be noted, first of all, the ability to effectively analyze the case of large and distant field sources in Q_{ext}, which generate the wave $U^i(g,t)$ ($\mathbf{U}^i(p,t)$) arriving onto the virtual boundary L, that was implemented without significant analytical transformations in the original EACs [12–15, 31]. The introduction of compact emitters and open resonators, so-called virtual waveguides (ports) into models of compact emitters and open resonators, became extremely important for an adequate mathematical description of real electrodynamic structures [12, 14, 15, 31]. These elements, represented in models by short segments of regular waveguides, locked on one side with a transverse metal wall and provided with accurate absorbing conditions on their virtual boundary with the domain Q_{int}, provide the same "electrodynamic picture" in the computational process as their real-world counterparts.

Problems equipped with EACs are solved in a bounded part of the space $\overline{Q_{int}}$, but the corresponding solutions, $U(g,t)$, $g \in \overline{Q_{int}}$ ($\mathbf{U}(p,t)$, $p \in \overline{Q_{int}}$) are easily recalculated to arbitrary points of the domain Q_{ext} using transport operators (exact radiation conditions) [11, 12, 14, 31]. The examples of such operators are the formulas (13), (54) and (63) in Sections 2 and 3. The latter allows, in particular, to calculate such an important for applied problems characteristic as the radiation pattern.

4. Compact Inhomogeneities of Two-Dimensional Space

4.1. Non-Local Condition for Problems in Polar Coordinates

The initial boundary value problem

$$\begin{cases} \left[-\varepsilon(g)\partial_t^2 - \sigma(g)\eta_0\partial_t + (\rho)^{-1}\partial_\rho\rho\,\partial_\rho + (\rho)^{-2}\partial_\phi^2\right]U(g,t) = F(g,t); \\ \quad g = \{\rho,\phi\} \in \mathrm{Q}, \quad t > 0 \\ U(g,0) = \varphi(g), \quad \partial_t U(g,0) = \psi(g); \quad g \in \bar{\mathrm{Q}} \\ \mathbf{E}_{\mathrm{tg}}(p,t)\big|_{p=\{x,y,z\}\in \mathrm{S}} = 0 \quad \text{and} \quad \mathbf{E}_{\mathrm{tg}}(p,t), \quad \mathbf{H}_{\mathrm{tg}}(p,t) \quad \text{are continuous} \\ \quad \text{when crossing} \quad \mathrm{S}^{\varepsilon,\sigma}, \quad U(\rho,\phi,t) = U(\rho,\phi+2\pi,t); \quad t \geq 0 \end{cases} \tag{64}$$

describes transient states of E-polarized ($\partial_x \equiv 0$, $E_\rho = E_\phi = H_x \equiv 0$, $U(g,t) = E_x(g,t)$) and H-polarized ($\partial_x \equiv 0$, $H_\rho = H_\phi = E_x \equiv 0$, $U(g,t) = H_x(g,t)$) fields, formed by sources and scattering objects concentrated in a limited part $\mathrm{Q}_{\mathrm{int}} = \{g = \{\rho,\phi\} \in \mathrm{Q} : \rho < L\}$ of the plane R^2 of polar coordinates ρ and ϕ (see Figure 4). In the domain $\mathrm{Q}_{\mathrm{ext}} = \mathrm{Q} \setminus (\mathrm{Q}_{\mathrm{int}} \cup \mathrm{L})$ the waves $U(g,t)$, which are generated in $\mathrm{Q}_{\mathrm{int}}$, are the outgoing waves crossing the virtual boundary $\mathrm{L} = \{g = \{\rho,\phi\} \in \mathrm{Q} : \rho = L\}$ only in one direction $\rho \to \infty$, and

$$\begin{cases} \left[-\partial_t^2 + (\rho)^{-1}\partial_\rho\rho\,\partial_\rho + (\rho)^{-2}\partial_\phi^2\right]U(g,t) = 0; \quad g \in \mathrm{Q}_{\mathrm{ext}}, \quad t > 0 \\ U(g,0) = \varphi(g), \quad \partial_t U(g,0) = \psi(g); \quad g \in \bar{\mathrm{Q}}_{\mathrm{ext}} \\ U(\rho,\phi,t) = U(\rho,\phi+2\pi,t); \quad t \geq 0. \end{cases} \tag{65}$$

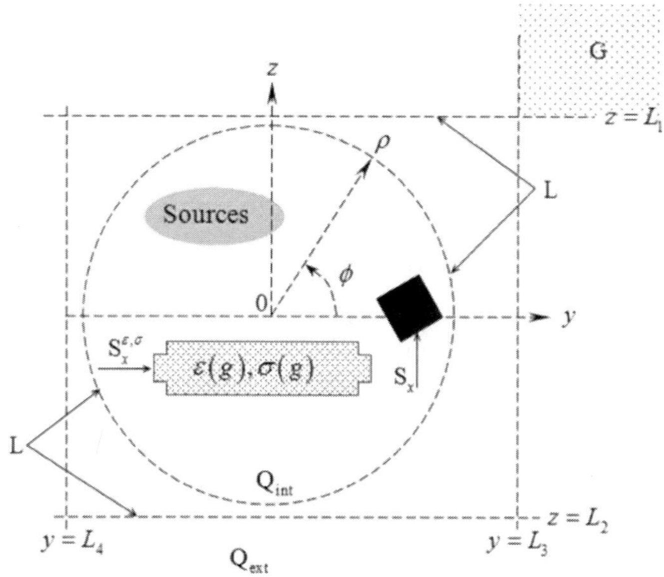

Figure 4. Geometry of the problem (64).

Separation of the variable ϕ in the problem (65) leads to the representation

$$U(\rho,\phi,t) = \sum_n u_n(\rho,t)\mu_n(\phi); \quad g \in \bar{Q}_{ext}, \quad t \geq 0, \tag{66}$$

where $\mu_n(\phi) = (2\pi)^{-1/2} \exp(in\phi)$, $n = 0, \pm 1, \pm 2, ...$ is complete (in $L_2(0 < \phi < 2\pi)$), orthonormal system of transverse functions, and space-time amplitudes $u_n(\rho,t)$ of the wave $U(\rho,\phi,t)$ are determined by the solutions of the following initial boundary value problem:

$$\begin{cases} \left[-\partial_t^2 + (\rho)^{-1}\partial_\rho \rho \partial_\rho - (\rho)^{-2} n^2\right] u_n(\rho,t) = 0; & \rho > L, \quad t > 0 \\ u_n(\rho,0) = 0, \quad \partial_t u(\rho,0) = 0; & \rho \geq L. \end{cases} \tag{67}$$

The expansion (66) puts the analysis in the space of complex functions $u_n(\rho,t)$. It is also possible (with insignificant complications in analytical representations) to continue the analysis in real space and thus preserve some of the computer resources diverted to work with complex numbers. In each approach, the derivation of absorbing conditions for the problem (64) is almost identical and is presented below (see also [14, 15]).

Multiplying (67) by $\chi(\rho-L)$ and using the Hankel transform over the variable ρ on the semi axis $\rho\geq 0$

$$\tilde{f}_n(\omega)=\int_0^\infty f_n(\rho)\rho J_{|n|}(\rho\omega)d\rho \quad\leftrightarrow\quad f_n(\rho)=\int_0^\infty \tilde{f}_n(\omega)\omega J_{|n|}(\rho\omega)d\omega \qquad (68)$$

we obtain for the images $\tilde{Z}_n(\omega,t)$ of the functions $Z_n(\rho,t)=u_n(\rho,t)\chi(\rho-L)$:

$$\begin{cases}\left[\partial_t^2+\omega^2\right]\tilde{Z}_n(\omega,t)=L\left[u_n(L,t)\partial_\rho J_{|n|}(\omega L)-\partial_\rho u_n(L,t)J_{|n|}(\omega L)\right];\\ \quad \omega>0, \quad t>0\\ \tilde{Z}_n(\omega,0)=\partial_t\tilde{Z}_n(\omega,0)=0; \quad \omega\geq 0.\end{cases} \qquad (69)$$

Here, $\partial_\rho u_n(L,t)=\partial_\rho u_n(\rho,t)\big|_{\rho=L}$, $\partial_\rho J_{|n|}(\omega L)=\partial_\rho J_{|n|}(\omega\rho)\big|_{\rho=L}$; and the following facts have been used in the derivation of (69): the well-known relation $-\omega^2\tilde{f}_n(\omega)\leftrightarrow\left[d_\rho^2+(\rho)^{-1}d_\rho-(\rho)^{-2}n^2\right]f_n(\rho)$ [18]; the sequence of obvious equalities

$$\begin{aligned}&\chi(\rho-L)\left[(\rho)^{-1}\partial_\rho\rho\,\partial_\rho\right]u_n(\rho,t)=\chi(\rho-L)\left[(\rho)^{-1}\partial_\rho+\partial_\rho^2\right]u_n(\rho,t)=\\ &=\left[(\rho)^{-1}\partial_\rho+\partial_\rho^2\right]Z_n(\rho,t)-\delta(\rho-L)\left[(\rho)^{-1}+\partial_\rho\right]u_n(\rho,t)-\\ &-\partial_\rho\left[\delta(\rho-L)u_n(\rho,t)\right],\end{aligned}$$

and the equality $\left(\partial^\alpha f,\gamma\right)=(-1)^{|\alpha|}\left(f,\partial^\alpha\gamma\right)$ defining a generalized derivative $\partial^\alpha f$ of a generalized function f [19].

The problem (69) are similar to the problem for the equation (48), considered in Subsection 3.1. Its solutions

$$\tilde{Z}_n(\omega,t)=\frac{L}{\omega}\int_0^t \sin\left[\omega(t-\tau)\right]\left[u_n(L,\tau)\partial_\rho J_{|n|}(\omega L)-\partial_\rho u_n(L,\tau)J_{|n|}(\omega L)\right]d\tau$$

after applying the inverse transform (68) lead to the representations

$$u_n(\rho,t)=L\int_0^t\left[u_n(L,\tau)\partial_r f_n(L,\rho,t-\tau)-\partial_\rho u_n(L,\tau)f_n(L,\rho,t-\tau)\right]d\tau; \quad (70)$$
$$\rho\geq L, \quad t\geq 0,$$

which allow us to trace all changes in amplitudes of the outgoing waves $U(\rho,\phi,t)$, propagating from the circle L to any circle with the radius $\rho>L$. Here,

$$f_n(r,\rho,t-\tau)=\int_0^\infty \sin\left[\omega(t-\tau)\right]J_{|n|}(\omega r)J_{|n|}(\omega\rho)d\omega \quad (71)$$

and $\partial_r f_n(L,\rho,t-\tau)=\partial_r f_n(r,\rho,t-\tau)\big|_{r=L}$. The integration in (71) is reduced to the determination of values of the Legendre functions of the first ($P_{|n|-1/2}(a)$) and second ($Q_{|n|-1/2}(-a)$) kinds with the argument $a_{r,\rho}=\left[r^2+\rho^2-(t-\tau)^2\right]/(2\rho r)$ [34]:

$$f_n(r,\rho,t-\tau)=\begin{cases}0; & 0<t-\tau<\rho-r\\ P_{|n|-1/2}(a_{r,\rho})/(2\sqrt{r\rho}); & \rho-r<t-\tau<\rho+r\\ Q_{|n|-1/2}(-a_{r,\rho})\cos(n\pi)/(\pi\sqrt{r\rho}); & \rho+r<t-\tau\end{cases}=$$
$$=\chi\left[(t-\tau)-(\rho-r)\right]Q_{|n|-1/2}(-a_{r,\rho})\cos(n\pi)/(\pi\sqrt{r\rho}); \quad 0<t-\tau.$$

The final transition in the above formula was made using the well-known properties of the Legendre functions [33]: $P_\nu(x)=\cos(\nu\pi)P_\nu(-x)-(2/\pi)\sin(\nu\pi)Q_\nu(-x)$.

Taking into account [19, 33], that the value of $Q_{|n|-1/2}(-a_{r,\rho})$ when $t-\tau=\rho-r$ is equal to $Q_{|n|-1/2}(-1)=\pi P_{|n|-1/2}(1)/2\cos(n\pi)=\pi/2\cos(n\pi)$, and $\partial_r\chi[(t-\tau)-(\rho-r)]=\delta[(t-\tau)-(\rho-r)]$, after performing the differentiation in (70) and summing up the results according to the representation (66), we obtain for $\rho\geq L$ and $t\geq 0$:

$$\begin{aligned} U(\rho,\phi,t)&=\frac{1}{2}\sqrt{\frac{L}{\rho}}U(L,\phi,t-\rho+L)+\\ &+\frac{1}{\pi}\sqrt{\frac{L}{\rho}}\sum_n(-1)^n\mu_n(\phi)\int_0^{t-(\rho-L)}\left\{\frac{u_n(L,\tau)}{2L}\left[\partial_x Q_{|n|-1/2}(-a_{L,\rho})\times\right.\right.\\ &\left.\left.\times\left(\frac{\rho^2-L^2-(t-\tau)^2}{L\rho}\right)-Q_{|n|-1/2}(-a_{L,\rho})\right]-\partial_\rho u_n(L,\tau)Q_{|n|-1/2}(-a_{L,\rho})\right\}d\tau. \end{aligned} \quad (72)$$

On the virtual boundary $\rho=L$, EAC resulting from (72) takes the form

$$U(L,\phi,t)=\frac{2}{\pi}\sum_n(-1)^n\mu_n(\phi)\left[\int_0^t\left[u_n(L,\tau)\xi_n(t-\tau)-\partial_\rho u_n(L,\tau)\eta_n(t-\tau)\right]d\tau\right]; \quad (73)$$
$$0\leq\phi\leq 2\pi,\quad t\geq 0.$$

Here, $\xi_n(t-\tau)=\left[2\partial_x Q_{|n|-1/2}(-a_{L,L})(a_{L,L}-1)-Q_{|n|-1/2}(-a_{L,L})\right](2L)^{-1}$, $\eta_n(t-\tau)= =Q_{|n|-1/2}(-a_{L,L})$ and $\partial_x Q_{|n|-1/2}(-a)=\partial_x Q_{|n|-1/2}(x)\big|_{x=-a}$. The conditions (72), (73) were first published in [11, 35], but similar conditions for subsurface sounding problems (conditions on semicircles $0\leq\phi\leq\pi$ terminated with a perfectly conducting surface in the plane $z=0$) were constructed earlier in [45].

4.2. Local Condition for Problems in Cartesian Coordinates

The problem (64) in the rectangular Cartesian coordinate system $\{x, y, z\}$, $\partial_x \equiv 0$ can be written in the following form:

$$\begin{cases} \left[-\varepsilon(g)\partial_t^2 - \sigma(g)\eta_0\partial_t + \partial_z^2 + \partial_y^2\right]U(g,t) = F(g,t); \quad g = (y,z) \in \mathrm{Q}, \quad t > 0 \\ U(g,0) = \varphi(g), \quad \partial_t U(g,0) = \psi(g); \quad g \in \bar{\mathrm{Q}} \\ \mathbf{E}_{\mathrm{tg}}(p,t)\big|_{p=\{x,y,z\}\in \mathrm{S}} = 0 \quad \text{and} \quad \mathbf{E}_{\mathrm{tg}}(p,t), \quad \mathbf{H}_{\mathrm{tg}}(p,t) \\ \quad \text{are continuous when crossing} \quad \mathrm{S}^{\varepsilon,\sigma}; \quad t \geq 0. \end{cases} \tag{74}$$

The functions $F(g,t)$, $\varphi(g)$, $\psi(g)$, $\sigma(g)$ and $\varepsilon(g)-1$ are finite in Q. Their supports at all considered moments of time $0 \leq t \leq T$ belong to the closure of the domain $\mathrm{Q}_{\mathrm{int}} = \{g = \{y,z\} \in \mathrm{Q}: L_4 < y < L_3; L_2 < z < L_1\}$ (see Figure 4). Above (below, to the right, to the left) of the virtual boundary $z = L_1$ ($z = L_2$, $y = L_3$, $y = L_4$), there are no sources and scattering objects. The function $U(g,t)$ here corresponds to an outgoing wave crossing the corresponding boundary only in one direction and satisfies the homogeneous problem (74) with $\varepsilon(g)-1 = \sigma(g) \equiv 0$. Applying the Fourier transform to this function

$$u_y(\lambda,z,t) = \frac{1}{2\pi}\int_{-\infty}^{\infty} U(y,z,t)\mathrm{e}^{i\lambda y}\,dy \leftrightarrow U(y,z,t) = \int_{-\infty}^{\infty} u_y(\lambda,z,t)\mathrm{e}^{-i\lambda y}\,d\lambda,$$

$$u_z(y,\mu,t) = \frac{1}{2\pi}\int_{-\infty}^{\infty} U(y,z,t)\mathrm{e}^{i\mu z}\,dz \leftrightarrow U(y,z,t) = \int_{-\infty}^{\infty} u_z(y,\mu,t)\mathrm{e}^{-i\mu z}\,d\mu$$

and using the approach to homogeneous problems for one-dimensional Klein-Gordon equations described in Subsection 2.1, we obtain:

$$\begin{aligned} &[\partial_t \pm \partial_z]u_y(\lambda,z,t) = -\lambda\int_0^t \frac{J_1(\lambda(t-\tau))}{t-\tau}u_y(\lambda,z,\tau)d\tau; \quad \begin{cases} z \geq L_1 \\ z \leq L_2 \end{cases}, \\ &\left[\partial_t \pm \partial_y\right]u_z(y,\mu,t) = -\mu\int_0^t \frac{J_1(\mu(t-\tau))}{t-\tau}u_z(y,\mu,\tau)d\tau; \quad \begin{cases} y \geq L_3 \\ y \leq L_4 \end{cases}. \end{aligned} \tag{75}$$

We now turn from non-local EACs (75), which are written in terms of Fourier amplitudes of the field $U(g,t)$ and reduce the analysis domain of the problem (74) to the band $L_2 < z < L_1$ or $L_4 < y < L_3$, to local relations in time and space. Following the approach from Subsection 3.2, we first use the integral Poisson formula [18] $J_1(x) = 2x(\pi)^{-1}\int_0^{\pi/2}\cos(x\cos\varphi)\sin^2\varphi d\varphi$, and then the substitutions

$$v_y = -\lambda\int_0^t \frac{\sin(\lambda(t-\tau)\cos\varphi)}{\cos\varphi}u_y\,d\tau, \quad \partial_t v_y = -\lambda^2\int_0^t \cos(\lambda(t-\tau)\cos\varphi)u_y\,d\tau \tag{76}$$

and the fundamental solution $G(a,t) = \chi(t)\sin(at)a^{-1}$ of the operator $D(a) \equiv \left[d^2/dt^2 + a^2\right]$ [12, 14] for the transition from the integral forms (76) to the equivalent differential form

$$\begin{cases} \left[\partial_t^2 + \lambda^2\cos^2\varphi\right]v_y = -\lambda^2 u_y; \quad t > 0 \\ \partial_t v_y\big|_{t=0} = v_y\big|_{t=0} = 0 \end{cases}.$$

The result is:

$$[\partial_t \pm \partial_z]u_y(\lambda,z,t) = \frac{2}{\pi}\int_0^{\pi/2}\partial_t v_y(\lambda,z,t,\varphi)\sin^2\varphi d\varphi; \quad t \geq 0, \quad \begin{cases} z \geq L_1 \\ z \leq L_2 \end{cases}, \tag{77a}$$

$$\begin{cases} \left[\partial_t^2 + \lambda^2\cos^2\varphi\right]v_y(\lambda,z,t,\varphi) = -\lambda^2 u_y(\lambda,z,t); \quad t > 0 \\ \partial_t v_y(\lambda,z,t,\varphi)\big|_{t=0} = v_y(\lambda,z,0,\varphi) = 0, \end{cases} \tag{77b}$$

$$\left[\partial_t \pm \partial_y\right] u_z(y,\mu,t) = \frac{2}{\pi}\int_0^{\pi/2} \partial_t v_z(y,\mu,t,\varphi)\sin^2\varphi d\varphi; \quad t \ge 0, \quad \begin{cases} y \ge L_3 \\ y \le L_4 \end{cases}, \tag{78a}$$

$$\begin{cases} \left[\partial_t^2 + \mu^2\cos^2\varphi\right] v_z(y,\mu,t,\varphi) = -\mu^2 u_z(y,\mu,t); \quad t > 0 \\ \partial_t v_z(y,\mu,t,\varphi)\big|_{t=0} = v_z(y,\mu,0,\varphi) = 0. \end{cases} \tag{78b}$$

Let $V_1(g,t,\varphi) = \int_{-\infty}^{\infty} v_y(\lambda,z,t,\varphi)\exp(-i\lambda y)d\lambda$ and $V_2(g,t,\varphi) = = \int_{-\infty}^{\infty} v_z(y,\mu,t,\varphi)\exp(-i\mu z)d\mu$. Then, after returning in (77), (78) to the originals (by applying the inverse Fourier transforms), local EACs, which reduce the analysis domain of the problem (74) to a strip in the plane $y0z$, are obtained in the form

$$[\partial_t \pm \partial_z] U(g,t) = \frac{2}{\pi}\int_0^{\pi/2} \partial_t V_1(g,t,\varphi)\sin^2\varphi d\varphi; \quad |y| \le \infty, \quad t \ge 0, \quad \begin{cases} z \ge L_1 \\ z \le L_2 \end{cases}, \tag{79a}$$

$$\begin{cases} \left[\partial_t^2 - \cos^2\varphi\, \partial_y^2\right] V_1(g,t,\varphi) = \partial_y^2 U(g,t); \quad |y| < \infty, \quad t > 0 \\ \partial_t V_1(g,t,\varphi)\big|_{t=0} = V_1(g,0,\varphi) = 0; \quad |y| \le \infty, \end{cases} \tag{79b}$$

$$\left[\partial_t \pm \partial_y\right] U(g,t) = \frac{2}{\pi}\int_0^{\pi/2} \partial_t V_2(g,t,\varphi)\sin^2\varphi d\varphi; \quad |z| \le \infty, \quad t \ge 0, \quad \begin{cases} y \ge L_3 \\ y \le L_4 \end{cases}, \tag{80a}$$

$$\begin{cases} \left[\partial_t^2 - \cos^2\varphi\, \partial_z^2\right] V_2(g,t,\varphi) = \partial_z^2 U(g,t); \quad |z| < \infty, \quad t > 0 \\ \partial_t V_2(g,t,\varphi)\big|_{t=0} = V_2(g,0,\varphi) = 0; \quad |z| \le \infty. \end{cases} \tag{80b}$$

Each of the four equations (79), (80) is EAC, which narrows the analysis domain to the half-plane $z < L_1$, $z > L_2$, $y < L_3$ or $y > L_4$. Closed differential problems (Cauchy problems) with respect to the functions $V_1(g,t,\varphi)$ (the coordinate z here plays the role of a numerical parameter) and $V_2(g,t,\varphi)$ (y is a numerical parameter) in these half-planes are formulated correctly. When narrowing an analysis domain to a rectangle, all four equations (79), (80) should be taken into account, and the closed differential problems should be supplemented with conditions at the ends of the intervals, namely at the points of intersection of the boundaries $z=const$ and $y=const$. There are several ways to solve the problem of corner points, which differ in the technique of analytical transformations. Let us stop on one of them, apparently, the most transparent.

We first take into consideration the equations with a plus sign from (79) and (80). The quadrant of the plane, where they simultaneously hold, is filled with dots in Figure 4. Select the domain $\mathrm{G}=\{g=\{y,z\}: L_3<y<L_3+2\pi;\ L_1<z<L_1+2\pi\}$ in this quadrant and use here the following representation of the function $f(g,t)\in \mathrm{W}_2^1(\mathrm{G}^\infty,\beta)=\{f(g,t): f(g,t)\exp(-\beta t)\in \mathrm{W}_2^1(\mathrm{G}^\infty)\}$, $\mathrm{Re}\, s\geq\beta\geq 0$:

$$f(g,t)=\frac{1}{2\pi i}\int\limits_{\alpha-i\infty}^{\alpha+i\infty}\sum_{n,m=-\infty}^{\infty}\tilde{f}(n,m,s)e^{i(ny+mz)+st}\,ds \leftrightarrow$$
$$\leftrightarrow \quad \tilde{f}(n,m,s)=\frac{1}{4\pi^2}\int\limits_0^\infty\int\limits_{L_1}^{L_1+2\pi}\int\limits_{L_3}^{L_3+2\pi} f(g,t)e^{-i(ny+mz)-st}\,dydzdt. \tag{81}$$

For the amplitudes $\tilde{u}=\tilde{u}(n,m,s)$, $\tilde{v}_j(\varphi)=\tilde{v}_j(n,m,s,\varphi)$ and $\tilde{w}_j(\varphi)=\tilde{w}_j(n,m,s,\varphi)$, $j=1,2$ of the functions $U(g,t)$, $V_j(g,t,\varphi)$ and $W_j(g,t,\varphi)=V_j(g,t,\varphi)\cos^2\varphi+U(g,t)$ the following relations are obtained:

$$(s+im)\tilde{u}=\frac{2s}{\pi}\int\limits_0^{\pi/2}\sin^2\varphi\,\tilde{v}_1 d\varphi,\quad (s+in)\tilde{u}=\frac{2s}{\pi}\int\limits_0^{\pi/2}\sin^2\varphi\,\tilde{v}_2 d\varphi,$$
$$\tilde{w}_1=\frac{s^2}{s^2+n^2\cos^2\varphi}\tilde{u},\quad \tilde{w}_2=\frac{s^2\tilde{u}}{s^2+m^2\cos^2\varphi},\quad \tilde{v}_1=-\frac{n^2\tilde{u}}{s^2+n^2\cos^2\varphi}, \tag{82}$$
$$\tilde{v}_2=-\frac{m^2\tilde{u}}{s^2+m^2\cos^2\varphi},\quad s^2+m^2+n^2=0.$$

Consider now the function

$$
\begin{aligned}
& in\tilde{w}_1(\varphi) = in\tilde{u}\frac{s^2}{s^2+n^2\cos^2\varphi} = \frac{s^2}{s^2+n^2\cos^2\varphi}\left[-s\tilde{u}+\frac{2s}{\pi}\int_0^{\pi/2}\sin^2\gamma\tilde{v}_2(\gamma)d\gamma\right] = \\
& = -s\tilde{u}\frac{s^2}{s^2+n^2\cos^2\varphi}\left[1+\frac{2\sin^2\varphi}{\pi}\int_0^{\pi/2}\frac{\sin^2\gamma}{\cos^2\varphi+\sin^2\varphi\cos^2\gamma}d\gamma\right]+ \\
& +\frac{2}{\pi}\int_0^{\pi/2}s\tilde{u}\frac{s^2}{s^2+m^2\cos^2\gamma}\times\frac{\sin^2\gamma}{\cos^2\varphi+\sin^2\varphi\cos^2\gamma}d\gamma = \\
& = -\frac{s}{\cos\varphi}\tilde{w}_1(\varphi)+\frac{2}{\pi}\int_0^{\pi/2}s\tilde{w}_2(\gamma)\frac{\sin^2\gamma}{\cos^2\varphi+\sin^2\varphi\cos^2\gamma}d\gamma.
\end{aligned}
\tag{83}
$$

In the transformations (83), the equations (82) and the equality

$$
\frac{s^2}{s^2+an^2}\times\frac{s^2}{s^2+bm^2} = \frac{a}{a+(1-a)b}\times\frac{s^2}{s^2+an^2}+\frac{b}{a+(1-a)b}\times\frac{s^2}{s^2+bm^2},
$$

valid for such s, m and n, that $s^2+m^2+n^2=0$ have been used. Returning in (83) to the originals, we get

$$
\left[\partial_t+\cos\varphi\,\partial_y\right]W_1(g,t,\varphi) = \frac{2\cos\varphi}{\pi}\int_0^{\pi/2}\frac{\sin^2\gamma\,\partial_t W_2(g,t,\gamma)}{\cos^2\varphi+\sin^2\varphi\cos^2\gamma}d\gamma;\quad z\ge L_1,\quad y\ge L_3.
$$

The upper boundaries $z=L_1+2\pi$ and $y=L_3+2\pi$ of the domain G, where this equation is valid, are not set here. Obviously, this is not necessary, since the dimensions of the domain G can be arbitrary.

The repetition of the above operations, but with the function $im\tilde{w}_2(\varphi)$ as the object of analysis, leads to the equation

$$
\left[\partial_t+\cos\varphi\,\partial_z\right]W_2(g,t,\varphi) = \frac{2\cos\varphi}{\pi}\int_0^{\pi/2}\frac{\sin^2\gamma\,\partial_t W_1(g,t,\gamma)}{\cos^2\varphi+\sin^2\varphi\cos^2\gamma}d\gamma;\quad z\ge L_1,\quad y\ge L_3.
$$

Relations that connect the auxiliary functions $W_1(g,t,\varphi)$ and $W_2(g,t,\varphi)$ in all four domains similar to G, solve the problem of corner points. They provide the possibility to correctly set closed initial boundary value problems with

EACs (79), (80) on the finite segments of the outer boundary L of the rectangular region Q_{int}. The corresponding complete system of equations, reducing the analysis domain Q of the open problem (74) to the bounded domain Q_{int}, takes the form:

$$\left[\partial_t \pm \partial_z\right]U(g,t) = \frac{2}{\pi}\int_0^{\pi/2} \partial_t V_1(g,t,\varphi)\sin^2\varphi d\varphi; \quad L_4 \le y \le L_3, \quad t \ge 0, \quad \begin{cases} z = L_1 \\ z = L_2 \end{cases}, \tag{84a}$$

$$\begin{cases} \left[\partial_t^2 V_1(g,t,\varphi) - \partial_y^2 W_1(g,t,\varphi)\right] = 0; \quad L_4 < y < L_3, \quad t > 0 \\ \partial_t V_1(g,0,\varphi) = V_1(g,0,\varphi) = 0; \quad L_4 \le y \le L_3, \end{cases} \tag{84b}$$

$$\left[\partial_t \pm \partial_y\right]U(g,t) = \frac{2}{\pi}\int_0^{\pi/2} \partial_t V_2(g,t,\varphi)\sin^2\varphi d\varphi; \quad L_2 \le z \le L_1, \quad t \ge 0, \quad \begin{cases} y = L_3 \\ y = L_4 \end{cases}, \tag{84c}$$

$$\begin{cases} \left[\partial_t^2 V_2(g,t,\varphi) - \partial_z^2 W_2(g,t,\varphi)\right] = 0; \quad L_2 < z < L_1, \quad t > 0 \\ \partial_t V_2(g,0,\varphi) = V_2(g,0,\varphi) = 0; \quad L_2 \le z \le L_1, \end{cases} \tag{84d}$$

$$\left.\begin{cases} \left[\partial_t \pm \cos\varphi\, \partial_y\right]W_1(g,t,\varphi) = \dfrac{2\cos\varphi}{\pi}\displaystyle\int_0^{\pi/2} \frac{\sin^2\gamma\, \partial_t W_2(g,t,\gamma)}{\cos^2\varphi + \sin^2\varphi\cos^2\gamma}d\gamma \\ \left[\partial_t \pm \cos\varphi\, \partial_z\right]W_2(g,t,\varphi) = \dfrac{2\cos\varphi}{\pi}\displaystyle\int_0^{\pi/2} \frac{\sin^2\gamma\, \partial_t W_1(g,t,\gamma)}{\cos^2\varphi + \sin^2\varphi\cos^2\gamma}d\gamma \end{cases}\right.; \quad t \ge 0, \tag{84e}$$

$$\begin{Bmatrix} + \\ + \end{Bmatrix} \to g = \{L_3, L_1\}, \quad \begin{Bmatrix} + \\ - \end{Bmatrix} \to \{L_3, L_2\}, \quad \begin{Bmatrix} - \\ + \end{Bmatrix} \to \{L_4, L_1\}, \quad \begin{Bmatrix} - \\ - \end{Bmatrix} \to \{L_4, L_2\}.$$

Only together, the formulas (84) determine local EAC for the entire coordinate virtual boundary L. The equations (84e) in this EAC play the role of boundary conditions in the closed initial boundary value problems (84b) and (84d). Here, the notation $\begin{Bmatrix} + \\ + \end{Bmatrix} \to g = \{L_3, L_1\}$ specify the rule for choosing signs in the upper and lower equations for different corner points $g = \{y, z\}$. EAC (84) was first published in [37].

5. Increasing Computational Efficiency. Approximation of Absorbing Conditions

The authors of [43] were the first to pay attention to the problem of efficient implementation of non-local EACs. To calculate their right-hand side at every moment of time t, it is necessary to store information about the field values on the virtual boundary at all previous moments of time $0 \le \tau \le t$. This leads to the growth of required memory volume with every time step. The simplest way to handle this is a truncation of the convolution integral by τ (history truncation), i.e., substitution of the lower limit $\tau = 0$ for $t - t_0$, where t_0 is a certain constant independent from t. Such a modification leads to approximate conditions, the correctness of which must be proved.

But, as was shown in [46], this approach does not lead to satisfactory results in the case of non-local EAC (14) due to a weak decrease of the Bessel function $J_0(t)$ for sufficiently large t. As an alternative, the algorithm of constant (on each time interval) updates of this EAC is proposed. With this approach, the history is stored only within the macro step, which could include from tens up to hundreds time steps, and EAC is updated via solving the problem (7) with non-uniform initial data. EAC, of course, become more complicated, integrals of unknown functions along the longitudinal coordinate z appear on segments of two macro-step length, but the number of arithmetic operations and computation errors does not increase with increasing observation time. But the paper [46] provides no information about the actual reduction of the required memory.

In the papers [47, 48], it is suggested to approximate the kernel $K(t, \lambda_n)$ of non-local EAC (18) by the function $\sum_{j=1}^{J(n)} c_j(\lambda_n) \exp(-d_j(\lambda_n)t)$. These results in a partial temporal localization of EAC: instead of a complete history of the space-time amplitudes, it is necessary to store information about some auxiliary functions $I_j(t, \lambda_n)$ at several previous moment of time only. This is a promising way to approximate EACs, providing the level of introduced relative errors is kept at the level of discretization's errors of the problem itself.

The paper [49] is dedicated to accurate discrete nonstationary radiation conditions in a spherical coordinate system. In numerical experiments discrete analogs of analytical radiation conditions are used, and sometimes it leads to difficulties in the control of mismatch of these discrete analogs with computational schemes used to solve the Maxwell's equations. Exact discrete nonstationary radiation conditions deal with this issue; their specific form is determined by the properties of computational scheme used to discretize the problem in the analysis domain bounded by a virtual boundary.

The transition to high-order finite-difference computational schemes demands the increase in accuracy of discrete analogs of absorbing conditions. In [24], this issue was analyzed within the framework of one simple waveguide problem with the analysis domain bounded by EAC (18). Numerical test confirmed the expected orders of accuracy and stability of computational schemes for moderate durations of simulations.

Our experience in implementing non-local and local EACs in solving a wide range of physical and applied problems of electrodynamics [11–15, 29, 50–78] allowed developing a number of important proposals for computational practice. Some of them are briefly described in the following subsections of this chapter.

5.1. History Truncation in Non-Local Conditions for Waveguide Nodes

Let us return to EACs (14), (16), (18) and rewrite them in this form:

$$u(z,t) = -\int_0^t J_0\left[\lambda(t-\tau)\right]\tilde{u}(z,\tau)\,d\tau, \tag{85}$$

$$(\lambda)^{-1}\left[\tilde{u}(z,t) + u^{(1)}(z,t)\right] = \int_0^t J_1\left[\lambda(t-\tau)\right]\tilde{u}(z,\tau)\,d\tau, \tag{86}$$

$$-(\lambda)^{-2}\left[\tilde{u}(z,t) + u^{(1)}(z,t)\right] = \int_0^t K_1(t-\tau)u(z,\tau)\,d\tau. \tag{87}$$

Here, we omitted the bottom index in λ_n, $u_n(z,\tau)$, and introduced the following notation: $\tilde{u}(z,t) = \pm\partial_z u(z,t)$ (the upper sign corresponds to the waves propagating in the direction of increasing values, the lower sign corresponds to the waves propagating in the opposite direction), $u^{(m)}(z,t) = \partial_t^m u(z,t)$, and $K_m(t) = J_m(\lambda t)(\lambda t)^{-m}$.

We have successfully implemented EAC (85) in our early programs, but the approximation of the first derivative with three point (instead of two) often led to the collapse of the finite-difference computational scheme. EACs (86), (87) are perfect. The only problem is the amount of computer resources required for their accurate implementation. The convolution values in these EACs are close to zero and much less than values of the derivatives $\tilde{u}(z,t)$ and $u^{(1)}(z,t)$. The further the spectrum of the signal $u(z,t)$ lies from the cutoff frequency of the relevant mode, the smaller the contribution of convolution is and the implementation of EACs (86), (87) is closer to the exact. But it is impossible to neglect the convolution magnitude, even in the cases when its contribution affects only the fifth or sixth decimal place. These conclusions are obtained via analyzing and solving test problems.

The kernel $K_1(t)$ of EAC (87) decreases with the growing t by an order of magnitude faster than the kernels of EACs (85), (86): $K_1(t) = O(t^{-1.5})$ for $t \to \infty$. Therefore, (87) seems more promising for history truncation. Twice differentiating (87) in time, we obtain EAC with the kernel decreasing as $O(t^{-2.5})$:

$$
\begin{aligned}
&-\left(3\lambda^4\right)^{-1}\left[\tilde{u}^{(2)}(z,t) + u^{(3)}(z,t)\right] - \left(6\lambda^2\right)^{-1}\left[2\tilde{u}(z,t) + 3u^{(1)}(z,t)\right] = \\
&= \int_0^t K_2(t-\tau)u(z,\tau)\,d\tau.
\end{aligned}
\tag{88}
$$

Apparently, the information from the distant history, for which the weighting factors were reduced in the convolution of (88), is moved outside of the integral with high-order derivatives. Differentiating two more times, we get

$$-(15\lambda)^{-6}\left[\tilde{u}^{(4)}(z,t)+u^{(5)}(z,t)\right]-\left(30\lambda^{4}\right)^{-1}\left[4\tilde{u}^{(2)}(z,t)+5u^{(3)}(z,t)\right]- \\ -\left(120\lambda^{2}\right)^{-1}\left[8\tilde{u}(z,t)+15u^{(1)}(z,t)\right]=\int_{0}^{t}K_{3}(t-\tau)u(z,\tau)d\tau. \quad (89)$$

It is EAC with the kernel decreasing as $O\left(t^{-3.5}\right)$. Unfortunately, (88), (89) and other similar EACs do not justify hopes caused by a rapid decrease of their kernels. EAC (88) fits correctly into the difference scheme, but breaks it up when there is an attempt of the history truncation; (89) leads to the collapse of a finite-difference scheme even without history truncation. EACs, which shown themselves as the best ones, have kernels decreasing as $O\left(t^{-1.5}\right)$. The only exception is EAC $-u(z,t)=\int_0^t K_1(t-\tau)\left[\tilde{u}(z,\tau)-u^{(1)}(z,\tau)\right]d\tau$ that follows from (87). It looks very elegant, but it crashes the computational scheme both with the history truncation and when it is fully preserved.

All these conclusions followed from the analysis of computational experiments. Let us touch one of these experiments in more details. Consider a segment $0\le z\le 5$ of a regular waveguide, the EACs being tested are employed on its virtual boundaries L_1 ($z=0$) and L_2 ($z=5$). The pulse wave $u^i(z,t)$ (the case of E-polarization is studied) arrives onto the boundary L_1, propagating in the direction of growing z (see Figure 5a). It should be noted that here and further on, $\tilde{f}(k)=\int_0^T f(t)\exp(ikt)dt$ and T is the observation time. The transverse eigenvalue is $\lambda=3$. The frequency band (of wave numbers k) covered by the pulse $u^i(z,t)$ includes the cut-off point of the waveguide, which is quite severe testing mode. The discretization steps are $h_z=0.005$ in space and $h_t=0.002$ in time.

Figure 5b shows the amplitudes $u(0,t)$ and $u(5,t)$ of the reflected and transmitted waves obtained using a well-adjusted absorbing condition (the amplitude of the reflection coefficient is about 10^{-6}). This is a reference solution. In Figures 6 and 7, the reference solution is compared with the solution obtained using the condition being tested.

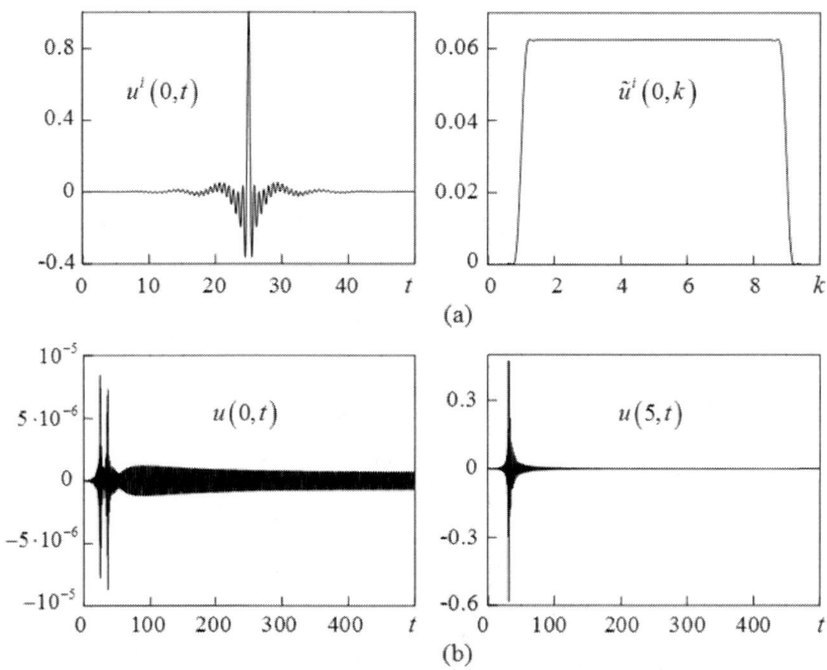

Figure 5. (a) The pulse $u^i(0,t)$ and its spectral amplitudes $\left|\tilde{u}^i(0,k)\right|$. (b) Reflected and transmitted pulses. Reference solution.

The use of EACs with kernels decreasing as $O\left(t^{-0.5}\right)$ and $O\left(t^{-1.5}\right)$ leads to solutions coinciding quite well with the reference one. Overall, 14 conditions were tested. When truncating the history, the amplitudes of the transmitted and reflected waves begin to increase (see Figure 6). On the boundary L_2, the difference between the reference solution and the solution obtained with the history truncation in the condition being tested may seem insignificant. However, on the boundary L_1, this difference is unacceptably large. Increasing the length of the stored history improves the situation, but not significantly. To get reliable results, one still needs to keep the entire history.

None of 15 EACs with kernels decreasing as $O\left(t^{-2.5}\right)$ was able to pass the test with the history truncation (see, for example, Figure 7). Moreover, some of these conditions led to the collapse of the standard computational scheme even with the untruncated history.

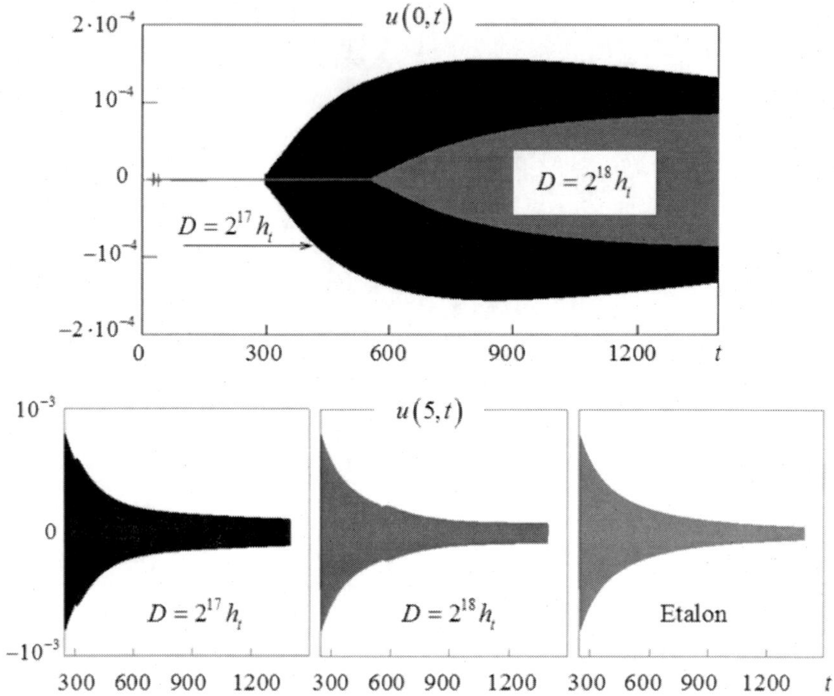

Figure 6. Reflected and transmitted pulses for different lengths D of the stored history: $D = 2^{17} h_t = 262.144$ and $D = 2^{18} h_t = 524.288$. EAC being tested is $-(\lambda)^{-1}\left[\tilde{u}(z,t)+u^{(1)}(z,t)\right]-u(z,t)=\int_0^t K_1(t-\tau)\left[\tilde{u}(z,\tau)-u^{(1)}(z,\tau)+\lambda u(z,\tau)\right]d\tau.$

5.2. History Thinning

A significant reduction of required computer resources could be achieved via saving the signal history not at every grid step, but only at each N-th. Accordingly, the convolution is calculated not at every time layer of the computational scheme, but only at each N-th. At intermediate times, the convolution value is restored by interpolation formulas. For small N (between 2 and 10) and λ, it works perfectly. For large transverse numbers λ (for higher modes), the Bessel functions in the kernels of non-local EACs strongly oscillate, and the signal's history thinning significantly affects the convolution accuracy. The whole computational scheme may collapse. At the same time, an interesting computational effect can be observed: the virtual boundary begins to radiate an increasing noise at high modes. A reasonable strategy to handle this is to process low modes with relatively large N, and to suppress high modes by moving the virtual boundary away into the waveguide to a distance of its several transverse dimensions.

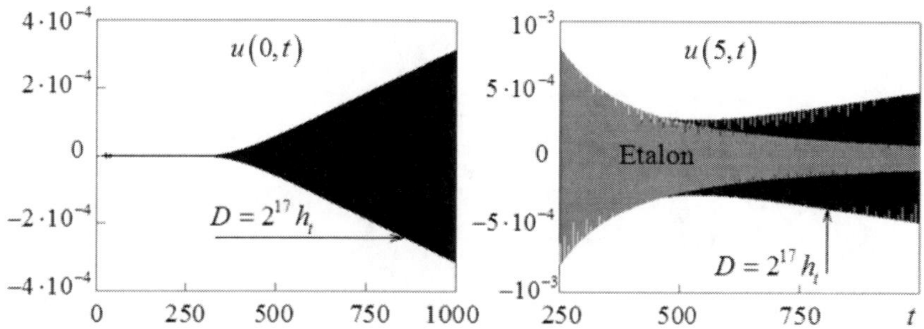

Figure 7. Reflected and transmitted pulses for the stored history length $D=2^{17}h_t=262.144$. Instability when truncating history in EAC $-(2\lambda^2)^{-1}\left[\tilde{u}^{(1)}(z,t)+u^{(2)}(z,t)\right]-u(z,t)=3\int_0^t K_2(t-\tau)\left[\tilde{u}(z,\tau)-u^{(1)}(z,\tau)\right]d\tau.$

We consider the details of the approach using EAC (87) and the case when the waveguide is filled with a medium with parameters $\varepsilon \geq 1$, $\mu \geq 1$ and $\sigma \geq 0$ [14], and the wave $u(z,t)$ propagates towards the growing z:

$$c\tilde{u}(z,t)+u^{(1)}(z,t)+su(z,t)=\int_0^t K(t-\tau)u(z,\tau)d\tau;$$
$$K(t)=-\eta\exp(-st)J_1(\eta t)(t)^{-1}, \quad c=(\varepsilon\mu)^{-1/2}, \quad s=\sigma\eta_0(2\varepsilon)^{-1}, \qquad (90)$$
$$\eta=\sqrt{\lambda^2(\varepsilon\mu)^{-1}-s^2}, \quad \lambda^2 \geq \varepsilon\mu s^2.$$

Let us turn in (90) to the grid functions $u(z_i,t_k)\to u\big|_i^k$, $K(t_k)\to K_k$, $z_i=ih_z$, $t_k=kh_t$ and use the following difference approximations of the first derivatives with respect to 4 and 5 points [20]:

$$\begin{aligned}\tilde{u}(z_i,t_k)&\to\frac{11u\big|_i^k-18u\big|_{i-1}^k+9u\big|_{i-2}^k-2u\big|_{i-3}^k}{6h_z},\\ u^{(1)}(z_i,t_k)&\to\frac{11u\big|_i^k-18u\big|_i^{k-1}+9u\big|_i^{k-2}-2u\big|_i^{k-3}}{6h_t},\end{aligned}\tag{91}$$

$$\begin{aligned}\tilde{u}(z_i,t_k)&\to\frac{25u\big|_i^k-48u\big|_{i-1}^k+36u\big|_{i-2}^k-16u\big|_{i-3}^k+3u\big|_{i-4}^k}{12h_z},\\ u^{(1)}(z_i,t_k)&\to\frac{25u\big|_i^k-48u\big|_i^{k-1}+36u\big|_i^{k-2}-16u\big|_i^{k-3}+3u\big|_i^{k-4}}{12h_t}.\end{aligned}\tag{92}$$

The formulas (91) have already been implemented in the actual codes. The formulas (92) are to be used with difference schemes of the 4-th order of accuracy.

Discretizing the integral in the right-hand side of (90) with the Bode's quadrature formula [20]

$$\int_{t_k}^{t_{k+4}} f(t)\,dt=h_t(45)^{-1}\left(14f_k+64f_{k+1}+24f_{k+2}+64f_{k+3}+14f_{k+4}\right)+O\left(h_t^7\right),$$

we obtain

$$\int_0^{t_k}K(t_k-\tau)u(z_i,\tau)\,d\tau\approx\sum_{m=0}^{k}u\big|_i^m M_{k-m},\tag{93}$$

where $M_k=h_t(45)^{-1}B_kK_k$, $B_0=14$, $B_{1,3,5,...}=64$, $B_{2,6,10,...}=24$, and $B_{4,8,12,...}=28$.

Using (91) and (92), we can write the difference analogues of EAC (90):

$$c\left(\frac{11\boxed{u|_i^k}-18u|_{i-1}^k+9u|_{i-2}^k-2u|_{i-3}^k}{6h_z}\right)+\frac{11\boxed{u|_i^k}-18u|_i^{k-1}+9u|_i^{k-2}-2u|_i^{k-3}}{6h_t}+$$
$$+s\boxed{u|_i^k}-\boxed{u|_i^k}M_0=\sum_{m=0}^{k-1}u|_i^m M_{k-m}, \tag{94}$$

$$c\left(\frac{25\boxed{u|_i^k}-48u|_{i-1}^k+36u|_{i-2}^k-16u|_{i-3}^k+3u|_{i-4}^k}{12h_z}\right)+$$
$$+\frac{25\boxed{u|_i^k}-48u|_i^{k-1}+36u|_i^{k-2}-16u|_i^{k-3}+3u|_i^{k-4}}{12h_t}+s\boxed{u|_i^k}-\boxed{u|_i^k}M_0=\sum_{m=0}^{k-1}u|_i^m M_{k-m}. \tag{95}$$

The results of numerical experiments showed that the sum in the right-hand parts of (94) and (95) is quite smooth (as function of k) and can be calculated with a time step larger than the one of the computational scheme. Considering this, we store the signal history on the virtual boundary at the time step Nh_t, N is an integer, and introduce the following notation $S_{N,k}=\sum_{m=0}^{k-1}u|_i^m M_{N,k-m}$, $M_{N,k}=Nh_t(45)^{-1}B_kK_{Nk}$. $S_{N,k}$ coincide with $S_{1,Nk}$ up to the error of integration. The missing values $S_{1,Nk+r}$ $(0<r<N)$ can be restored by the interpolation formula [20] $S_{1,Nk+r}=A_{-2}S_{N,k-2}+A_{-1}S_{N,k-1}+A_0S_{N,k}+A_1S_{N,k+1}+A_2S_{N,k+2}$, where $A_{-2}=(r^2-1)r(r-2)/24$, $A_{-1}=-(r-1)r(r^2-4)/6$, $A_0=(r^2-1)(r^2-4)/4$, $A_1=-(r+1)r(r^2-4)/6$, $A_2=(r^2-1)r(r+2)/24$.

Collecting similar terms in (94) and (95), we finally get:

$$\begin{aligned}
&u|_i^k=C_0S_{1,k}+C_1u|_{i-1}^k+C_2u|_{i-2}^k+C_3u|_{i-3}^k+C_4u|_i^{k-1}+C_5u|_i^{k-2}+C_6u|_i^{k-3};\\
&C_0=6h_zh_tZ,\quad C_1=18ch_tZ,\quad C_2=-9ch_tZ,\quad C_3=2ch_tZ,\quad C_4=18h_zZ,\\
&C_5=-9h_zZ,\quad C_6=2h_zZ,\quad Z=\left[11(h_z+ch_t)+6h_zh_t(s-M_0)\right]^{-1}
\end{aligned} \tag{96}$$

when the 4 points approximation is used for the first derivatives, and

$$
\begin{aligned}
u\big|_i^k &= C_0 S_{1,k} + C_1 u\big|_{i-1}^k + C_2 u\big|_{i-2}^k + C_3 u\big|_{i-3}^k + C_4 u\big|_{i-4}^k + C_5 u\big|_i^{k-1} + C_6 u\big|_i^{k-2} + C_7 u\big|_i^{k-3} + \\
&+ C_8 u\big|_i^{k-4}; \quad C_0 = 12h_z h_t Z, \quad C_1 = 48ch_t Z, \quad C_2 = -36ch_t Z, \quad C_3 = 16ch_t Z, \\
&C_4 = -3ch_t Z, \quad C_5 = 48h_z Z, \quad C_6 = -36h_z Z, \quad C_7 = 16h_z Z, \quad C_8 = -3h_z Z, \\
&Z = \left[25\left(ch_t + h_z\right) + 12h_z h_t\left(s - M_0\right)\right]^{-1}
\end{aligned}
\tag{97}
$$

when the 5 points approximation is used.

5.3. Fast Calculation of the Convolution

History thinning (see Subsection 5.2), using the fast Fourier transform (FFT) [15, 19], storing large blocks of information on a hard disk and reading them as necessary (see below) are the operations that allow to calculate the convolutions in non-local EACs with almost no loss of accuracy, very quickly, using small and almost non-increasing amounts of RAM. The implementation of these operations made it possible to solve problems with tens to hundreds million time steps. Without them, even relatively simple problems, such as analysis of circular direct-flow electromagnetic energy compressors [15, 64], which require one to two million time steps, were solved at the limit of available personal computers.

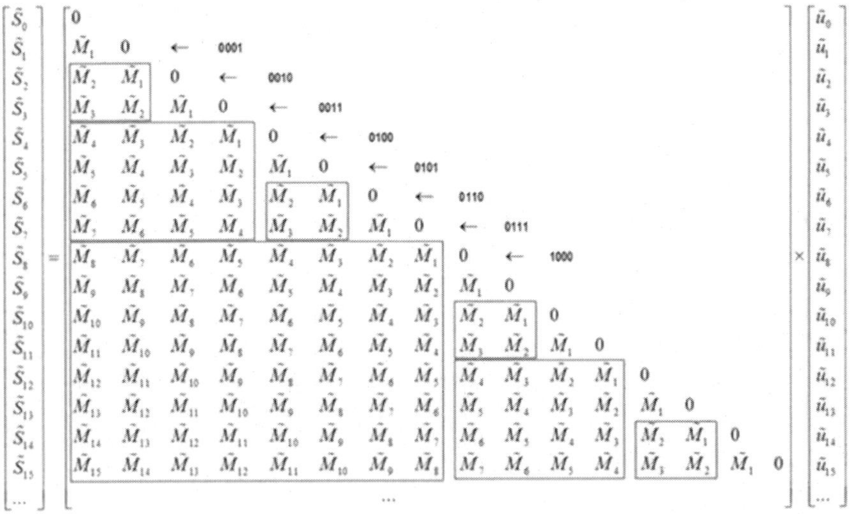

Figure 8. Matrix representation of the convolution.

Let us return to the difference convolution $S_{N,k} = \sum_{m=0}^{k-1} u\big|_i^m M_{N,k-m}$. It can be represented as the product $\tilde{\mathbf{S}} = \tilde{\mathbf{M}}\tilde{\mathbf{U}}$ of the lower triangular Toeplitz matrix $\tilde{\mathbf{M}}$ and the column vector $\tilde{\mathbf{U}}$ composed of the values $u\big|_i^m$, see Figure 8. Here, for the sake of convenience, we have introduced the notation $\tilde{u}_k = u\big|_i^k$, $\tilde{M}_k = M_{N,k}$ and $\tilde{S}_k = S_{N,k}$.

We divide the lower triangular matrix into square blocks with power of two sides. Each block in the upper right corner contains the element $\tilde{M}_1$, and its size $s \times s$ is uniquely determined by the number of time layer, k. If in the binary representation, the k-th rightmost unit (bold numbers near the main diagonal in Figure 8) is in the low order position, then $s = 1$, if the second from the right, then, $s = 2$, etc. Multiplication in the matrix representation of the convolution is carried out block-by-block (see details in [15, 29]). Small blocks (less than 128×128) are multiplied using the common row–by-column rule, large blocks are multiplied using the convolution theorem and FFT. In the latter case, the product of a square block and a fragment of a column vector

$$\begin{bmatrix} \tilde{S}_k \\ \tilde{S}_{k+1} \\ \cdots \\ \tilde{S}_{k+s-1} \end{bmatrix} = \begin{bmatrix} \tilde{M}_s & \cdots & \tilde{M}_2 & \tilde{M}_1 \\ \tilde{M}_{s+1} & \cdots & \tilde{M}_3 & \tilde{M}_2 \\ \cdots & \cdots & \cdots & \cdots \\ \tilde{M}_{2s-1} & \cdots & \tilde{M}_{s+1} & \tilde{M}_s \end{bmatrix} \times \begin{bmatrix} \tilde{u}_{k-s} \\ \tilde{u}_{k-s+1} \\ \cdots \\ \tilde{u}_{k-1} \end{bmatrix} \tag{98}$$

is reduced to the cyclic convolution $W_n = (\mathbf{M} * \mathbf{U})_n$, $n = 0,1,...,N-1$ of two vectors $\mathbf{U} = \{\tilde{u}_{k-s}, \tilde{u}_{k-s+1},...,\tilde{u}_{k-1},0,...,0\}$ and $\mathbf{M} = \{\tilde{M}_1, \tilde{M}_2,...,\tilde{M}_{2s-1},0\}$ of the size $N = 2s$ according the rule $\tilde{S}_{k+p} = W_{s-1+p}$, $p = 0,1,...,s-1$. W_n are found using the convolution theorem [79]

$$\mathbf{W} = DFT^{-1}\big[DFT[\mathbf{M}]DFT[\mathbf{U}]\big]. \tag{99}$$

Here, $\mathbf{W} = \{W_0, W_1, W_2, ..., W_{N-1}\}$, DTF and DTF^{-1} mean direct and inverse discrete Fourier transform, respectively:

$$y_n = DFT[\mathbf{X}] \equiv \sum_{m=0}^{N-1} x_m \exp\left(-\frac{2\pi i}{N} mn\right) \leftrightarrow$$
$$\leftrightarrow \quad x_m = DTF^{-1}[\mathbf{Y}] \equiv \frac{1}{N} \sum_{n=0}^{N-1} y_n \exp\left(\frac{2\pi i}{N} mn\right). \qquad (100)$$

Let now $s \times s$ is the maximum size of a square block which still allows to use FFT regardless of the memory restrictions. Then within $0 < k < s$, multiply blocks in the triangular fragment as described above. For $k = s$, multiply one $s \times s$ block. Then, in the interval $s < k < 2s$, multiply again blocks in the triangular fragment. For $k = 2s$, multiply two $s \times s$ blocks, and then again blocks in the triangular fragment, etc. The partition of the matrix into such blocks is shown in Figure 9.

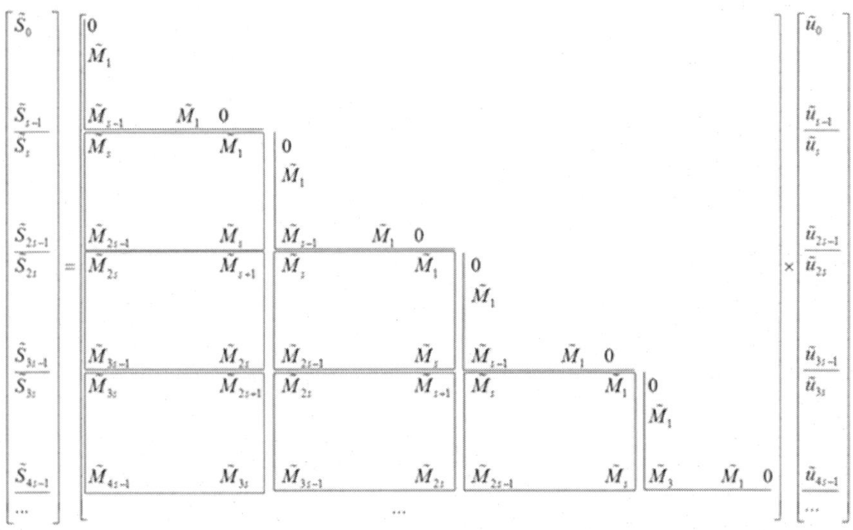

Figure 9. Block-wise calculation of the convolution with maximum square block size $s \times s$.

Upon reaching the time layer $k = Ms$, M block multiplications is performed. The key to building an efficient implementation of the algorithm is in storing the spectra of square $s \times s$ blocks of the kernel and the history column vector fragments of the length s not in RAM but on a hard disk. This information is rarely required, only at each s-th time step, and therefore, storage on a hard disk leads to significant memory saving with minimal loss in the computation.

The algorithms for accelerated computation of convolutions in problems with non-local EACs were tested in [29, 66]. In [29], it was applied to EAC (14) and analogs of EACs (62), (63) on a hemispherical boundary. With the growing number of time steps, M, the computation time grows as $O(M^2)$ and $O(M \log^2 M)$ for the standard convolution calculations and the calculations using the block technique and FFT, respectively. This means that only in the latter case the computation efforts associated with EACs have negligible effect on the total computational time. In [66], the convolution in EAC (18) calculated directly using the trapezium quadrature formula was compared with the one calculated using the block technique described above. For $M = 10^5$, the difference in computation time was about two orders of magnitude, and it increased to four orders of magnitude for $M = 10^6$.

6. Thinning of History and Its Storage on a Hard Disk. How It Works

The previous section is devoted to proper approximation and implementation of non-local EACs, which lead to significant reduction of required computer resources. In this section, we will illustrate the statements that non-local EAC (87) is very suitable for history truncation by solving several specific problems and, in particular, by results which could not be obtained using other (non EACs-based) approaches.

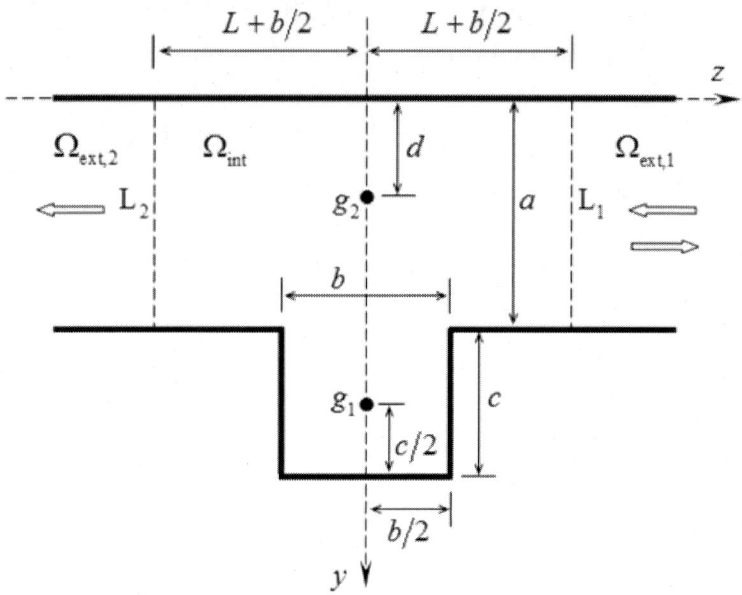

Figure 10. Geometry of the test problem.

6.1. Storing History on a Hard Disk

Consider the problem (4) with geometry shown in Figure 10. Let $a=1.0$, $b=0.25$, $c=0.4$, $d=0.5$, and assume there are no current and instant sources. The pulse TEM -wave runs on the virtual boundary L_1 :

$$
\begin{gathered}
U^i(g,t)=u_0^i(z,t)\mu_0(y): u_0^i(L+b/2,t)=S(t)\sin\left[\Delta k\left(t-\tilde{T}\right)\right]\left(t-\tilde{T}\right)^{-1}\cos\left[\tilde{k}\left(t-\tilde{T}\right)\right]= \\
=F_1(t);\quad S(t)=4s^2(t)\left[3-2s(t)\right],\quad s(t)=\left(t-t_0\right)/\left(\tilde{T}-t_0\right)\quad \text{for}\quad t_0\le t\le\tilde{T}, \\
s(t)=\left(t-t_1\right)/\left(\tilde{T}-t_1\right)\quad \text{for}\quad \tilde{T}<t\le t_1,\quad \text{and}\quad s(t)=0\quad \text{for}\quad t<t_0,\quad t>t_1,
\end{gathered}
\tag{101}
$$

$t_0=0$, $t_1=5$, $\tilde{T}=2.5$, $\tilde{k}=4.5$, $\Delta k=4.5$. Its time profile and the standard amplitude-frequency characteristics are shown in Figure 11.

Here, $R_{n0}(k)$ and $T_{n0}(k)$ are the portions of energy reflected into the waveguide $\mathrm{Q}_{\text{ext},1}$ and transmitted into the waveguide $\mathrm{Q}_{\text{ext},2}$ TM_{0n} -waves. The cutoff points of these waves are $k_n=n\pi$, and the conservation of energy law requires that $\eta(k)=1$, where $\eta(k)=\sum_n\left[R_{n0}(k)+T_{n0}(k)\right]$ [14]. Within the frequency range covered by the pulse (101), the input and output waveguides support without attenuation up to four TM_{0n} -waves. The data presented in Figure 11 allows us to observe the effect of resonant total reflection of the incident wave ($R_{00}(k)=1$), threshold effects at the points of appearance of new propagating waves, effects of partial transformation of the incident wave into higher reflected and transmitted waves.

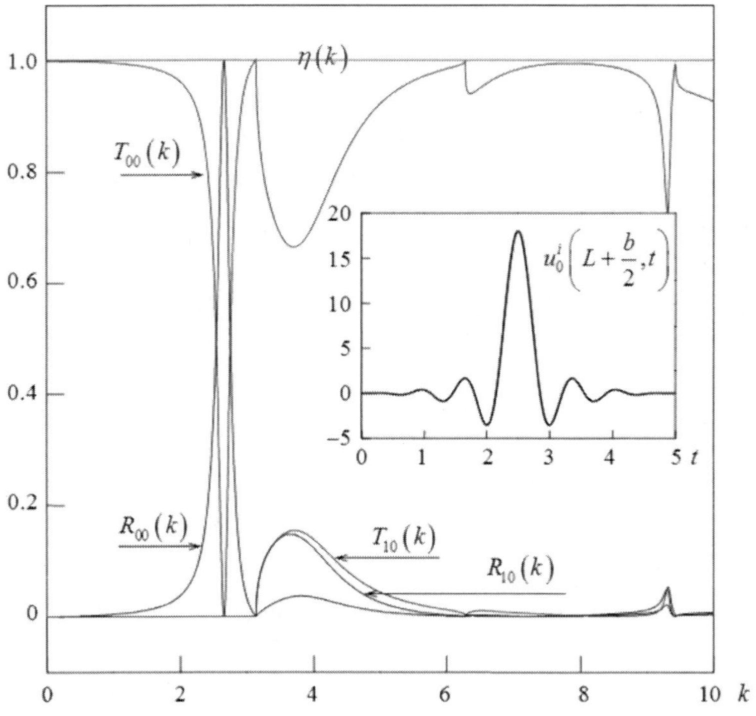

Figure 11. Characteristics of the incident wave (inner fragment) and the waves generated by the waveguide node.

The difference scheme's discretization steps are $h_s=0.005$ (space step) and $h_t=0.002$ (time step). With this space step, 200 Yee cells fit in the cross-section of the waveguides, 20 cells fit within one spot of the tenth waveguide mode, and 125 cells fit in the minimum wavelength of the studied frequency range

$0 < k \leq 10$. All this allows us to consider the approximation's accuracy sufficient for the accurate EACs testing.

First we set $L = 1$ (Figure 10), which corresponds to a small computational domain. On the virtual boundaries L_1 and L_2, EAC (87) is enforced and it is set to take into account the first ten modes $(TEM, TM_{01} \div TM_{09})$ of the pulse waves $U(g,t) - U^i(g,t)$ and $U(g,t)$ propagating into the domains $\mathrm{Q}_{\text{ext},1}$ and $\mathrm{Q}_{\text{ext},2}$. Computation time is $T = 1000$ ($5 \cdot 10^5$ time steps). The maximum size of a square $s \times s$ block of the matrix $\tilde{\mathbf{M}}$ is determined by the value $s = 2^{12} = 4096$.

36 binary files are stored on a hard disk, they contain the history of 9 modes ($TM_{01} \div TM_{09}$) of the outgoing waves on each virtual boundary and the Fourier transforms of the corresponding convolution kernels (square blocks). At time steps that are multiples of $4096h_t = 8.192$, the stored files are copied into RAM, supplemented and returned back to a hard disk. By the end of computation, the size of each file with history is 7 997 344 bytes, and the size of each file with convolutional kernels is 8 062 516 bytes. The total size of all 36 files is 289 077 480 bytes.

The amount of RAM allocated for arrays with grid functions is 4.136 MB. For 20 handlers (processors) of non-local EACs 5.689 MB were allocated. The entire program occupies 15.5 MB of RAM. Some space is reserved for all other required information such as the function $u_0^i(L + b/2, t)$ and its normal derivative, a packed map of the computation domain, total wave amplitudes and field values at observation points, file names, graphic arrays, etc.

So, it was possible to move a megabyte to a hard disk, leaving 15.5 MB in RAM. This is a good result, as for more serious problems, even two-dimensional, the total volume of history and kernels stored on a hard disk could reach tens of gigabytes.

The outcome of the small domain numerical experiment is the functions $H_x(g_1,t)$ and $H_x(g_2,t)$, Figure 12a. They are fully affected by the introduction of the virtual boundaries into the computational scheme, it will be used below during tests.

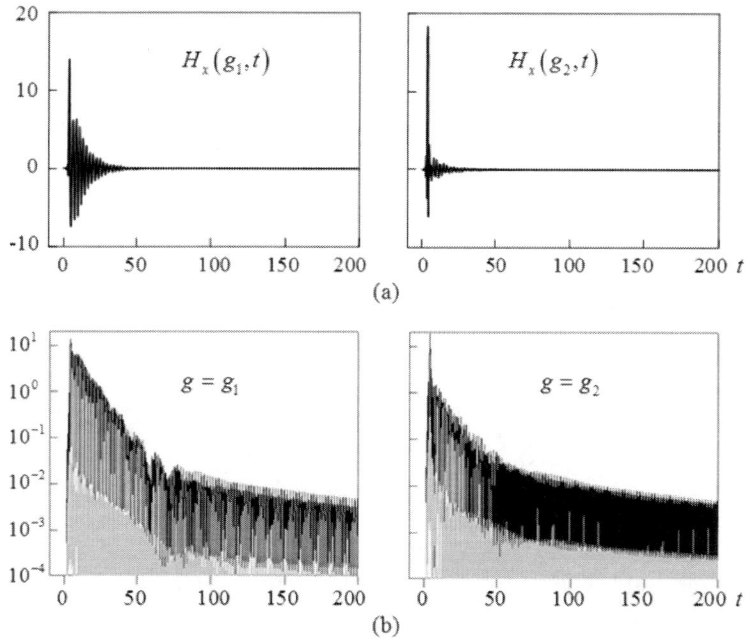

Figure 12. (a) The field $U(g,t)=H_x(g,t)$ in the points $g=g_1$ and $g=g_2$: $L=1$. (b) On the estimation of the error introduced into the difference solution by absorbing boundaries.

Let now $L=101$ (large computational domain). On the virtual boundaries L_1 and L_2, EAC (87) is enforced and it is set to take into account the first five pulse waves modes ($TEM, TM_{01} \div TM_{04}$) propagating into the domains $Q_{ext,1}$ and $Q_{ext,2}$. Higher (evanescent) modes at such large distances from the virtual boundaries could be ignored. Computation time is $T=300$, the maximum size of a square $s \times s$ block of the matrix $\tilde{\mathbf{M}}$ is determined by the value $s=2^{18}=262\,144$, the entire computational scheme is placed in RAM.

The outcome of the large domain numerical experiment is again the functions $H_x(g_1,t)$ and $H_x(g_2,t)$. Their values $H_x(g_1,t+100)=$ $=H_x^{\text{etalon}}(g_1,t)$ and $H_x(g_2,t+100)=H_x^{\text{etalon}}(g_2,t)$ are taken as the referenced ones for the interval $0 \le t \le 200$. It is possible because they are not affected by the presence of the virtual boundaries in the computational scheme, and at the time moment $t=100$, the incident *TEM* -wave arrives on the boundary L_1 of the small computational domain without any distortions. But there are discretization errors in the reference solution. Therefore, the difference $H_x^{\text{etalon}}(g_j,t)-H_x(g_j,t)$, $j=1,2$, where $H_x(g_j,t)$ is the small domain result, reflects actually the influence of the virtual boundaries, namely, the level of errors introduced into the solution by limiting the computational domain using EACs. Since the functions $H_x^{\text{etalon}}(g_j,t)$ and $H_x(g_j,t)$ are visually indistinguishable, we use a logarithmic scale to represent the result (see Figure 12b, $\left|H_x^{\text{etalon}}(g_j,t)\right|$ is in black, $\left|H_x^{\text{etalon}}(g_j,t)-H_x(g_j,t)\right|$ is in grey). The absolute error introduced into the numerical solution by EACs is two to three orders of magnitude smaller than this solution. This result allows us to make the following conclusion: non-local EAC (87) and the approach of storing history and convolutional kernels on a hard disk work well enough.

6.2. How Works the History Thinning

Consider the problem (4) with almost the same geometry as in the previous subsection (see Figure 10). Let $a=1.0$, $b=0.35$, $c=0.3$, $d=0.25$, but the waveguide node is now excited by the current source $F(g,t)=\eta_0 G(g)\partial_t j_x(t)$. Here, the function $G(g)$ determines the source support, it is 0.02×0.02 square with its center at the point $g=g_1$, and the function $j_x(t)=F_1(t)$ (see (101)) is an external current density with $t_0=0$, $t_1=10$, $\tilde{T}=5$, $\tilde{k}=6.4$, $\Delta k=3$. The source's time profile and spectral amplitudes are depicted in Figure 13a. The discretization steps of the difference scheme are $h_s=0.005$ and $h_t=0.002$. The value of the function $U(g,t)=E_x(g,t)$ is calculated at the point $g=g_2$.

The first two waveguide modes (TE_{01} and TE_{02}) carry the major part of the source energy into both arms. Some energy is pumped at frequencies below the cutoff frequency of the first mode, $k_1=\pi$, and a little – at frequencies above the cutoff frequency of the third mode, $k_3=3\pi$. But the first two modes dominate. When testing EAC, 10 modes are taken into account at each virtual boundary. Accordingly, $TE_{0,11}$-mode with the cutoff frequency $k_{11} = 11\pi \approx 34.56$ is reflected from the virtual boundaries. The data presented in the right fragment of Figure 13a allows one to estimate the level of signals at frequencies larger than k_{11}.

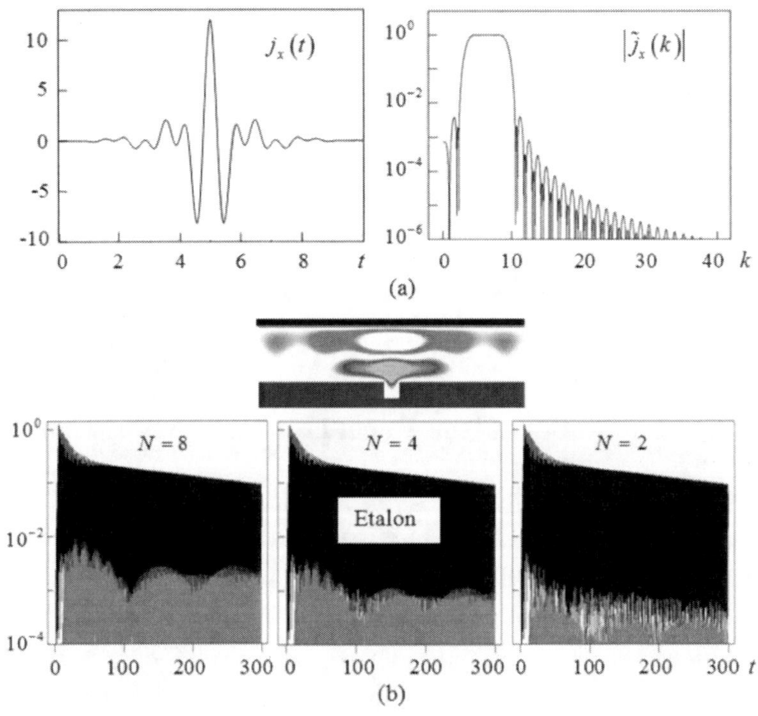

Figure 13. (a) Characteristics of the current source. (b) To the estimation of the error introduced into the difference solution by absorbing boundaries and history thinning.

Constructing the reference solution $E_x^{\text{etalon}}(g_2,t)$, we assume $L=153$ and $T=300$. For this length and time, the perturbation reaches the virtual boundaries and gets partially reflected by them, but the reflected signal does not have time to return to the observation point $g=g_2$. In Figure 13b the spatial distribution of $E_x^{\text{etalon}}(g,t)$, $g\in \text{Q}_{\text{int}}$, at the moment of time $t=100$ is also presented: a sufficiently high-Q free oscillation of the field is excited in the waveguide node, and that is why the function $E_x^{\text{etalon}}(g_2,t)$ decays so slowly over time t.

Then we assume $L=3$ and calculate the dependencies $E_x^N(g_2,t)$ corresponding to different values of the history thinning parameter N ($N=2,4,8$; the signal history on the virtual boundaries L_1 and L_2 is saved with the time step Nh_t). Further (see Figure 13b), we compare $\left|E_x^{\text{etalon}}(g_2,t)\right|$ (black graph) and $\left|E_x^{\text{etalon}}(g_2,t)-E_x^N(g_2,t)\right|$ (grey graphs). The error introduced by EACs is two to three orders of magnitude smaller than the calculated value. The obtained data suggests that it is possible to thin out the history, a reasonable choice is $N=2\div 4$. It seems that the gain is not too big, but when tens of gigabytes are stored on a hard disk, even a twofold reduction can play a significant role.

6.3. Unique Capabilities of Non-Local Exact Absorbing Conditions

Different EACs manifest themselves in various ways when solving fairly complex problems. The differences are mainly related to the maximum value of observation time T, for which the corresponding computational scheme still remains stable. EAC (87) is unique in this regard, the possibilities of its effective implementation were discussed above, let us give here one confirming example.

Consider the problem of excitation of a chain of three coupled waveguide resonators of the same size, Figure 14. Let $a=1.0$, $b=0.8$, $c=0.02$, $d=0.1$, and a pulsed TE_{01}-wave covers the frequency band $\pi<k<2\pi$. Only the basic monochromatic wave can propagate without attenuation in the waveguide's regular sections. Enforcing EAC (87) on the virtual boundaries L_1 and L_2, we take into account the modes $TE_{01}\div TE_{06}$.

During the computational experiment, which results are presented in Figure 14 (see also [77]), about 100 million time layers were "scanned." On a hard disk, 12 files with the history were stored, each of them was 796 920 800 bytes by the end of calculations, and 12 files with the convolutional operators kernels, 801 114 836 bytes each. In total, on a hard disk 17.859 GB were occupied. In

RAM, 8.268 MB were allocated for the grid functions and 240.041 MB for 12 handlers (processors) of EACs. The maximum size of a square $s \times s$ block of the matrix $\tilde{\mathbf{M}}$ was determined by the value $s = 2^{18} = 262\ 144$, and during the history thinning, every second computed set of data ($N = 2$) was stored.

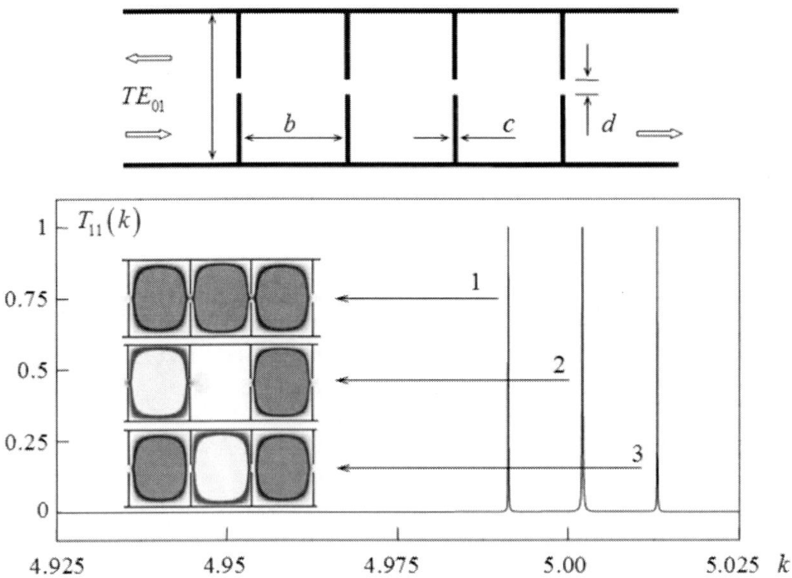

Figure 14. Triplet in a chain of three coupled resonators.

Why are such large resources required for an accurate numerical solution of the problem? The fact is that within the entire studied frequency range $\pi < k < 2\pi$, the energy transmission coefficient $T_{11}(k)$ is almost zero ($T_{11}(k)$ is between 10^{-15} and 10^{-9}), with the exception of very small neighborhoods of three frequencies $k \approx 4.991\,326$, $k \approx 5.002\,118\,3$, $k \approx 5.013\,005\,5$, where $T_{11}(k) = 1.0$ (Figure 14, triplet). These frequencies coincide with the real part $\operatorname{Re}\bar{k}$ of the complex eigenfrequencies $\bar{k} = \operatorname{Re}\bar{k} + i\operatorname{Im}\bar{k}$ corresponding to the high-quality ($|\operatorname{Im}\bar{k}| << 1$) free oscillations 1, 2, and 3 [14, 51, 61], their field patterns are shown in Figure 14. The quality factors of these oscillations, which are determined by the half-width of the spectral lines, are $Q \approx 83\,000$ for the oscillation 1, $Q \approx 41\,000$ for the oscillation 2, and $Q \approx 81\,000$ for the oscillation 3.

This and other examples of successfully solving fairly complex problems allow us to conclude that using EAC (87), we can confidently handle resonances with Q-factor $Q \approx 10^6$, which requires up to 10^9 time layers. The correct implementation of this EAC allows one to transfer to a hard disk amounts of data which are tens and hundreds of times larger than kept in RAM. Tuning the history thinning parameters and maximum sizes of square blocks of the matrix $\tilde{\mathbf{M}}$, one can flexibly control the computation process and obtain a desired result with required accuracy.

7. The Method of Exact Absorbing Conditions: Testing Algorithms and a Brief Review of Physical Results

7.1. Test Problems

In this subsection, we consider only two of many tests performed, which aim to evaluate the efficiency and accuracy of the algorithms of the EACs method, to determine the real capabilities of the method in analyzing space-time and spatial-frequency transformations of electromagnetic field in high-Q resonance structures. Test results can be found in our previously published papers [11, 12, 14, 45]. And in [80], a comparative analysis of ABCs, PMLs, and EACs (18) and (25) was carried out for problems of the electrodynamic theory of grating.

The first test problem

$$\begin{cases} \left[-\partial_t^2 + \partial_z^2 - \pi^2\right] u(z,t) = 0; \quad 0 < z < L, \quad t > 0 \\ u_n(z,0) = 0, \quad \partial_t u_n(z,t)\big|_{t=0} = 0; \quad 0 \le z \le L \\ D\left[u(z,t) - u^i(z,t)\right]\Big|_{z=0} = 0 \quad \text{and} \quad D\left[u(z,t)\right]\Big|_{z=L} = 0 \end{cases} \tag{102}$$

describes the propagation of one of the eigen pulsed waves of a hollow regular waveguide through its segment of the length $L = 100$. Its numerical solution allows one to test EACs both in the pure absorption mode (the wave $u(z,t)$ crosses the virtual boundary $z=L$ only in one direction $z\to\infty$), and in the mode when on the virtual boundary $z=0$, the primary wave encounters the reflected wave $u(z,t)-u^i(z,t)$ generated by this boundary. Here, $u(z,t)$ is the space-time amplitude of the total field generated by the virtual waveguide node, $u^i(z,t)$ is the amplitude of the primary wave arriving on the boundary $z=0$, and the operator $D[...]$ corresponds to EAC (87).

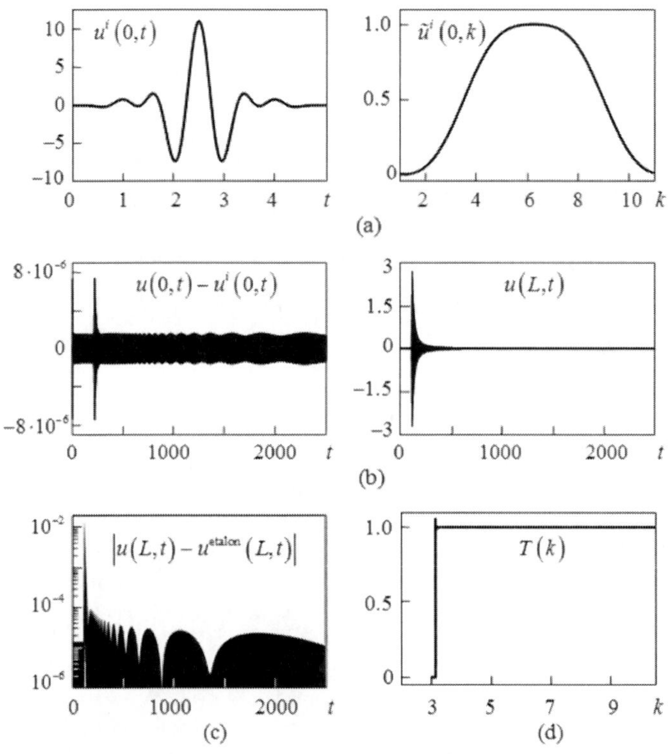

Figure 15. (a) Time profile of the primary pulse and its amplitude spectrum on the virtual boundary $z=0$. (b) Time profiles of the reflected ($u(0,t)-u^i(0,t)$) and transmitted through the waveguide segment ($u(L,t)$) pulses. (c) The absolute error of the calculation of the profile of the transmitted pulse. (d) Transmitted energy coefficient for the considered waveguide section.

The discretization steps in (102) are $h_z = 2h_t = 0.002$, the computational time is $T = 25000$. The result of its solution is presented in Figure 15. The time profile of the primary pulse is shown in Figure 15a. The amplitude spectrum of this pulse occupies the frequency band, part of which lies below the cut-off point $k = \pi$ of the corresponding monochromatic wave. The time profile $u(0,t) - u^i(0,t)$ of the pulse reflected from $z = 0$ is shown in Figure 15b. The exact value of this function is obviously zero. That is why $u(0,t) - u^i(0,t)$ represents the cumulative error of the difference scheme, including the approximation error of the equation of the problem and its boundary conditions on the virtual boundaries $z = 0$ and $z = L$. The absolute values of this error are five to six orders of magnitude smaller than $u^i(0,t)$, thus the reflection from the boundary $z = 0$ can be considered negligible.

The absolute error $\left|u(L,t) - u^{\text{etalon}}(L,t)\right|$ is obtained calculating $u(L,t)$, and $u^{\text{etalon}}(L,t)$ is determined using the transport operators in the time domain (see [81–86]) or in the frequency domain. This error is two to three orders of magnitude smaller than $u(L,t)$ itself, Figure 15c. Interestingly, the maximum values of this error fall on the leading front of the pulse $u(L,t)$, where its high-frequency components are concentrated. This indirectly indicates that the analyzed error is caused mostly by discretizing the equation of the problem (102) rather than its boundary conditions.

The function $T(k)$ represents the part of the energy that goes through the boundary $z = L$ in the direction of increasing z. As expected, it is equal to zero at frequencies below the critical and close to unity for $k > \pi$ ($\left|1 - T(k)\right| < 10^{-5}$), see Figure 15d. The only exception is a small vicinity of the point $k = \pi$ at which the deviation of $T(k)$ from the unit is more noticeable. As the computational time T increases, this deviation decreases.

In general, the test problem results presented in Figure 15 allow us to state that EAC (87) can be correctly and efficiently embedded in finite-difference schemes and ensure that the results are obtained with the required accuracy.

Computational schemes with EAC (87) remain stable well beyond 10^7 time steps, we have successfully considered problems even with $5 \cdot 10^8$ time steps.

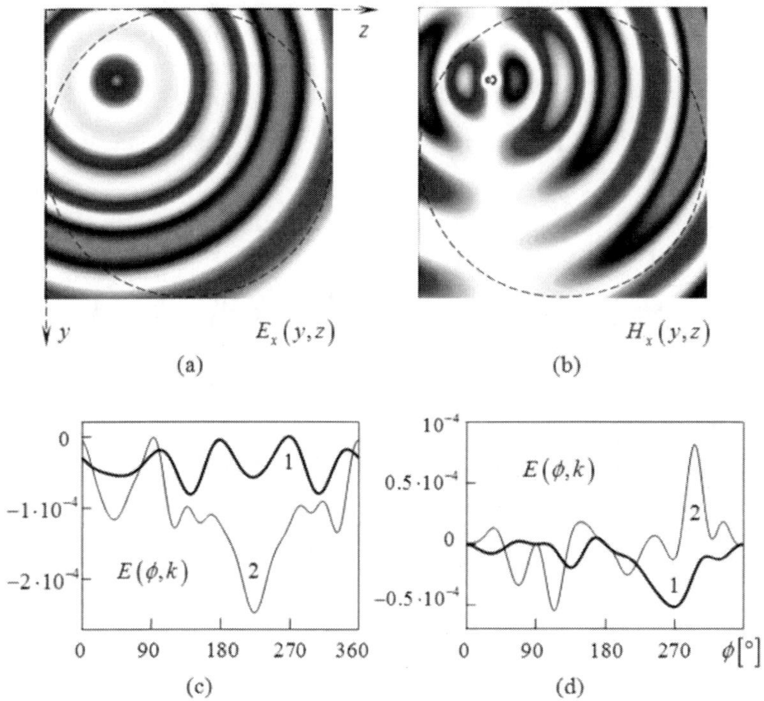

Figure 16. (a) and (b) Fields of electric dipoles oriented along the axes x and y. (c) and (d) The absolute error $D^{\text{etalon}}(\phi,k)-D(\phi,k)$ in the calculation of dipole patterns (curves 1 are for $k=4$, curves 2 are for $k=6$).

The second test is associated with the problem (74) ($\varepsilon(g)\equiv1$, $\sigma(g)=\varphi(g)=\psi(g)\equiv0$), which computation domain is limited by EAC (84). The radiation of an electric dipole oriented either along the x axis (E -polarization of the field, $F(g,t)=\eta_0\partial_t j_x(t)\delta(g-g_0)$) or along the y axis (H -polarization of the field, $F(g,t)=\partial_z j_y(t)\delta(g-g_0)$) is simulated (see Figure 16).

The pulses of the external currents $j_x(t)$ and $j_y(t)$ are set by the function $F_1(t)$ (see (101)) with the parameters $t_0=0$, $t_1=20$, $\tilde{T}=10$, $\tilde{k}=6$, $\Delta k=4$ and occupy the frequency band $2\leq k\leq 10$. The size of the computation domain is 2.0×2.0, the discretization steps are $h_s=2h_t=0.002$. The center of a circle with radius 0.99 is superposed with the geometric center of the domain Q_{int}. Using the values of the field $U(g,t)$ calculated on this circle, the normalized radiation patterns $D(\phi,k)$ were calculated [13–15], the dipole position in Q_{int} is given by the coordinates $g_0=\{0.25,0.25\}$.

The exact value of the pattern functions for the first dipole is $D^{\text{etalon}}(\phi,k)=1$, and for the second one is $D^{\text{etalon}}(\phi,k)=\sin^2\phi$ [87]. The patterns error $E(\phi,k)=D^{\text{etalon}}(\phi,k)-D(\phi,k)$ within the entire frequency range does not exceed 10^{-3} (see Figures 16c and 16d). The considered test verifies the computational scheme of the EACs method for the problem (74) and the method of calculating the radiation patterns of emitters, which can be studied using this problem.

7.2. Physical Results of the Method of Exact Absorbing Conditions. Short Review

Studies of radiation, propagation and scattering of pulsed electromagnetic waves are in the focus of modern theoretical radiophysics. First of all, because the design of promising communication, electronics, and radar devices requires reliable information about space-time and space-frequency field transformations in fairly complex electrodynamic structures, and the possibilities of the frequency domain approaches in this regard are significantly limited. The method of exact absorbing conditions provides researchers and engineers with algorithms which allow to study, with no additional errors, transient states of the field, whose careful analysis requires long observation times and huge number of time steps. Essentially, the EACs method algorithms analytically regularize problems with high-Q free field oscillations in open periodic, waveguide and compact resonant structures. Such problems play an important role in design and development of new devices for resonant quasi-optics, high-power electronics, communication, and radar technology.

Obviously, to study the physics of resonant wave scattering, results obtained using the EACs method itself need to be effectively and quickly converted into standard amplitude-frequency characteristics widely used in practice and characteristics that are usually ignored by researchers because of problems with their calculation, but playing an important role in understanding the dynamics of space-time transformations of electromagnetic waves. Corresponding tools (designing pulse diagrams of emitters, determining complex eigenfrequencies of free field oscillations and their configuration, calculating instantaneous efficiencies of energy storage in resonators of active compressors of electromagnetic power, etc.) were developed in parallel with algorithms of the EACs method [12–15, 31, 50, 51, 56, 59, 64] and are incorporated into all our programs.

We briefly list some basic physical results obtained using the EACs method.

Effective approaches to solving problems of analysis and synthesis of compact open resonators with conventional and dispersive (gratings) mirrors were developed and implemented [11, 12, 14, 50, 51, 61].

For the first time, information on the basic electrodynamic characteristics (amplitude-frequency and pulse) of canonical axially symmetric emitters of monochromatic and pulse TE_0- and TM_0-waves (monopoles, mirror and resonant antennas of various types) was widely presented [14, 15, 56].

Discovered and studied: effects of mode splitting of ultra-wideband signals by stepwise and smooth connections of waveguides of different cross-sections; effects of frequency mode splitting by conical plugs in circular and coaxial waveguides [14, 53, 60].

For the first time, slot resonances (resonances on TEM-wave in segments of narrow radial and coaxial waveguides) are investigated in detail. Their implementation significantly expands the set of practically interesting scattering modes for quasi-monochromatic and broadband TM_0-signals in simple axially symmetric waveguide units. It is proved that the excitation of gap resonances can be accompanied by the effects of complete locking of a coaxial waveguide, a stepwise widening of a circular waveguide and a U-turn of a "circular waveguide – coaxial circular waveguide"; effects of almost complete transformation of TM_{01}-wave of a narrow waveguide into TM_{02}-wave of a wide waveguide (stepwise expansion), full transformation of TM_{01}-wave of a circular waveguide into TEM-wave of a coaxial circular waveguide (U-turn) [14, 54, 58, 63].

For the first time, slot resonances which have a significant impact on the basic characteristics of axially symmetrical emitters of TM_0-wave of standard configurations have been studied. The following effects have been discovered and analyzed: effects of antenna blinding by resonances in thin transverse and longitudinal shorted slits in the external and internal conductors of a feeding coaxial line; effects of complete transition of the input energy into free space through narrow gaps in the closed end of a coaxial waveguide and in its thick outer conductor [14, 57].

For the first time, a rigorous electrodynamic analysis of the forming line of an emitter of ultrashort electromagnetic video pulses was carried out. Effects accompanying the propagation of ultra-wideband signals in structures of this type are considered, the conclusions regarding the influence of this effect on basic characteristics of forming lines have been done. A new scheme for the formation of short pulses allowing a trapezoidal pulse to be converted into a bipolar stepped pulse with low losses is proposed [14, 55].

Developed and implemented rigorous model synthesis schemes for active electromagnetic power compressors with accumulators and locks of various types [14, 15, 59, 64, 65, 71].

New physical results allowing characterizing the formation of forbidden zones in spatially limited 2-D photonic crystals, the processes of energy accumulation in resonant cavities of such structures, the propagation, and emission of pulsed and monochromatic waves in compact photonic crystals, whose defects play the role of various classical functional elements, have been obtained [78].

The characteristics of one-dimensionally periodic gratings placed in the field of homogeneous and inhomogeneous plane waves are studied in detail. For the first time, using the results of spectral theory, we studied the deformations of pulses propagating and scattering in homogeneous and inhomogeneous Floquet channels (virtual wave-guiding channels), which can be extracted due to the periodicity of scattering structures. The effects of quasi-complete transmission and reflection of narrow-band Gaussian pulses on semitransparent gratings, the effects of bifurcation of the amplitude center of such pulses, the effects of capturing and accumulating signal energy in the near field of an inhomogeneity or source, the phenomenon of pulse focusing with phase modulation, and others were discovered and investigated [11, 12, 13, 62, 73, 83].

The physics of diffraction radiation processes, which is the wave analogs of radiation, in systems "finite grating – an open surface wave transmission line" has been studied in detail; and conclusions and recommendations important for the theory and technique of antennas are made [14, 15, 68, 69, 72].

CONCLUSION

The properties of computational schemes that implement grid-based methods for solving initial boundary value problems of electrodynamics, and, consequently, the quality of the numerical results obtained with their help, greatly depend on the type of embedded absorbing conditions. It is important that absorbing conditions preserve the correctness classes of open problems when they are transformed into closed ones, do not violate the stability and convergence of computational schemes, and ensure the predicted accuracy of final results. In principle, all these requirements are met by the so-called exact absorbing conditions. Virtual boundaries with such conditions are absolutely transparent for waves passing through them in one direction, regardless of angles of arrival, configurations of fronts, etc. For some of practically used conditions, it is possible to prove the equivalence of original open and modified closed problems. In theory everything is perfect. But in practice, the possibility of effective and successful implementation of exact absorbing conditions must be analyzed in details, and the results of this analysis must be reliable and unambiguous, as mistakes made at this stage may be too expensive. There are quite a lot of exact absorbing conditions that have already been developed and researchers have plenty of versions to choose from and decide on which basis the modern computational electrodynamics will be built in the following years.

REFERENCES

[1] Engquist, B. and Majda, A. "Absorbing Boundary Conditions for the Numerical Simulation of Waves." *Mathematics of Computation* 31, no. 139 (1977): 629–651.

[2] Mur, G. "Absorbing Boundary Conditions for the Finite Difference Approximation of the Time-Domain Electromagnetic Field Equations." *IEEE Transactions on Electromagnetic Capability* 23, no. 4 (1981): 377–382.

[3] Fang, J. "Absorbing Boundary Conditions Applied to Model Wave Propagation in Microwave Integrated Circuits." *IEEE Transactions on Microwave Theory and Techniques* 42, no. 8 (1994): 1506–1513.

[4] Mei, K. and Fang, J. "Superabsorbtion – a Method to Improve Absorbing Boundary Conditions." *IEEE Transactions on Antennas and Propagation* 40, no. 9 (1992): 1001–1010.

[5] Betz, V. and Mittra, R. "A Boundary Condition to Absorb Both Propagating and Evanescent Waves in a FDTD Simulation." *IEEE Microwave and Guided Wave Letters* 3, no. 6 (1993): 182–184.

[6] Berenger, J. P. "A Perfectly Matched Layer for the Absorption of Electromagnetic Waves." *Journal of Computational Physics* 114, no. 1 (1994): 185–200.

[7] Berenger, J. P. "Three-Dimensional Perfectly Matched Layer for Absorption of Electromagnetic Waves." *Journal of Computational Physics* 127, no. 2 (1996): 363–379.

[8] Sacks, Z., Kingsland, D., Lee, R., and Lee, J. "A perfectly Matched Anisotropic Absorber for use as an Absorbing Boundary Condition." *IEEE Transactions on Antennas and Propagation* 43, no. 12 (1995): 1460–1463.

[9] Taflove, A. and Hagness, S. *Computational Electrodynamics: The Finite-Difference Time-Domain Method*. Boston: Artech House, 2005.

[10] Jin, J. *The Finite Element Method in Electromagnetics*. New York: John Wiley & Sons, 2002.

[11] Sirenko, Y. *Simulation and Analysis of Transient Processes in Open Periodic, Waveguide, and Compact Resonators*. Kharkiv: EDENA, 2003.

[12] Sirenko, Y., Strom, S., and Yashina, N. *Modeling and Analysis of Transient Processes in Open Resonant Structures. New Methods and Techniques*. New York: Springer, 2007.

[13] Sirenko, Y. and Strom, S., eds. *Modern Theory of Gratings. Resonant Scattering: Analysis Techniques and Phenomena*. New York: Springer, 2010.

[14] Kravchenko, V., Sirenko, Y., and Sirenko, K. *Electromagnetic Wave Transformation and Radiation by the Open Resonant Structures. Modelling and Analysis of Transient and Steady-State Processes.* Moscow: Fizmathlit, 2011.

[15] Sirenko, Y. and Velychko, L., eds. *Electromagnetic Waves in Complex Systems: Selected Theoretical and Applied Problems*. New York: Springer, 2016.

[16] Maikov, A., Sveshnikov, A., and Yakunin, S. "A Difference Scheme for the Non-Stationary Maxwell Equations in Waveguide Systems." *USSR Computational Mathematics and Mathematical Physics* 26, no. 3 (1986): 130–138.

[17] Ladyzhenskaya, O. *The Boundary Value Problems of Mathematical Physics.* New York: Springer-Verlag, 1985.

[18] Korn, G. and Korn, T. *Mathematical Handbook for Scientists and Engineers*. New York: McGraw-Hill, 1961.

[19] Vladimirov, V. *Equations of Mathematical Physics*. New York: Dekker, 1971.

[20] Abramowitz, M. and Stegun, I., eds. *Handbook of Mathematical Functions*. New York: Dover, 1972.

[21] Bateman, H. and Erdelyi, A. *Tables of Integral Transforms. Vol.1*. New York: McGraw-Hill, 1954.

[22] Perov, A. and Sirenko, Y. "Nonstationary Model Problems of the Electrodynamic Theory of Gratings." *Radiofizika i Elektronika* 2, no. 2 (1997): 66–86.

[23] Sirenko, Y. "Exact 'Absorbing' Conditions in Outer Initial Boundary Value Problems of the Electrodynamics of Nonsinusoidal Waves. Part 2: Waveguide Units and Periodic Structures." *Telecommunications and Radio Engineering* 57, no. 12 (2002): 1–31.

[24] Sofronov, I., Dovgilevich, L., and Krasnov, N. "On High-Order Approximation of Transparent Boundary Conditions for the Wave Equation." *Komp'yuternye Isledovaniya i Modelirovaniye* 6, no. 1 (2014): 45–56.

[25] Sofronov, I. "Conditions of Complete Transparency on the Sphere for the Three-Dimensional Wave Equation." *Doklady Akademii Nauk (RAN)* 326, no. 6 (1992): 953–957.

[26] Sirenko, Y. "Exact 'Absorbing' Conditions in Outer Initial Boundary Value Problems of the Electrodynamics of Nonsinusoidal Waves. Part 5: Plausible Extensions and Unresolved Problems. *Telecommunications and Radio Engineering* 59, no. 5-6 (2003): 1–27.

[27] Sirenko, K. "Transport Operators in the Axially-Symmetrical Problems of the Electrodynamics of Pulsed Waves." *Elektromagnitnye Volny i Elektronnye Sistemy* 11, no. 11 (2006): 15–26.

[28] Sirenko, K. and Sirenko, Y. "Exact 'Absorbing' Conditions in the Initial Boundary Value Problems of the Theory of Open Waveguide Resonators." *Computational Mathematics and Mathematical Physics* 45, no. 3 (2005): 490–506.

[29] Sirenko, K., Pazynin, V., Sirenko, Y., and Bagci, H. "An FFT-Accelerated FDTD Scheme with Exact Absorbing Conditions for Characterizing Axially Symmetric Resonant Structures." *Progress in Electromagnetics Research* 111 (2011): 331–364.

[30] Maikov, A., Poezd, A., Sveshnikov, A., and Yakunin, S. "Difference Scheme of Initial Boundary Value Problems for Maxwell Equations in Unlimited Domain." *Zhurnal Vychislitel'noy Matematiki i Matematicheskoy Fiziki* 29, no.2 (1989): 239–250.

[31] Shafalyuk, O., Sirenko, Y., and Smith, P. "Simulation and analysis of transient processes in open axially-symmetrical structures: Method of exact absorbing boundary conditions." In *Electromagnetic Waves* edited by V. Zhurbenko, 99–116. Rijeka: InTech, 2011.

[32] Bateman, H. and Erdelyi, A. *Higher Transcendental Functions. Vol. 2.* New York: McGraw-Hill, 1953.

[33] Gradshteyn, I. and Ryzhik, I. *Table of Integrals, Series, and Products*. San Diego, London: Academic Press, 2000.

[34] Prudnikov, A., Brychkov, Y., and Marichev, O. *Integrals and Series. Vol.1*. New York: Gordon & Breach, 1986.

[35] Sirenko, Y. "Exact 'Absorbing' Conditions in Outer Initial Boundary Value Problems of the Electrodynamics of Nonsinusoidal Waves. Part 3: Compact Inhomogeneties in Free Space." *Telecommunications and Radio Engineering* 59, no. 1-2 (2003): 1–31.

[36] Vyaz'mitinova, A., Pazynin, V., and Sirenko, Y. "Simulation of Radiation of Pulsed Waves of a Coaxial Circular Waveguide." *Radiofizika i Elektronika* 6, no. 1 (2001): 22–29.

[37] Vyaz'mitinova, A., Pazynin, V., Perov, A., Sirenko, Y., Akdogan, H., and Yaldiz, E. Pulse Antennas: Accurate Solution of Two-Dimensional Model Problems for Structures with Gratings as Dispersive Elements." *Telecommunications and Radio Engineering* 55, no. 8 (2001): 1–14.

[38] Sofronov, I. "Artificial Boundary Conditions of Absolute Transparency for Two- and Three-Dimensional External Time-Dependent Scattering Problems." *European Journal of Applied Mathematics*, 9, no. 6 (1998): 561–588.

[39] Sofronov, I. "About Using Transparent Boundary Conditions in Aeroacoustics." *Matematicheskoye Modelirovaniye* 19, no. 8 (2007): 105–112.

[40] Hagstrom, T. "Radiation Boundary Conditions for the Numerical Simulation of Waves." *Acta Numerica. Cambridge University Press* 8 (1999): 47–106.

[41] Olivier, J. "On the Synthesis of Exact Free Space Absorbing Boundary Conditions for the FDTD Method." *IEEE Transactions on Antennas and Propagation* 40, no. 4 (1992): 456–460.

[42] Grote, M. *Local Nonreflecting Boundary Condition for Maxwell's Equations*. Basel: Preprint No. 2004-15, Department of Mathematics University of Basel, 2004.

[43] Maikov, A., Sveshnikov, A., and Yakunin, S. "Nonlocal Radiation Conditions for the Non-Stationary System of Maxwell Equations." *Zhurnal Vychislitel'noy Matematiki i Matematicheskoy Fiziki* 30, no. 12 (1990): 1785–1796.

[44] Shafalyuk, O., Smith, P., and Velychko, L. "Rigorous Substantiation of the Method of Exact Absorbing Conditions in Time-Domain Analysis of Open Electrodynamic Structures." *Progress in Electromagnetics Research B* 41 (2012): 231–249.

[45] Perov, A., Sirenko, Y., and Yashina, N. "Explicit Conditions for Virtual Boundaries in Initial Boundary Value Problems in the Theory of Wave Scattering." *Journal of Electromagnetic Waves and Applications* 13, no. 10 (1999): 1343–1371.

[46] Maikov, A., Poezd, A., and Yakunin, S. "An Economical Method for Calculating Non-Stationary Non-Local-Time Radiation Conditions for Waveguide Systems." *Zhurnal Vychislitel'noy Matematiki i Matematicheskoy Fiziki* 30, no. 8 (1990): 1267–1271.

[47] Maikov, A. "On Approximate Conditions on the Open Boundary for the Wave Equation and the Klein-Gordon Equation." *Vychislitel'nye Metody i Programirovaniye* 6, no. 1 (2005): 290–303.

[48] Maikov, A. "On Approximate Conditions on the Open Boundary for the Klein-Gordon Equation and their Efficiency over Large Time Intervals." *Vychislitel'nye Metody i Programirovaniye* 13, no. 1 (2012): 139–148.

[49] Poezd, A., Sveshnikov, A., and Yakunin, S. "Exact Discrete Nonstanionary Radiation Conditions for Spherical System of Coordinates in Absence of Azimuthal Symmetry." *Matematicheskiye Modeli Yavleniy i Protsesov* 3, no. 2 (1991): 30–41.

[50] Sirenko, Y., Velychko, L., and Erden, F. "Time-Domain and Frequency-Domain Methods Combined in the Study of Open Resonance Structures of Complex Geometry." *Progress in Electromagnetics Research* 44 (2004): 57–79.

[51] Velychko, L., Sirenko, Y., and Velychko, O. "Time-Domain Analysis of Open Resonators. Analytical Grounds." *Progress in Electromagnetics Research* 61 (2006): 1–26.

[52] Pazynin, V. and Sirenko, K. "The Strong Approach to Analysis of Transients in the Axially Symmetrical Waveguide Units." *Telecommunication and Radio Engineering* 65, no. 1 (2006): 1–18.

[53] Pazynin, V. and Sirenko, K. "Transformation of Pulsed TE_{0n} - and TM_{0n} - Waves by Axially Symmetric Waveguide Nodes. Simple Inhomogeneities." *Elektromagnitnye Volny i Elektronnye Sistemy* 10, no. 8 (2005): 29–39.

[54] Pazynin, V. and Sirenko, K. "Transformation of TE_{0n} - and TM_{0n} -Waves by Axially-Symmetrical Waveguide Units. Slot Resonances." *Elektromagnitnye Volny i Elektronnye Sistemy* 10, no. 10 (2005): 21–26.

[55] Melezhyk, P., Pazynin, V., and Sirenko, K. "Transformation of Pulsed TE_{0n} - and TM_{0n} -Waves by Axially Symmetric Waveguide Nodes. Special Tasks." *Elektromagnitnye Volny i Elektronnye Sistemy* 11, no. 2-3 (2006): 22–29.

[56] Sirenko, K. and Pazynin, V. "Axially-Symmetrical Radiators of Pulsed and Monochromatic TE_{0n} - and TM_{0n} -Waves." *Uspehi Sovremennoy Radioelektroniki* no. 4 (2006): 52–69.

[57] Sirenko, K. "Slot Resonances in Axially Symmetric Radiators of Pulse-Modulated and Monochromatic TM_{0n} -Modes." *Telecommunication and Radio Engineering* 66, no. 1 (2007): 9–21.

[58] Chernobrovkin, R., Ivanchenko, I., Korolev, A., Popenko, N., and Sirenko, K. "The Novel Microwave Stop-Band Filter." *Active and Passive Electronic Components* 2008, ID 745368 (2008).

[59] Kuzmitchev, I., Melezhyk, P., Pazynin, V., Sirenko, K., Sirenko, Y., Shafalyuk, O., and Velychko, L. "Model Synthesis of Energy Compressors." *Radiofizika i Elektronika* 13, no. 2 (2008): 166–172.

[60] Sirenko, K. "Splitting of Super-Broadband Pulses by Simple Inhomogeneities of Circular and Coaxial Waveguide." *Telecommunication and Radio Engineering* 67, no. 16 (2008): 1415–1428.

[61] Velychko, L. and Sirenko, Y. Controlled Changes in Spectra of Open Quasi-Optical Resonators." *Progress in Electromagnetics Research B* 16 (2009): 85–105.

[62] Sirenko, K., Sirenko, Y., and Yashina, N. "Modeling and Analysis of Transients in Periodic Gratings. II. Resonant Wave Scattering." *Journal of the Optical Society of America A* 27, no. 3 (2010): 544–552.

[63] Yevdokymov, A., Kryzhanovskiy, V., Pazynin, V., and Sirenko, K. "Ka-Band Waveguide Rotary Joint." *IET Microwaves, Antennas & Propagation* 7, no. 5 (2011): 365–369.

[64] Sirenko, K., Pazynin, V., Sirenko, Y., and Bagci, H. "Compression and Radiation of High-Power Short Radio Pulses. I. Energy Accumulation in Direct-Flow Waveguide Compressors." *Progress in Electromagnetics Research* 116 (2011): 239–270.

[65] Sirenko, K., Pazynin, V., Sirenko, Y., and Bagci, H. "Compression and Radiation of High-Power Short Radio Pulses. II. A Novel Antenna Array Design with Combined Compressor/Radiator Elements." *Progress in Electromagnetics Research* 116 (2011): 271–296.

[66] Pazynin, V. "Compression of frequency-Modulated Electromagnetic Pulses in Sections of Regular Waveguides." *Telecommunications and Radio Engineering* 71, no. 29 (2012): 1833–1857.

[67] Granet, G., Melezhik, P., Sirenko, K., and Yashina, N. "Time-and-Frequency Domains Approach to Data Processing in Multiwavelength Optical Scatterometry of Dielectric Gratings. *Journal of the Optical Society of America A* 30, no. 3 (2013): 427–436.

[68] Sautbekov, S., Sirenko, Y., and Vertiy, A. "Rigorous Simulation of Diffraction Radiation Effect: Electrodynamics of Periodic Structures Places in the Surface Wave Field of Dielectric Waveguides." *Telecommunications and Radio Engineering* 72, no. 14 (2013): 1263–1278.

[69] Vertiy, A., Sautbekov, S., Sirenko, Y., and Yashina, N. "Effects of Diffraction Radiation in Finite Planar and Axially Symmetric Periodic Structures." *Fizicheskie Osnovy Priborostroeniya* 2, no. 4 (2013): 36–52.

[70] Sautbekov, S., Sirenko, Y., Yashina, N., and Vertiy, A. "Rigorous 2-D Model for Study of Pulsed and Monochromatic Waves Propagation near the Earth's Surface." *International Journal of Antennas and Propagation* 2014, Article ID 158297 (2014).

[71] Burambayeva, N., Sautbekov, S., Sirenko, Y., and Vertiy, A. "Compact Open Resonator as the Power-Storage Unit for a Microwave

Compressor." *Telecommunications and Radio Engineering* 74, no. 1 (2015): 29–40.

[72] Sautbekov, S., Sirenko, K., Sirenko, Y., and Yevdokimov, A. "Diffraction Radiation Effects: A Theoretical and Experimental Study." *Antennas and Propagation Magazine, IEEE* 57, no. 5 (2015): 73–93.

[73] Balatay, A., Begimova, A., Duisengali, G., Yeleusinova, G., Kasabayeva, D., Maimurynova, A., Tlegen, M., and Sirenko, Y. "Polarization Sensitivity of Reflection and Semi-Transparent Periodic Structures." *Telecommunications and Radio Engineering* 74, no. 14 (2015): 1241–1256.

[74] Melezhyk, P., Ney, M., Sautbekov, S., Sirenko, Y., Vertiy, A., and Yashina, N. "Directional Antenna Based on Vavilov-Cherenkov Radiation." *Telecommunications and Radio Engineering* 75, no. 12 (2016): 1051–1056.

[75] Pazynin, V. and Maiboroda, M. "Electromagnetic Pulse Compression in Sections of Helically Coiled Waveguides." *Telecommunication and Radio Engineering* 76, no. 3 (2017): 209–225.

[76] Pazynin, V. and Maiboroda, M. "Compression of Electromagnetic Pulses with Piecewise Linear Laws of Amplitude and Frequency Modulation." *Telecommunication and Radio Engineering* 76, no. 4 (2017): 277–284.

[77] Pazynin, V. "Multiplet Properties in the Spectrum of Oscillations of Coupled Waveguide Resonators Chain." *Radiofizika i Elektronika* 8(22), no. 1 (2017): 3–14.

[78] Ney, M., Sirenko, K., Sirenko, Y., Sliusarenko, H., and Yashina, N. "2-D Photonic Crystals: Electromagnetic Models of the Method of Exact Absorbing Conditions." *Telecommunication and Radio Engineering* 76, no. 3 (2017): 185–207.

[79] Oppenheim, A. and Schafer, R. *Discrete-Time Signal Processing.* Englewood Cliffs, NJ: Prentice-Hall, 2009.

[80] Pazynin, V., Sautbekov, S., Sirenko, K., Sirenko, Y., Vertiy, A., and Yashina, N. "Comparison of Exact and Approximate Absorbing Conditions for Initial Boundary Value Problems of the Electromagnetic Theory of Gratings." *Telecommunications and Radio Engineering* 77, no. 18 (2018): 1581–1595.

[81] Sirenko, Y., Shestopalov, V., and Yashina, N. "New Methods in the Dynamic Linear Theory of Open Waveguide Resonators." *Computational Mathematics and Mathematical Physics* 37, no. 7 (1997): 845–853.

[82] Sirenko, Y. and Yashina, N. "Nonstationary Model Problems for Waveguide Open Resonator Theory." *Electromagnetics* 19, no. 5 (1999): 419–442.

[83] Sirenko, Y., Sukharevskiy, I., Sukharevskiy, O., and Yashina, N. *Fundamental and Applied Problems in the Scattering Theory of Electromagnetic Waves*. Kharkiv: Krok, 2000.

[84] Sirenko, Y. and Yashina, N. "Time Domain Theory of Open Waveguide Resonators: Canonical Problems and a Generalized Matrix Technique." *Radio Science* 38, no. 2 (2003): VIC 26-1–VIC 26-12.

[85] Kravchenko, V., Sirenko, K., and Sirenko, Y. "Transport Operators and Exact Absorbing Conditions in the Plane Problems of the Electrodynamics of Pulsed Waves for Compact Open Resonators with the Waveguide Feeder Line." *Elektromagnitnye Volny i Elektronnye Sistemy* 14, no. 1 (2009): 4–19.

[86] Kravchenko, V., Pazynin, V., Sirenko, K., and Sirenko, Y. "Planar Problems of Non-Sinusoidal Electrodynamics for Compact Open Resonators with a Waveguide Feeding Line. Grating as Diagram-Forming Structures." *Elektromagnitnye Volny i Elektronnye Sistemy* 14, no. 6 (2009): 4–20.

[87] Nikol'skiy, V. *Electrodynamics and Radio Waves Propagation*. Moscow: Nauka, 1973.

In: A Closer Look at Boundary Value Problems ISBN: 978-1-53617-857-9
Editor: Mustafa Avci

Chapter 4

DIFFRACTION BOUNDARY VALUE PROBLEMS FOR ELECTROMAGNETIC THEORY OF INHOMOGENEOUS MULTILAYERED MEDIA: RICCATI EQUATION METHOD

Petr N. Melezhik[1], Anatoliy Ye. Poyedinchuk[1], Yuriy K. Sirenko[1,2] and Nataliya P. Yashina[1,*]

[1]Department of Diffraction Theory and Diffraction Electronics,
O.Ya. Usikov Institute for Radiophysics and Electronics,
Kharkiv, Ukraine

[2]Department of Applied Mathematics,
V.N. Karazin Kharkiv National University, Kharkiv, Ukraine

ABSTRACT

This chapter is devoted to the construction of the solutions to boundary value problems of an electromagnetic wave diffraction by layered inhomogeneous media with constitutive parameters depending on the single spatial coordinate. The new results related to the development of the robust numerical-analytical approach to

* Corresponding Author's Email: yashinanp@gmail.com.

an analysis of such problems, basing on the algorithms for constructing solutions to the Cauchy problem for the Riccati equation, are presented. The computational efficiency of the approach has been demonstrated by a number of problems of electromagnetic wave diffraction by dielectric structures: a transversely inhomogeneous isotropic and anisotropic dielectric layers; a radially inhomogeneous circular cylinder, a nonlinear dielectric layer. The principal possibility of applying the developed approach to solving the spectral problems of electromagnetic theory of layered inhomogeneous media is shown and verified.

Keywords: electromagnetic theory, electrodynamics of layered inhomogeneous media, boundary value problems of wave diffraction, numerical-analytical methods, Cauchy problem, Riccati equation, spectral problems

1. INTRODUCTION

For more than a dozen of years, researchers have been actively engaged in the study of the propagation and scattering of acoustic and electromagnetic waves in media with constitutive parameters depending on a single spatial variable [1–7]. Wave processes in such media are described by boundary value problems for Maxwell's equations with variable coefficients that admit an analytical solution only in a few special cases [7, 8]. Under certain restrictions on functions modeling inhomogeneous media (for example, in the case of smoothly changing material parameters), several asymptotic methods can be used: the ray method, the parabolic equation method, the "standard solution" method, etc. [9]. The construction and application of various numerical-analytical and numerical methods for solving model problems of the electromagnetic theory of layered inhomogeneous media significantly expanded the range of already solved problems [10–18]. The methods based on approximation of the material parameters of layered inhomogeneous media by piecewise constant functions are the most common and easiest to use among the corresponding set of methods. It is clear in this case that in order to achieve a sufficiently high accuracy of the solution it is necessary to consider a media with a very large number of homogeneous layers that cannot be accurately carried out without accumulating significant calculation errors [10].

A new numerical-analytical approach for solving the problems of electromagnetic wave diffraction by the objects with permittivity and permeability depending on a single spatial variable is presented in this chapter. This is a rigorous method based on the use of special solutions to the Riccati equation [13–17]. The method is not limited by any significant restrictions on the constitutive parameters of the considered layered inhomogeneous media; it is sufficient to require the permittivity and permeability of these media to be described by piecewise continuous functions of one spatial variable.

Outline the chapter structure: the problems of plane wave diffraction by transversely inhomogeneous layers of finite thickness – isotropic and anisotropic -are considered in Sections 2 and 4; Section 3 is devoted to solving the problem of a plane wave diffraction by a radially inhomogeneous circular dielectric cylinder; in Sections 5 and 7 the nonlinear and spectral (homogeneous) problems of the electromagnetic theory of layered inhomogeneous media are considered, and in Section 6 the solution to the problem of plane wave diffraction by a perfectly conducting strip grating backed with a transversely inhomogeneous dielectric layer is discussed.

The dimensions of physical quantities in the text of chapter are omitted; they all correspond to the international system of units SI.

2. Transversely Inhomogeneous Dielectric Layer

At the beginning let us consider the problem of a plane monochromatic electromagnetic wave diffraction by a flat inhomogeneous magnetoelectric layer of the thickness h. The geometry of the problem is depicted in Figure 1. Let the layer take part $\mathrm{D}=\{p=\{x,y,z\}:-h\le z<0,-\infty<x,y<\infty\}$ of the space R^3; x, y, z are the Cartesian coordinates. It is supposed that the relative permittivity and permeability of the layer depend only on the coordinate z: $\varepsilon_1=\varepsilon_1(z)$, $\mu_1=\mu_1(z)$. The half spaces $z>0$ and $z<-h$ are filled with homogeneous magnetodielectric media characterized by $\varepsilon=\mu=1$ and $\varepsilon_2=const$, $\mu_2=const$.

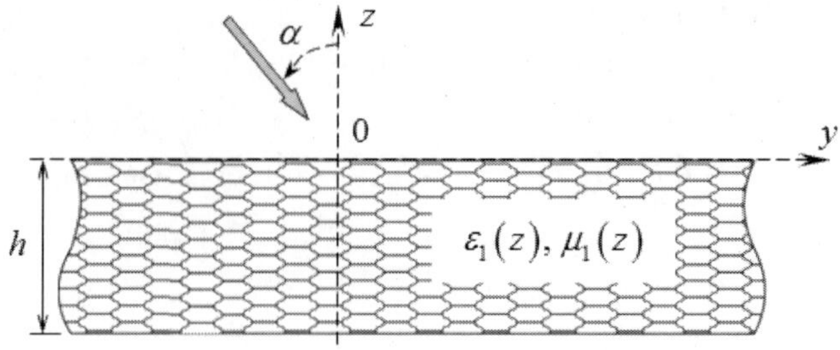

Figure 1. To the problem of electromagnetic wave diffraction by a transversely inhomogeneous dielectric layer.

A monochromatic plane wave with electric field vector parallel to the axis x, propagates in the half space $z>0$, wave vector $\vec{k}$ lays in the plane $y0z$ and makes an angle α with the axis z. The non-zero components of the electromagnetic field of this wave have the form

$$E_x^0 = \exp\left[ik\left(y\sin\alpha - z\cos\alpha\right) - i\omega t\right], \quad H_y^0 = -\cos\alpha E_x^0, \quad H_z^0 = -\sin\alpha E_x^0, \tag{1}$$

where $k=\omega/c$ and c is the speed of light in a medium filling half-space $z>0$. Subsequently, the time factor $\exp(-i\omega t)$ in all representations is omitted. The problem is to find out the electromagnetic field resulting from the diffraction of the wave (1) by a plane layer. As follows from Maxwell's equations, this problem is reduced to finding out $u(y,z)$, satisfying everywhere outside the boundaries of the layer the Helmholtz equation

$$\left[\mu(z)\frac{\partial}{\partial z}\left(\frac{1}{\mu(z)}\frac{\partial}{\partial z}\right) + \frac{\partial^2}{\partial y^2} + k^2 n(z)\right]u(y,z) = 0, \tag{2}$$

and the following boundary conditions:

$$\left(u^0+u\right)\Big|_{z=0+0}=u\Big|_{z=0-0},\quad \left.\frac{\partial\left(u^0+u\right)}{\partial z}\right|_{z=0+0}=\gamma_1\left.\frac{\partial u}{\partial z}\right|_{z=0-0}, \tag{3}$$

$$u\Big|_{z=-h+0}=u\Big|_{z=-h-0},\quad \left.\frac{\partial u}{\partial z}\right|_{z=-h+0}=\gamma_2\left.\frac{\partial u}{\partial z}\right|_{z=-h-0}. \tag{4}$$

Here, $u^0=E_x^0$, $\gamma_1=\mu_1^{-1}(0)$, $\gamma_2=\mu_1(-h)/\mu_2$, and functions $n(z)$ and $\mu(z)$ are defined by formulas

$$n(z)=\begin{cases}1; & z>0\\ \varepsilon_1(z)\mu_1(z); & -h<z<0,\\ \varepsilon_2\mu_2; & z<-h\end{cases}\quad \mu(z)=\begin{cases}1; & z>0\\ \mu_1(z); & -h<z<0.\\ \mu_2; & z<-h\end{cases}$$

The solution $u(y,z)$ to the problem (2)–(4) must satisfy in half-spaces $z>0$ and $z<-h$ the radiation conditions, meaning that the field corresponding to the function $u(y,z)$, should not include waves coming from the infinity [19].

It is easy to show that the non-zero components of the diffraction field (secondary field) are determined via $u(y,z)$:

$$E_x=u(y,z),\quad H_y=\frac{1}{ik\mu(z)}\frac{\partial u(y,z)}{\partial z},\quad H_z=-\frac{1}{ik\mu(z)}\frac{\partial u(y,z)}{\partial y}.$$

Suppose $\mu_1(z)=const$ (this restriction does not significantly affect the generality of the results obtained below) and introduce dimensionless variables $\bar{z}=z/h$, $\bar{y}=y/h$, in which equation (2) can be written as,

$$\left[\frac{\partial^2}{\partial\bar{y}^2}+\frac{\partial^2}{\partial\bar{z}^2}+\kappa^2\bar{n}(\bar{z})\right]\bar{u}(\bar{y},\bar{z})=0;\quad \kappa=kh. \tag{5}$$

Here, $\bar{u}(\bar{y},\bar{z})=u(\bar{y}h,\bar{z}h)$ and $\bar{n}(\bar{z})=n(\bar{z}h)$.

Separating the variables in the equation (5), we obtain the following representation for $\bar{u}(\bar{y},\bar{z})$:

$$\bar{u}(\bar{y},\bar{z})=\begin{cases} R(\kappa,\alpha)\exp\left[i\kappa(\bar{y}\sin\alpha+\bar{z}\cos\alpha)\right]; & \bar{z}\ge 0 \\ \left[A_+(\kappa,\alpha)u_+(\bar{z})+A_-(\kappa,\alpha)u_-(\bar{z})\right]\exp(i\kappa\bar{y}\sin\alpha); & -1\le\bar{z}\le 0 \\ T(\kappa,\alpha)\exp\left[i\kappa\left(\bar{y}\sin\alpha-(\bar{z}+1)\sqrt{\varepsilon_2\mu_2-\sin^2\alpha}\right)\right]; & \bar{z}\le -1. \end{cases} \quad (6)$$

Here, R, $A_\pm$, and T are unknown complex values, and $u_\pm(\bar{z})$ are linearly independent solutions of an ordinary second-order differential equation

$$\left[\frac{d^2}{d\bar{z}^2}+\kappa^2\left(\bar{\varepsilon}_1(\bar{z})\mu_1-\sin^2\alpha\right)\right]u_\pm(\bar{z})=0;\quad \bar{\varepsilon}_1(\bar{z})=\varepsilon_1(\bar{z}h). \quad (7)$$

Now suppose that the linearly independent solutions $u_+(\bar{z})$ and $u_-(\bar{z})$ of the equation (7) are constructed, at $\bar{z}=-1$ the conditions

$$u_\pm(-1)=1,\quad \frac{du_\pm(-1)}{d\bar{z}}=\pm ip;\quad p\ne 0,\quad p=\gamma_2\kappa\sqrt{\varepsilon_2\mu_2-\sin^2\alpha} \quad (8)$$

hold. Then, substituting (6) into conditions (4) at the lower layer's boundary $\bar{z}=-1$, we obtain $A_+=0$ and $A_-=T$. Therefore, inside the layer (for $-1\le\bar{z}\le 0$) function $\bar{u}(\bar{y},\bar{z})$ has a form $T(\kappa,\alpha)u_-(\bar{z})\exp(i\kappa\bar{y}\sin\alpha)$.

Now applying the conditions (3) at the upper boundary of the layer $\bar{z}=0$ the following representations for R and T

$$R=\frac{i\kappa\mu_1u_-(0)\cos\alpha+du_-(\bar{z})/d\bar{z}\big|_{\bar{z}=0}}{i\kappa\mu_1u_-(0)\cos\alpha-du_-(\bar{z})/d\bar{z}\big|_{\bar{z}=0}}, \quad (9)$$

$$T=-\frac{2i\kappa\mu_1\cos\alpha}{i\kappa\mu_1u_-(0)\cos\alpha-du_-(\bar{z})/d\bar{z}\big|_{\bar{z}=0}}$$

are derived. The formulas (6)–(9) give the solution to the diffraction problem (2)–(4).

It is clear now that the key to constructing this solution is to find solutions of equation (7) satisfying conditions (8). Note that the solutions to the ordinary differential equation (7) in the general case are not expressed in terms of elementary or special functions. However, in a number of special cases (see, for example, [1, 7, 8]) a solution can be obtained in an analytical form. To construct solutions to the equation (7) (with insignificant restrictions on the function $\bar{\varepsilon}_1(\bar{z})$), it is possible to use one of numerical-analytical approaches [12]: the power series method, the method of successive approximations, the method of polynomial operators, the method of approximation by polynomials, the grid method, projection methods, etc. Below we propose a new algorithm for calculating the functions $u_{\pm}(\bar{z})$; for its founded application it is sufficient to assume that the function $\bar{\varepsilon}_1(\bar{z})$ is a piecewise continuous one.

So, it is required to construct a solution to the following problem

$$\begin{cases} \left[\dfrac{d^2}{d\,\bar{z}^2}+\kappa^2 q(\bar{z})\right]u(\bar{z})=0; \quad -1<\bar{z}<0 \\ u(-1)=1, \quad \dfrac{du(-1)}{d\bar{z}}=-ip, \quad p\neq 0, \end{cases} \tag{10}$$

where $q(\bar{z})=\bar{\varepsilon}_1(\bar{z})\mu_1-\sin^2\alpha$. Let us introduce a new unknown function $V(\bar{z})$ using the representation $u(\bar{z})=\exp\left(\int_{-1}^{\bar{z}}V(\tau)d\tau\right)$. Substituting this representation in the boundary conditions for $u(\bar{z})$ and considering $du/d\,\bar{z}=Vu$, we reduce (10) to the Cauchy problem for the Riccati equation:

$$\begin{cases} \dfrac{dV(\bar{z})}{d\,\bar{z}}+V^2(\bar{z})+\kappa^2 q(\bar{z})=0; \quad -1<\bar{z}<0 \\ V(-1)=-ip, \quad p=\gamma_2\kappa\sqrt{\varepsilon_2\mu_2-\sin^2\alpha}. \end{cases} \tag{11}$$

Construct an approximate solution to this problem using the following algorithm. Let us approximate $V(\bar{z})$ with the grid function $V_n \approx V(\bar{z}_n)$, $\bar{z}_n = -1 + \delta(n-1)$, $n = 1,2,...,N$, $\delta = (N-1)^{-1}$, N is an integer positive number. Integrating the equation in (11) over $\bar{z}$ from $\bar{z}_n$ to $\bar{z}_{n+1}$, we obtain

$$V_{n+1} - V_n + \int_{\bar{z}_n}^{\bar{z}_{n+1}} V^2(\bar{z})d\bar{z} + \kappa^2 \int_{\bar{z}_n}^{\bar{z}_{n+1}} q(\bar{z})d\bar{z} = 0; \quad n = 1,2,...,N-1. \tag{12}$$

Assuming that N is a sufficiently large number and approximating the integrals in (12) by the quadrature sum of the two-point trapezoidal formula [20], we arrive to

$$V_{n+1} - V_n + \frac{\delta}{2}\left(V_{n+1}^2 + V_n^2\right) + \kappa^2 \frac{\delta}{2}\left(q_{n+1} + q_n\right) = 0; \quad q_n = q(\bar{z}_n). \tag{13}$$

Now let us represent (13) in the form $\delta V_{n+1}^2/2 + V_{n+1} + F_n = 0$, here $F_n = -V_n + \delta V_n^2/2 + \kappa^2\delta(q_{n+1} + q_n)/2$, and find out $V_{n+1} = \left(-1 \pm \sqrt{1 - 2\delta F_n}\right)/\delta$. In this representation we single out the value V_{n+1}, tending to V_n, at $\delta \to 0$ (at $N \to \infty$) and as a result obtain

$$V_{n+1} = \frac{-2F_n}{1 + \sqrt{1 - 2\delta F_n}}; \quad n = 1,2,...,N-1. \tag{14}$$

The relation (14) allows to sequentially calculate V_n, if $V_1 = -ip$ is known. It can be shown (see, for example, [20]) that the error in these calculations will be of order δ^3. Thus, formula (14) determines the algorithm for calculating approximate values of the solutions to the Cauchy problem (11) $\{V_n\}_{n=1}^N$ on the grid $\{\bar{z}_n\}_{n=1}^N$, covering the interval $-1 < \bar{z} < 0$.

Reflection and transmission coefficients (9) of a plane E-polarized electromagnetic wave (1) incident on a transversely inhomogeneous dielectric layer, in terms of functions $V(\overline{z})$ can be represented in the following form:

$$R=\frac{i\kappa\mu_1\cos\alpha+V(0)}{i\kappa\mu_1\cos\alpha-V(0)},\quad T=-\frac{2i\kappa\mu_1\cos\alpha}{i\kappa\mu_1\cos\alpha-V(0)}\exp\left(-\int_{-1}^{0}V(\overline{z})d\,\overline{z}\right). \qquad (15)$$

Here, $V(0)\approx V_N$; the exponential factor is easily calculated using the recurrence formula

$$\exp\left(-\int_{-1}^{0}V(\overline{z})d\,\overline{z}\right)\approx B_N;\quad B_1=1,\quad B_{n+1}=B_n\exp\left(-\frac{\delta}{2}(V_n+V_{n+1})\right),$$
$$n=1,2,...,N-1.$$

In the case of H-polarization (the vector of the magnetic field of the incident wave is parallel to the axis x) the formulas similar to (15) can be written in the form

$$R=\frac{i\kappa\cos\alpha+V(0)}{i\kappa\cos\alpha-V(0)},\quad T=\frac{2i\kappa\cos\alpha}{i\kappa\cos\alpha-V(0)}\exp\left(-\int_{-1}^{0}\overline{\varepsilon}_1(\overline{z})V(\overline{z})d\,\overline{z}\right).$$

As a result, the Cauchy problem for finding out the function $V(\overline{z})$ takes the form

$$\begin{cases}\dfrac{dV(\overline{z})}{d\,\overline{z}}+\overline{\varepsilon}_1(\overline{z})V^2(\overline{z})+\kappa^2\left(\mu_1(\overline{z})-\dfrac{\sin^2\alpha}{\varepsilon_1(\overline{z})}\right)=0;\quad -1<\overline{z}<0\\[2ex] V(-1)=-i\kappa\dfrac{\sqrt{\varepsilon_2\mu_2-\sin^2\alpha}}{\varepsilon_2}.\end{cases}$$

The constructed algorithm for solving the diffraction problem allows to obtain an analytical representation of the reflection coefficient R from (14) in the limit case $\kappa \to 0$ (the long-wave approximation). Let us assume that a permeability of the layer is a continuous function, and this function is a piecewise continuous function of the coordinate $\bar{z}$ (discontinuities of the first kind are allowed). At the next step let us integrate the equation of problem (11) and introduce a new unknown function $W(\bar{z}) = V(\bar{z})/\kappa$. After these rearrangements the Cauchy problem (11) is reduced to the equivalent integral equation

$$W(\bar{z}) + ip + \kappa \int_{-1}^{\bar{z}} W^2(\tau)\,d\tau + \kappa \int_{-1}^{\bar{z}} q(\tau)\,d\tau = 0;$$
$$p = \frac{\mu_1}{\mu_2}\sqrt{\varepsilon_2\mu_2 - \sin^2\alpha}, \quad q(\tau) = \bar{\varepsilon}_1(\tau)\mu_1 - \sin^2\alpha. \tag{16}$$

We seek the solution to (16) in the form of an asymptotic series in powers of the small parameter κ:

$$W = a_0 + \kappa a_1 + \kappa^2 a_2 + \ldots \tag{17}$$

Substituting (17) into (16) and retaining the term containing κ^2, we obtain

$$a_0 = -ip, \quad a_1 = ip(\bar{z}+1) - \int_{-1}^{\bar{z}} q(\tau)\,d\tau, \quad a_2 = -p^2(\bar{z}+1)^2 - 2ip\int_{-1}^{\bar{z}} d\xi \int_{-1}^{\xi} q(\tau)\,d\tau,$$

and

$$R = \frac{i\mu_1\cos\alpha + \bar{V}}{i\mu_1\cos\alpha - \bar{V}}; \quad \bar{V} = -ip + \kappa\left[ip - \int_{-1}^{0} q(\tau)\,d\tau\right] - \kappa^2 p\left[p + 2i\int_{-1}^{0} d\xi \int_{-1}^{\xi} q(\tau)\,d\tau\right]. \tag{18}$$

In Figure 2, the absolute values of reflection coefficients calculated using the approximate formula (18) (dashed lines) and using the above-described algorithm for numerical solution of the diffraction problem (solid lines) are presented. The permittivity of the layer is $\bar{\varepsilon}_1(\bar{z}) = 4(\bar{z}+1)^q + 1$, and $\mu_1 = 1$, $\alpha = 0$. Permittivity and permeability of the media, filling half-spaces $\bar{z} > 0$ and $\bar{z} < -1$ are assumed equal to unity. Results with graphic accuracy coincide up to $\kappa \approx 0.5$.

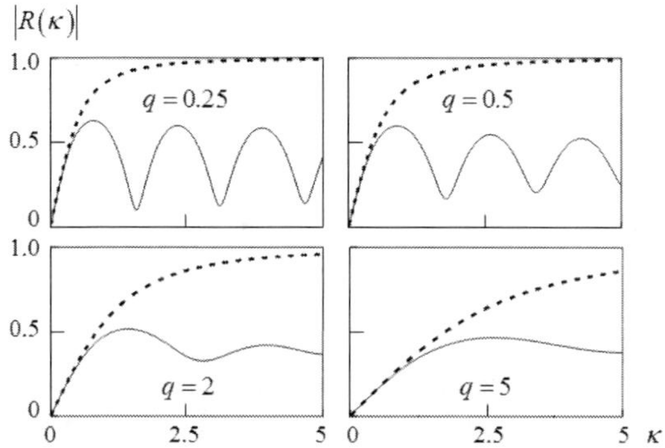

Figure 2. E-polarization. To the verification of the long-wave approximation for the calculation of $|R(\kappa)|$.

An analysis of similar results for various types of dependences on $\bar{z}$ of the constitutive parameters of the layer allowed us to draw the following conclusion. If the relations $\left|\kappa\sqrt{\mu_1 \int_{-1}^{0} \bar{\varepsilon}_1(\bar{z})\, d\,\bar{z}}\right| < 0.5$ and $\left|\kappa\sqrt{\varepsilon_2 \mu_2}\right| < 0.5$ hold, then the relative error in calculating the reflection coefficient by the approximate formula (18) is less than 1% for any angle of incidence of a plane wave (1). It can also be noted that with $\kappa \to 0$ reflection coefficient (18) tends to the reflection coefficient for a half-space with constitutive parameters ε_2 and μ_2, in particular, when $\varepsilon_2 = \mu_2 = 1$ we have $R \to 0$. In the parameter variation region $\kappa\sqrt{\max\limits_{-1<\bar{z}<0} |\bar{\varepsilon}_1(\bar{z})|} \approx 1$, the formula (18) arrives to rather large errors in R. The principal error in the algorithm for calculation of R and T appears due to the approximation of the integrals in (12) by the quadrature sum of the two-point trapezoidal formula. Analytically, the error of such approximation is estimated as $O\left((\kappa\delta)^2\right)$.

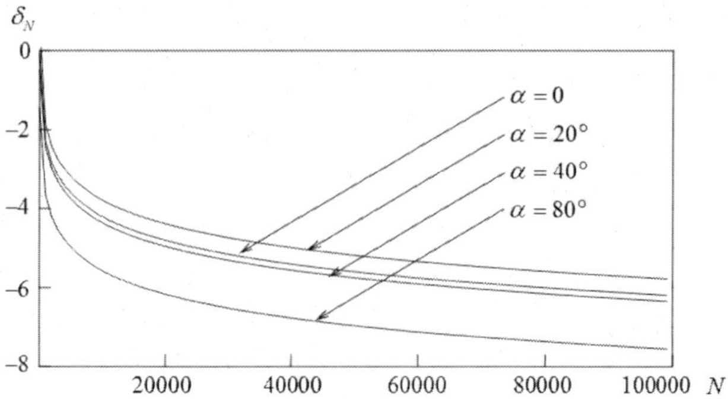

Figure 3. E -polarization. δ_N as a function of the discretization parameter N for various angles of incidence, $\kappa = 10$.

In Figure 3 some numerical results estimating the efficiency of the constructed algorithm are presented. As a parameter characterizing the algorithm's rate of convergence, the value $\delta_N = \ln\left|\left(R_N - R_{2N}\right)/R_{2N}\right|$ is used; here R_N and R_{2N} are the reflection coefficients calculated within the interval $-1 < \overline{z} < 0$ at sampling steps $\delta = \left(N-1\right)^{-1}$ and $\delta = \left(2N-1\right)^{-1}$. The calculations have been carried out for an inhomogeneous layer with the permittivity $\overline{\varepsilon}_1\left(\overline{z}\right) = 5 + \sin 2\pi q\left(\overline{z}+1\right) + i0.01$, $q = 10$. With parameter values $q >> 1$ function $\overline{\varepsilon}_1\left(\overline{z}\right)$ is a fast oscillating function of $\overline{z}$. The application of well-known approximate methods for such a type of inhomogeneous layer is practically impossible. When using suggested here algorithm, δ_N rapidly decrease with increasing number of sampling points N. The calculation time of reflection coefficient R_N at fixed values of the frequency parameter κ and angle of incidence α is a few milliseconds on a computer with a conventional processor.

The capabilities of the algorithm can also be estimated by the numerical results obtained for the case when the permittivity of a layer is a piecewise constant function (a discrete-layered medium). In this case, the explicit solution of the problem can be compared with its numerical solution and thus the exact value of the error of the implemented approach can be determined. In the test problem, we considered a medium formed by four homogeneous dielectric layers:

$$\bar{\varepsilon}_1(\bar{z}) = \begin{cases} 2.0 + i0.001; & -1.0 < \bar{z} < -0.75 \\ 3.0 + i0.001; & -0.75 < \bar{z} < -0.5 \\ 5.0 + i0.001; & -0.5 < \bar{z} < -0.25 \\ 8.0 + i0.001; & -0.25 < \bar{z} < 0. \end{cases}$$

In Figure 4 the results for $|R(\kappa)|$ obtained by the input impedance method [4] (explicit solution) and in the framework of the algorithm described above are presented. As one can see, even with a small number of sampling points $N = 100$ ($\delta \approx 10^{-2}$), this algorithm leads to a fairly good result within the frequency parameter region limited by $\kappa \leq 6$. At $N \geq 1000$ the exact and approximate solutions, coincide with graphic accuracy in the entire considered range of frequency parameter variation. The time required for calculation of these solutions is also the same.

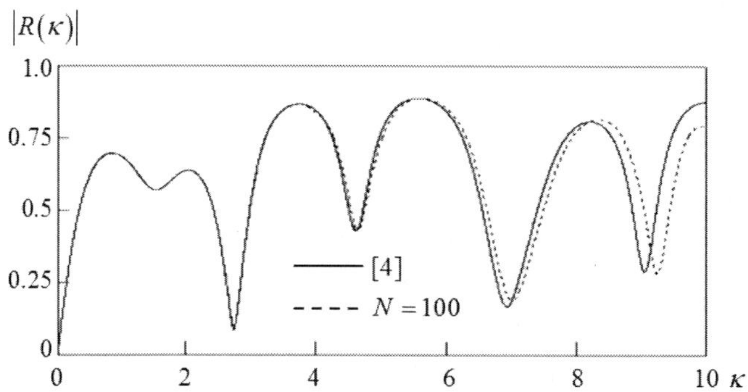

Figure 4. E -polarization. Comparison of calculated $|R(\kappa)|$ with results [4], $\alpha = 0$.

Two quite different examples illustrate the abilities and advantages of the proposed method for the numerical solution to problem (2)–(4). The comment on the physical content of the obtained here solutions are out of the scope of the study. The first example corresponds to plasma-like medium with frequency dispersion. The permittivity of the medium is modeled by the function:

$$\bar{\varepsilon}_1(\bar{z},\omega)=1-\frac{\omega_p^2 N(\bar{z})}{\omega(\omega+i\nu)}.$$

Here the dimensionless function $0 \le N(\bar{z}) \le 1$ determines the distribution of electrons over the layer thickness, ω_p is the plasma frequency corresponding to the maximum value of the electron density of the layer. For the linear, parabolic, and exponential profiles of normalized density $N(\bar{z})$ the exact solutions to the problem (2)–(4) [3] have been obtained. We have chosen $N(\bar{z})=a+b\cos\left[2\pi q(\bar{z}+1)\right]$, where $a=0.6$ и $b=0.4$, and parameter q can take any values. The layer with $\mu_1=1$ is located in half space with $\varepsilon_2(\omega)=1-\omega_p^2/(\omega(\omega+i\nu))$ and $\mu_2=1$. In Figure 5a, the numerical results for $\left|R(\omega/\omega_p)\right|$ are presented. In the frequency range $\omega<\omega_p$ $|R|\approx 1$, and for $\omega>\omega_p$ the manifestation of resonant behavior of $\left|R(\omega/\omega_p)\right|$ is possible. In Figure 5b illustrated the behavior of $|R(\alpha)|$. If $\omega>\omega_p$, then there is a range $\cos\alpha \le \omega_p/\omega$ of incidence angles α, where $|R(\alpha)|\approx 1$.

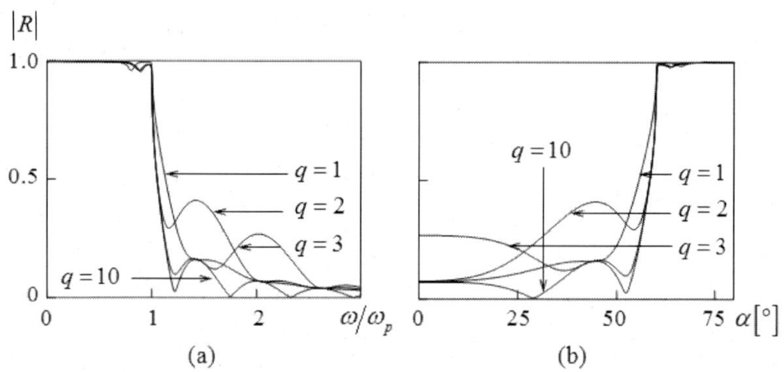

Figure 5. E-polarization, $\omega_p h/2\pi c=5$, $\nu/\omega_p=0.01$. (a) The absolute value of reflection coefficient $\left|R(\omega/\omega_p)\right|$, $\alpha=0$. (b) The absolute value of reflection coefficient $|R(\alpha)|$, $\omega/\omega_p=2$.

Let us now consider the case of parameters $\bar{\varepsilon}_1(\bar{z}) = \varepsilon\left(1+\sin^2 20\pi(\bar{z}+1)\right)+i0.01$, $\varepsilon = 5$; $\bar{\varepsilon}_1(\bar{z})$ is a periodic fast oscillating function of a spatial variable $\bar{z}$. Here $\bar{\varepsilon}_1(\bar{z})$ does not depend on the frequency of the incident plane wave. Figure 6 shows a typical behavior of $|R(\kappa)|$, indicating the existence of a finite interval of frequency parameter variation, where $|R(\kappa)| \approx 1$ (the so-called "forbidden zones"). This behavior of the reflection coefficient is an analog of the Bragg reflection of waves propagating in a periodic medium. Outside the Bragg reflection ranges, the function $|R(\kappa)|$ is subject to oscillations, which are caused by multiple reflection from the boundaries of the inhomogeneous layer. The period of these oscillations is determined by 1.5ε that is the real part of the dielectric constant $\bar{\varepsilon}_1(\bar{z})$ averaged over the thickness of the layer.

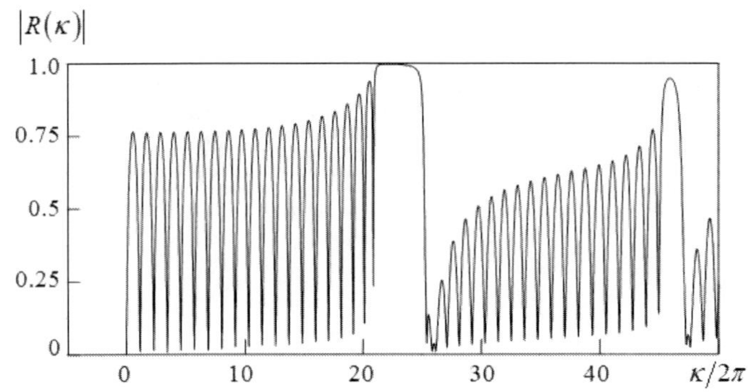

Figure 6. E -polarization. Reflection of a plane wave from a layer with permittivity that is a periodic rapidly oscillating function of a variable z.

3. Radial Inhomogeneous Circular Dielectric Cylinder

Let a cylindrical inhomogeneity occupy the part $D = \{\{r,\phi,z\} : 0 \le r < a, 0 \le \phi \le 2\pi, -\infty < z < +\infty\}$ of the space $\mathbf{R}^3$, where r, ϕ, z are cylindrical coordinates, a is a radius of the outer circular cylindrical surface (Figure 7). Permittivity and permeability inside the region D depend on the single coordinate r: here $\varepsilon_1 = \varepsilon_1(r)$ and $\mu_1 = \mu_1(r)$ are the continuous functions. The space outside the domain D is filled with a homogeneous isotropic medium with $\varepsilon = \mu = 1$.

Let a plane homogeneous monochromatic wave of unit amplitude, having electric field vector parallel to the axis z and a wave vector $\vec{k}$ opposite in direction to axis x, propagates in the region $r > a$. The non-zero field components of this wave have the form:

$$E_z^0 = \exp(-ikr\cos\phi), \quad H_r^0 = E_z^0 \sin\phi, \quad H_\phi^0 = E_z^0 \cos\phi. \tag{19}$$

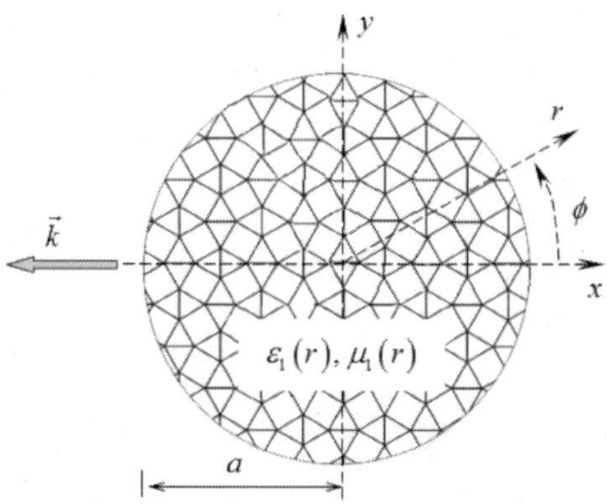

Figure 7. Cross section of a radially inhomogeneous dielectric cylinder.

The problem is to determine the electromagnetic field resulting from the diffraction of the wave (19) by a radially inhomogeneous cylinder D. This problem is reduced to the search of a twice continuously differentiable function $u(r,\phi)$, satisfying the equations

$$\begin{gathered}\left[\Delta + k^2\right] u(r,\phi) = 0; \quad r > a, \quad 0 \le \phi \le 2\pi \quad \text{and} \\ \left[\Delta - \frac{\mu'(r)}{\mu(r)}\frac{\partial}{\partial r} + k^2 \varepsilon_1(r)\mu_1(r)\right] u(r,\phi) = 0; \quad r < a, \quad 0 \le \phi \le 2\pi\end{gathered} \tag{20}$$

($\Delta = \partial^2/\partial r^2 + r^{-1}\partial/\partial r + r^{-2}\partial^2/\partial\phi^2$), boundary conditions

$$\left(u^0 + u\right)\Big|_{r=a+} = u\Big|_{r=a-}, \quad \left(\frac{\partial u^0}{\partial r} + \frac{\partial u}{\partial r}\right)\Bigg|_{r=a+} = \left(\frac{1}{\mu_1(r)}\frac{\partial u}{\partial r}\right)\Bigg|_{r=a-}; \quad u^0 = E_z^0, \tag{21}$$

and Sommerfeld radiation conditions

$$\frac{\partial u}{\partial r} - iku = o\left(\frac{1}{\sqrt{r}}\right) \text{ and } u = O\left(\frac{1}{\sqrt{r}}\right) \text{ for } r \to \infty. \tag{22}$$

All the nonzero components of the diffraction field (secondary field) are determined via the function $u(r,\phi)$ using the following representations:

$$E_z = u(r,\phi), \quad H_\phi = \frac{i}{k}\begin{cases} \dfrac{\partial u}{\partial r}; & r > a \\ \dfrac{1}{\mu_1(r)}\dfrac{\partial u}{\partial r}; & r < a \end{cases}, \quad H_r = -\frac{i}{kr}\begin{cases} \dfrac{\partial u}{\partial \phi}; & r > a \\ \dfrac{1}{\mu_1(r)}\dfrac{\partial u}{\partial \phi}; & r < a \end{cases}. \tag{23}$$

Introducing the dimensionless variable $\rho = r/a$, separating the variables ρ and ϕ in (20) we obtain

$$\bar{u}(\rho,\phi) = \begin{cases} \sum\limits_{m=0}^{+\infty} A_m(\kappa)\cos(m\phi); & \rho \ge 1 \\ \sum\limits_{m=0}^{+\infty} T_m(\kappa)u_m(\rho)\cos(m\phi); & \rho \le 1. \end{cases} \tag{24}$$

Here, $\bar{u}(\rho,\phi) = u(\rho a,\phi)$, $\kappa = ka$, $A_m = a_m(-i)^m J_m(\kappa\rho) + R_m H_m^{(1)}(\kappa\rho)$, $a_m = 2 - \delta_0^m$, δ_0^m is Kronecker symbol, R_m and T_m are sought coefficients, $J_m(...)$ and $H_m^{(1)}(...)$ are Bessel and Hankel functions, and $u_m(\rho)$ is a solution to the differential equation

$$\left[\frac{d^2}{d\rho^2}+\left(\frac{1}{\rho}-\frac{\bar{\mu}_1'(\rho)}{\bar{\mu}_1(\rho)}\right)\frac{d}{d\rho}+\left(\kappa^2\bar{\varepsilon}_1(\rho)\bar{\mu}_1(\rho)-\frac{m^2}{\rho^2}\right)\right]u_m(\rho)=0; \tag{25}$$
$$u_m(0)<\infty,\ \ m=0,1,2,\dots \text{ and } \bar{\varepsilon}_1(\rho)=\varepsilon_1(\rho a),\ \bar{\mu}_1(\rho)=\mu_1(\rho a).$$

The boundary conditions (21) for representations (24) lead to the system of equations

$$\begin{cases} T_m u_m(1)-R_m H_m^{(1)}(\kappa)=(-i)^m a_m J_m(\kappa) \\ \dfrac{1}{\kappa\bar{\mu}_1(1)}T_m u_m'(1)-R_m H_m^{(1)\prime}(\kappa)=(-i)^m a_m J_m'(\kappa) \end{cases};\ \ m=0,1,2\dots$$

(here and further on $f'(b)=df(x)/dx\big|_{x=b}$), solutions R_m and T_m can be represented as

$$R_m=a_m(-i)^m F_m,\ \ T_m=a_m(-i)^m G_m, \tag{26}$$

where

$$F_m=\frac{J_m(\kappa)u_m'(1)-\kappa\bar{\mu}_1(1)J_m'(\kappa)u_m(1)}{\kappa\bar{\mu}_1(1)u_m(1)H_m^{(1)\prime}(\kappa)-u_m'(1)H_m^{(1)}(\kappa)} \text{ and}$$
$$G_m=\frac{2i\bar{\mu}_1(1)}{\pi\left(\kappa\bar{\mu}_1(1)u_m(1)H_m^{(1)\prime}(\kappa)-u_m'(1)H_m^{(1)}(\kappa)\right)}.$$

From the above it follows that the key point in solving the diffraction problem under consideration is the finding solutions $u_m(\rho)$ of the differential equations (25) that are bounded at zero.

New unknown functions $V_m(\rho)$ are introduced via relations $u_m(\rho)=(\rho/2)^m\exp\left(\int_0^\rho\bar{\mu}_1(\tau)V_m(\tau)d\tau\right)$. From (25) and particular solutions

$u_m(0)=\delta_0^m$, $u_m'(0)=\delta_0^m\bar{\mu}_1(0)V_0(0)+0.5\delta_1^m$ it follows that $V_m(\rho)$ must satisfy the ordinary differential equations:

$$\begin{aligned}&(\rho V_m(\rho))'+2mV_m(\rho)+\rho\bar{\mu}_1(\rho)V_m^2(\rho)+\kappa^2\rho\bar{\varepsilon}_1(\rho)-m\frac{\bar{\mu}_1'(\rho)}{\bar{\mu}_1(\rho)}=0;\\&V_m(0)<\infty,\quad m=0,1,2,...\end{aligned}\tag{27}$$

Supposing

$$V_m(0)=\frac{m\bar{\mu}_1'(0)}{(2m+1)\bar{\mu}_1^2(0)},\quad \bar{\mu}_1(0)\neq 0,\tag{28}$$

we arrive at the Cauchy problem (27), (28) for the Riccati equation.

The equality $u_m'(1)=[m+\bar{\mu}_1(1)V_m(1)]u_m(1)$, valid for all integer $m\geq 0$, follows from the relation introducing the functions $V_m(\rho)$. Substituting this equality into (26), we obtain

$$R_m=a_m(-i)^m\bar{F}_m,\quad T_m=a_m(-i)^m\bar{G}_m,\tag{29}$$

where

$$\bar{F}_m=\frac{J_m(\kappa)[m+\bar{\mu}_1(1)V_m(1)]-\kappa\bar{\mu}_1(1)J_m'(\kappa)}{\kappa\bar{\mu}_1(1)H_m^{(1)'}(\kappa)-[m+\bar{\mu}_1(1)V_m(1)]H_m^{(1)}(\kappa)}\quad\text{and}$$

$$\bar{G}_m=\frac{2^{m+1}i\bar{\mu}_1(1)\exp\left(-\int_0^1\bar{\mu}_1(\rho)V_m(\rho)d\rho\right)}{\pi\left[\kappa\bar{\mu}_1(1)H_m^{(1)'}(\kappa)-[m+\bar{\mu}_1(1)V_m(1)]H_m^{(1)}(\kappa)\right]}.$$

From the representations (29) it is clear that for the final solution of the problem (20)–(22) it is necessary to calculate $V_m(1)$ and $\exp\left(-\int_0^1 \bar{\mu}_1(\rho)V_m(\rho)d\rho\right)$. To this end, we return to the problem (27), (28) and construct its approximate solution. For this, we introduce in the interval $0<\rho<1$ the grid functions $V_{mn} \approx V_m(\rho_n)$, $\varepsilon_n = \bar{\varepsilon}_1(\rho_n)$ and $\mu_n = \bar{\mu}_1(\rho_n)$; $\rho_n = (n-1)\delta$, $\delta = 1/N-1$, $n = 1,2,...,N$. It follows from (28), that values $V_{m1} = V_m(0)$ are known. To determine V_{m2}, we integrate equation (27) in the range from ρ_1 to ρ_2. These rearrangements result in

$$\begin{aligned}&\rho_2 V_{m2} - \rho_1 V_{m1} + 2m\int_{\rho_1}^{\rho_2} V_m(\rho)d\rho + \\&+\int_{\rho_1}^{\rho_2} \rho\bar{\mu}_1(\rho)V_m^2(\rho)d\rho + \kappa^2\int_{\rho_1}^{\rho_2}\rho\bar{\varepsilon}_1(\rho)d\rho - m\int_{\rho_1}^{\rho_2}\frac{\bar{\mu}_1'(\rho)}{\bar{\mu}_1^2(\rho)}d\rho = 0.\end{aligned} \tag{30}$$

Let us approximate the first three integrals in (30) by the quadrature sum of the two-point trapezoidal formula [20]

$$\begin{aligned}&\rho_2 V_{m2} - \rho_1 V_{m1} + m\delta\left(V_{m1} + V_{m2}\right) + \frac{\delta}{2}\left(\rho_1\mu_1 V_{m1}^2 + \rho_2\mu_2 V_{m2}^2\right) + \\&+\frac{\kappa^2\delta}{2}\left(\rho_1\varepsilon_1 + \rho_2\varepsilon_2\right) + m\left(\frac{1}{\mu_2} - \frac{1}{\mu_1}\right) \approx 0,\end{aligned} \tag{31}$$

consider $\rho_1 = 0$, $\rho_2 = \delta$, and reduce (31) to the following equation for V_{m2}:

$$\frac{\mu_2\delta}{2}V_{m2}^2 + (m+1)V_{m2} + \frac{\kappa^2\varepsilon_2\delta}{2} + m\left(\alpha_m + \frac{1/\mu_2 - 1/\mu_1}{\delta}\right) \approx 0;\quad \alpha_m = \frac{m\bar{\mu}_1'(0)}{(2m+1)\bar{\mu}_1^2(0)}.$$

Choosing a solution to this equation, “close” to V_{m1} at $\delta \to 0$, we obtain

$$V_{m2} = -\frac{\kappa^2 \varepsilon_2 \delta + 2m\left(\alpha_m + \dfrac{1/\mu_2 - 1/\mu_1}{\delta}\right)}{m+1+\sqrt{D_m}}; \tag{32}$$

$$D_m = (m+1)^2 - \kappa^2 \varepsilon_2 \mu_2 \delta^2 - 2m\mu_2\left(\alpha_m \delta + 1/\mu_2 - 1/\mu_1\right).$$

Determine now V_{mn} for $n = 2,3,...,N$. Transform the equations (27) to the form

$$V_m'(\rho) + \frac{2m+1}{\rho} V_m(\rho) + \bar{\mu}_1(\rho) V_m^2(\rho) + \kappa^2 \bar{\varepsilon}_1(\rho) - m\frac{\bar{\mu}_1'(\rho)}{\bar{\mu}_1^2(\rho)\rho} = 0 \tag{33}$$

and integrate (33) in the range from ρ_n to ρ_{n+1}. These result in the following relation

$$V_{m,n+1} - V_{mn} + (2m+1)\int_{\rho_n}^{\rho_{n+1}} \frac{V_m(\rho)}{\rho} d\rho + \int_{\rho_n}^{\rho_{n+1}} \bar{\mu}_1(\rho) V_m^2(\rho)\, d\rho +$$
$$+\kappa^2 \int_{\rho_n}^{\rho_{n+1}} \bar{\varepsilon}_1(\rho)\, d\rho - m\int_{\rho_n}^{\rho_{n+1}} \frac{\bar{\mu}_1'(\rho)}{\bar{\mu}_1^2(\rho)\rho} d\rho = 0. \tag{34}$$

Let us represent the last integral term in (34) as

$$-\int_{\rho_n}^{\rho_{n+1}} \frac{\bar{\mu}_1'(\rho)}{\bar{\mu}_1^2(\rho)\rho} d\rho = \int_{\rho_n}^{\rho_{n+1}} \frac{1}{\bar{\mu}_1(\rho)\rho^2} d\rho + \frac{1}{\rho_{n+1}\bar{\mu}_1(\rho_{n+1})} - \frac{1}{\rho_n \bar{\mu}_1(\rho_n)}$$

and, assuming N being sufficiently large, we approximate the first three integrals in (34) and the integral term in the last relation by the quadrature sum of the two-point trapezoidal formula. As a result, we obtain

$$\frac{\delta\mu_{n+1}}{2}V^2_{m,n+1}+\left(1+\frac{2m+1}{2n}\right)V_{m,n+1}+F_{mn}=0;$$
$$F_{mn}=\left(\frac{2m+1}{2(n-1)}-1\right)V_{mn}+\frac{\delta\mu_n}{2}V^2_{mn}+\frac{\kappa^2\delta}{2}\left(\varepsilon_n+\varepsilon_{n+1}\right)+ \tag{35}$$
$$+\frac{m}{\delta}\left(\frac{1}{\mu_{n+1}n}-\frac{1}{\mu_n(n-1)}\right)+\frac{m}{2\delta}\left(\frac{1}{\mu_n(n-1)^2}-\frac{1}{\mu_{n+1}n^2}\right).$$

From (35) the following relation

$$V_{m,n+1}=\frac{-\left[1+(2m+1)/2n\right]\pm\sqrt{\bar{D}_{mn}}}{\delta\mu_{n+1}};\quad \bar{D}_{mn}=\left(1+\frac{2m+1}{2n}\right)^2-2\delta\mu_{n+1}F_{mn} \tag{36}$$

is derived. Choosing in (36) the value $V_{m,n+1}$ which at $N\to\infty$ (at $\delta\to 0$) tends to V_{mn}, after simple transformations, we obtain

$$V_{m,n+1}=-\frac{2F_{mn}}{1+(2m+1)/2n+\sqrt{\bar{D}_{mn}}};\quad n=2,3,\ldots,N-1,\quad m=0,1,2,\ldots \tag{37}$$

It can be shown that the error in the values of R_m and T_m defined by formulas (29), (32), (37) is $O(\delta^2)$.

The integral characteristics of the field generated by an incident plane wave in the region $r>a$ are calculated via R_m: $\sigma_s=(4a/\kappa)\left(|R_0|^2+0.5\sum_{m=1}^{\infty}|R_m|^2\right)$ is the total scattering cross section; $\sigma_B=(4a/\kappa)\left|\sum_{m=0}^{\infty}(-i)^m R_m\right|^2$ is the back scattering cross section and etc.

The solution of the diffraction problem in the case of H-polarization of the field (the vector of the magnetic field is parallel to the axis z) can be obtained using a similar scheme.

Now we present several numerical results allowing the evaluation of the proposed approach efficiency. As a parameter characterizing the convergence of the numerical algorithm (37), we choose $\delta_N = \ln\left|\left(V_{mN} - V_{m,2N}\right)/V_{m,2N}\right|$. Remind now that V_{mN} and $V_{m,2N}$ are approximate solutions $V_m(1)$ of the functions $V_m(\rho)$, calculated at the sampling steps $\delta = (N-1)^{-1}$ and $\delta = (2N-1)^{-1}$. In Figures 8 and 9 the numerical results for δ_N calculated for two different radially inhomogeneous dielectric cylinders ($\bar{\mu}_1(\rho) = 1$) are presented. For the first of them (fragments (a) of Figures) $\bar{\varepsilon}_1(\rho) = 3.5\sin(8\pi\rho) + 4.5 + i0.0001$, for the second (fragments (b) of Figures) – $\bar{\varepsilon}_1(\rho) = 3.5\rho^4 + 4.5 + i0.0001$. For fixed N and m, with increasing frequency parameter κ, the value of δ_N increases. The opposite result is obtained by increasing the index m for fixed values N and κ. This behavior of δ_N allows, for given values m and κ, to select the number of sampling points N so that the quantity of δ_N does not exceed a given value. In most cases analyzed, the corresponding optimal value of N was well described by such a simple relation: $N \approx 1 + \kappa\sqrt{\tilde{\varepsilon}}10^{\gamma}$. Here, γ is certain positive value providing validity of $\delta_N \leq -\gamma$, and $\tilde{\varepsilon} = 2a^{-2}\int_0^a \left|\varepsilon_1(r)\right| r dr$ is averaged (over the cross-section of a radially inhomogeneous cylinder) permittivity $\varepsilon_1(r)$.

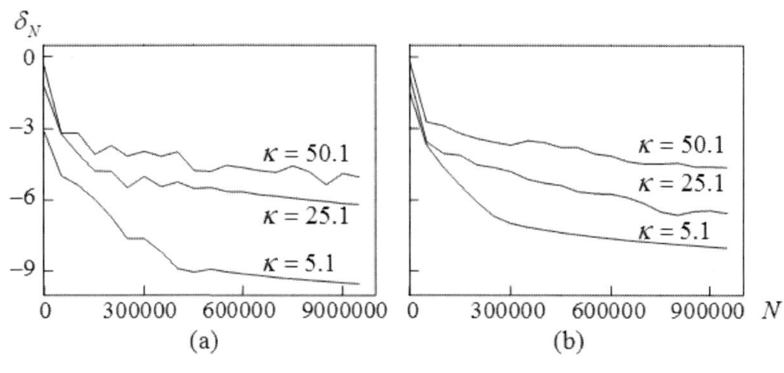

Figure 8. δ_N for two different radially inhomogeneous cylinders, $m = 1$.

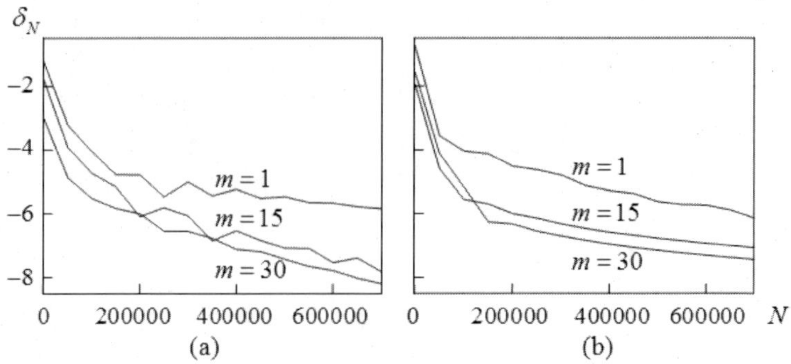

Figure 9. δ_N for two different radially inhomogeneous cylinders, $\kappa = 25$.

4. Transversely Inhomogeneous Anisotropic Layer

Let us consider the problem of a plane homogeneous H-polarized monochromatic electromagnetic wave ($H_y^0 = H_z^0 = E_x^0 = 0$, $H_x^0 = \exp\left[-ik\left(z\cos\alpha - y\sin\alpha\right)\right]$, $E_y^0 = H_x^0 \cos\alpha$, $E_z^0 = H_x^0 \sin\alpha$) diffraction by a flat transverse inhomogeneous anisotropic layer characterized by the permittivity tensor

$$\bar{\bar{\varepsilon}}(z) = \begin{Vmatrix} \varepsilon(z) & 0 & 0 \\ 0 & \varepsilon_1(z) & i\varepsilon_2(z) \\ 0 & -i\varepsilon_2(z) & \varepsilon_1(z) \end{Vmatrix}.$$

The permeability of the layer is equal to unity.

Let the layer occupies the part $\mathrm{D} = \left\{p = \{x, y, z\} : -h \le z < 0, -\infty < x, y < \infty\right\}$ of the space R^3 (Figure 1). We assume the function $\varepsilon(z)$ is piecewise continuous in the interval $[-h, 0]$, and the functions $\varepsilon_j(z)$, $j = 1,2$ are continuously differentiable in the interval $(-h, 0)$. The medium filling the half-spaces $z > 0$ and $z < -h$ is isotropic and homogeneous. Here the relative permittivity and permeability are equal to unity.

The diffraction problem is reduced to determining a function $u(y,z)$ satisfying the equations

$$\begin{aligned}&\left[\Delta+k^2\right]u(y,z)=0;\quad z>0 \ \text{ and } \ z<-h\\&\left[\varepsilon_\perp\Delta-\frac{d\varepsilon_\perp}{dz}\frac{\partial}{\partial z}+i\varepsilon_\perp^2\frac{d}{dz}\left(\frac{\varepsilon_\parallel}{\varepsilon_\perp}\right)\frac{\partial}{\partial y}+k^2\varepsilon_\perp^2\right]u(y,z)=0;\quad -h<z<0\end{aligned}\qquad(38)$$

($\Delta=\partial^2/\partial y^2+\partial^2/\partial z^2$), boundary conditions

$$\begin{aligned}&\left(u^0+u\right)\Big|_{z=0+0}=u\Big|_{z=0-0},\quad \left(\frac{\partial u^0}{\partial z}+\frac{\partial u}{\partial z}\right)\Bigg|_{z=0+0}=\frac{1}{\varepsilon_\perp}\left(\frac{\partial u}{\partial z}+i\varepsilon_\parallel\frac{\partial u}{\partial y}\right)\Bigg|_{z=0-0},\\&u\Big|_{z=-h+0}=u\Big|_{z=-h-0},\quad \frac{1}{\varepsilon_\perp}\left(\frac{\partial u}{\partial z}+i\varepsilon_\parallel\frac{\partial u}{\partial y}\right)\Bigg|_{z=-h+0}=\frac{1}{\varepsilon_s}\frac{\partial u}{\partial z}\Bigg|_{z=-h-0},\end{aligned}\qquad(39)$$

and conventional radiation conditions at infinity [19]. Here, $u^0=H_x^0$, $\varepsilon_\perp(z)=\left[\varepsilon_1^2(z)-\varepsilon_2^2(z)\right]/\varepsilon_1(z)$ and $\varepsilon_\parallel(z)=\varepsilon_2(z)/\varepsilon_1(z)$ are effective transverse and longitudinal permittivity of the medium occupying the region D. The non-zero components of the diffraction field are determined by the relations $H_x=u(y,z)$,

$$E_y=\frac{i}{k}\begin{cases}\dfrac{\partial u}{\partial z}; & z>0 \ \text{ and } \ z<-h\\ \dfrac{1}{\varepsilon_\perp(z)}\left(\dfrac{\partial u}{\partial z}+i\varepsilon_\parallel(z)\dfrac{\partial u}{\partial y}\right); & -h<z<0,\end{cases}\qquad\text{and}$$

$$E_z=-\frac{i}{k}\begin{cases}\dfrac{\partial u}{\partial y}; & z>0 \ \text{ and } \ z<-h\\ \dfrac{1}{\varepsilon_\perp(z)}\left(\dfrac{\partial u}{\partial y}-i\varepsilon_\parallel(z)\dfrac{\partial u}{\partial z}\right); & -h<z<0.\end{cases}$$

Introducing the dimensionless variables $\bar{y} = y/h$ and $\bar{z} = z/h$, separating the variables $\bar{y}$ and $\bar{z}$ in (38), as a result we arrive to

$$\bar{u}(\bar{y},\bar{z}) = \begin{cases} R(\kappa)\exp\left[i\kappa(\bar{y}\sin\alpha + \bar{z}\cos\alpha)\right]; & \bar{z} > 0 \\ \left[A_+u_+(\bar{z}) + A_-u_-(\bar{z})\right]\exp(i\kappa\bar{y}\sin\alpha); & -1 < \bar{z} < 0 \\ T(\kappa)\exp\left[i\kappa(\bar{y}\sin\alpha - (z+1)\cos\alpha)\right]; & \bar{z} < -1. \end{cases} \quad (40)$$

Here, $\bar{u}(\bar{y},\bar{z}) = u(\bar{y}h, \bar{z}h)$, $\kappa = kh$; R, $A_\pm$ and T are unknown complex values and the functions $u_\pm(\bar{z})$ are the linearly independent solutions to the ordinary differential equation of the second order

$$\left[\frac{d^2}{d\bar{z}^2} - \frac{1}{\bar{\varepsilon}_\perp(\bar{z})}\frac{d\bar{\varepsilon}_\perp(\bar{z})}{d\bar{z}}\frac{d}{d\bar{z}} + \kappa^2\bar{\varepsilon}_\perp(\bar{z}) - \right. \\ \left. -\left(\kappa\bar{\varepsilon}_\perp(\bar{z})\sin\alpha\frac{d}{d\bar{z}}\left(\frac{\bar{\varepsilon}_\parallel(\bar{z})}{\bar{\varepsilon}_\perp(\bar{z})}\right) + (\kappa\sin\alpha)^2\right)\right]u_\pm(\bar{z}) = 0. \quad (41)$$

Here, $\bar{\varepsilon}_\perp(\bar{z}) = \varepsilon_\perp(\bar{z}h)$ and $\bar{\varepsilon}_\parallel(\bar{z}) = \varepsilon_\parallel(\bar{z}h)$ ($\bar{\varepsilon}_1(\bar{z}) = \varepsilon_1(\bar{z}h), \bar{\varepsilon}_2(\bar{z}) = \varepsilon_2(\bar{z}h)$).

Now suppose that the linearly independent solutions $u_+(\bar{z})$ and $u_-(\bar{z})$ to the equations (41) satisfying the following conditions

$$u_\pm(-1) = 1, \quad \left.\frac{du_\pm(\bar{z})}{d\bar{z}}\right|_{\bar{z}=-1} = \pm ip; \quad p = \kappa\bar{\varepsilon}_\perp(-1)\left(\cos\alpha + i\sin\alpha\frac{\bar{\varepsilon}_\parallel(-1)}{\bar{\varepsilon}_\perp(-1)}\right) \neq 0$$

at $\bar{z} = -1$, are constructed. Substituting (40) into the conditions (39) in the boundary $z = -h$ ($\bar{z} = -1$), we obtain $A_+ = 0$, $A_- = T$. This means that inside the layer $\bar{u}(\bar{y},\bar{z}) = Tu_-(\bar{z})\exp(i\kappa\bar{y}\sin\alpha)$. From the boundary conditions in layer's upper boundary $z = 0$ ($\bar{z} = 0$), it is possible to obtain the representations for R and T:

$$R=\frac{\kappa u_{-}(0)\left[\bar{\varepsilon}_{\perp}(0)\cos\alpha+\bar{\varepsilon}_{\|}(0)\sin\alpha\right]-iu_{-}'(0)}{\kappa u_{-}(0)\left[\bar{\varepsilon}_{\perp}(0)\cos\alpha-\bar{\varepsilon}_{\|}(0)\sin\alpha\right]+iu_{-}'(0)},\tag{42}$$

$$T=\frac{2\kappa\bar{\varepsilon}_{\perp}(0)\cos\alpha}{\kappa u_{-}(0)\left[\bar{\varepsilon}_{\perp}(0)\cos\alpha-\bar{\varepsilon}_{\|}(0)\sin\alpha\right]+iu_{-}'(0)}.\tag{43}$$

Thus, the solution to the original diffraction problem (38), (39) in the same way as in the case of the isotropic layer is reduced to finding out the function $u_{-}(\bar{z})$. Let us put $u_{-}(\bar{z})=\exp\left(\int_{-1}^{\bar{z}}\bar{\varepsilon}_{\perp}(\xi)V(\xi)d\xi\right)$, substitute this representation in (41) and into the boundary conditions corresponding to the function $u_{-}(\bar{z})$. We arrive to the Cauchy problem for the Riccati equation for unknown function $V(\bar{z})$:

$$\frac{dV(\bar{z})}{d\bar{z}}+\bar{\varepsilon}_{\perp}(\bar{z})V^{2}(\bar{z})++\kappa^{2}-\kappa\sin\alpha\frac{d}{d\bar{z}}\left(\frac{\bar{\varepsilon}_{\|}(\bar{z})}{\bar{\varepsilon}_{\perp}(\bar{z})}\right)-\frac{(\kappa\sin\alpha)^{2}}{\bar{\varepsilon}_{\perp}(\bar{z})}=0\tag{44}$$

with initial condition

$$V(-1)=-i\bar{p};\quad \bar{p}=p/\bar{\varepsilon}_{\perp}(-1).\tag{45}$$

Problem (44), (45) is solved numerically in the same way as in the case of an isotropic transversely inhomogeneous dielectric layer. Relations (42), (42) in terms of function $V(\bar{z})$ take the following form:

$$\begin{gathered}R=\frac{(i\kappa\cos\alpha+V(0))f(0)-\kappa\bar{\varepsilon}_{2}(0)\sin\alpha}{(i\kappa\cos\alpha-V(0))f(0)+\kappa\bar{\varepsilon}_{2}(0)\sin(\alpha)};\quad f(\bar{z})=\bar{\varepsilon}_{1}^{2}(\bar{z})-\bar{\varepsilon}_{2}^{2}(\bar{z}),\\ T=\frac{2i\kappa f(0)\cos\alpha}{(i\kappa\cos\alpha-V(0))f(0)+\kappa\sin\alpha\bar{\varepsilon}_{2}(0)}\exp\left(-\int_{-1}^{0}\bar{\varepsilon}_{\perp}(\xi)V(\xi)d\xi\right).\end{gathered}\tag{46}$$

The exponential factor in (46) can be calculated using the following recurrence formula:

$$\exp\left(-\int_{-1}^{0}\bar{\varepsilon}_{\perp}(\xi)V(\xi)d\xi\right)\approx B_N,\quad B_1\equiv 1,\quad B_{n+1}=B_n\exp\left(-\frac{\delta}{2}\left(\varepsilon_{\perp n}V_n+\varepsilon_{\perp n+1}V_{n+1}\right)\right);$$
$$n=1,2,\ldots,N-1,\quad \delta=(N-1)^{-1},\quad \varepsilon_{\perp n}=\bar{\varepsilon}_{\perp}(\bar{z}_n),\quad \bar{z}_n=-1+\delta(n-1).$$

The following analytical representation for R

$$R=\sum_{n=1}^{\infty}c_n\kappa^n;\quad c_1=\frac{qW_1(0)}{2\cos\alpha}\quad\text{and}$$
$$c_n=\frac{q}{2\cos\alpha}\left[W_n(0)+\sum_{p=1}^{n-1}W_{n-p}(0)c_p\right]\quad\text{for}\quad n\geq 2 \tag{47}$$

is obtained [14] at $\kappa\to 0$ from (44)–(46). Here, $W(\bar{z})=V(\bar{z})/i\kappa q=\sum_{n=0}^{\infty}W_n(\bar{z})\kappa^n$, $q=\cos\alpha+i\sin\alpha\left[\bar{\varepsilon}_{\parallel}(-1)/\bar{\varepsilon}_{\perp}(-1)\right]$, and

$$W_0(0)=\frac{i}{q}S(0)\sin\alpha-1,$$
$$W_1(0)=\frac{i}{q}-\frac{i\sin^2\alpha}{q}\int_{-1}^{0}\frac{d\xi}{\bar{\varepsilon}_{\perp}(\xi)}-iq\int_{-1}^{0}\bar{\varepsilon}_{\perp}(\xi)\left(1-\frac{i}{q}S(\xi)\sin\alpha\right)^2d\xi,$$
$$W_n(0)=-iq\sum_{p=0}^{n-1}\int_{-1}^{0}\bar{\varepsilon}_{\perp}(\xi)W_p(\xi)W_{n-p-1}(\xi)d\xi\quad\text{for}\quad n\geq 2.$$

Functions $W_n(\bar{z})$ are defined by the relations:

$$W_0(\bar{z})=\frac{1}{q}S(\bar{z})\sin\alpha-1;\quad S(\bar{z})=\frac{\bar{\varepsilon}_{\parallel}(-1)}{\bar{\varepsilon}_{\perp}(-1)}-\frac{\bar{\varepsilon}_{\parallel}(\bar{z})}{\bar{\varepsilon}_{\perp}(\bar{z})},$$
$$W_1(\bar{z})=\frac{i}{q}(1+\bar{z})-\frac{i\sin^2\alpha}{q}\int_{-1}^{\bar{z}}\frac{d\xi}{\bar{\varepsilon}_{\perp}(\xi)}-iq\int_{-1}^{\bar{z}}\bar{\varepsilon}_{\perp}(\xi)\left(1-\frac{i}{q}S(\xi)\sin\alpha\right)^2d\xi,$$
$$\text{and}\quad W_n(\bar{z})=-iq\sum_{p=0}^{n-1}\int_{-1}^{\bar{z}}\bar{\varepsilon}_{\perp}(\xi)W_p(\xi)W_{n-p-1}(\xi)d\xi\quad\text{for}\quad n\geq 2.$$

From (47), in particular, it follows that in the long-wave approximation the reflection coefficient R is practically independent of the anisotropic properties of the medium filling the domain D .

The described above algorithm, providing the numerical solution to the problems of plane wave diffraction by a transversely inhomogeneous anisotropic layer, has been used to analyze practically interesting situations, with the permittivity tensor $\bar{\bar{\varepsilon}}(z)$,

$$\bar{\varepsilon}_1(\bar{z}) = \varepsilon\left(1 - \frac{\kappa_p^2 f(\bar{z})(\kappa + i\nu)}{\kappa\left[(\kappa + i\nu)^2 - \kappa_h^2\right]}\right), \quad \bar{\varepsilon}_2(\bar{z}) = -\varepsilon\frac{\kappa_p^2 f(\bar{z})\kappa_h}{\kappa\left[(\kappa + i\nu)^2 - \kappa_h^2\right]}. \quad (48)$$

Representation (48) for $\bar{\bar{\varepsilon}}(z)$ simulates a semiconductor stratum in a constant magnetic field, with intensity parallel to the axis x. Here $f(\bar{z})$ is the semiconductor electron density distribution normalized to the maximum value, $\kappa_p = \omega_p h/c$, $\kappa_h = \omega_h h/c$, $\nu = \bar{\nu} h$, ω_p and ω_h are plasma and cyclotron frequencies, $\bar{\nu}$ is an electron collision frequency, ε is the permittivity of the semiconductor.

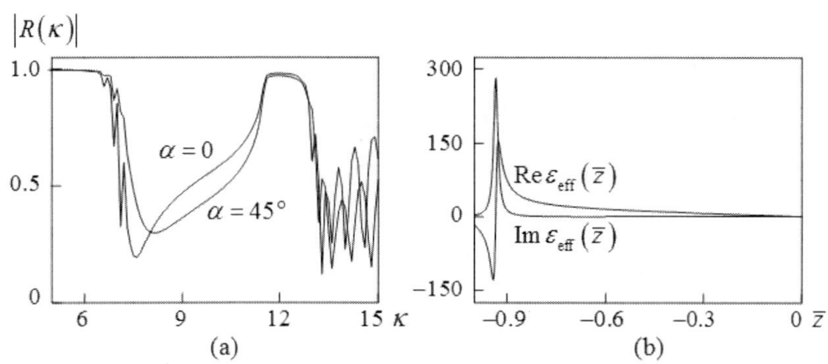

Figure 10. H -polarization. (a) $|R(\kappa)|$ for different incident angles and (b) distribution of $\varepsilon_{\mathrm{eff}}(\bar{z})$ along the thickness of the anisotropic layer at $\kappa = \kappa_r = 7.6$.

The calculations have been carried out for the model problem with the following values of parameters: $\kappa_p = 10$, $\kappa_h = \sqrt{0.5}\kappa_p$, $\nu = 0.05$, $\varepsilon = 12.53$ and $f(\bar{z}) = -4\bar{z}(\bar{z}+1)$. Already for $N = 25000$ (discretization parameter of the problem; see previous sections) the relative error in calculating $|R(\kappa)|$ for $\kappa = 25$ did not exceed 0.01%, and for $\kappa = 50$ it was less 0.1%.

The layer's permittivity tensor components (48) depend both on the frequency κ and $\bar{z}$. In Figure 10a, the curves $|R(\kappa)|$ for the case of normal and oblique ($\alpha = 45°$) incidence are presented. At low values of κ, almost complete reflection takes place $|R(\kappa)| \approx 1$. The real part of the effective permittivity of the layer $\varepsilon_{\text{eff}}(\bar{z}) = \left[\bar{\varepsilon}_1^2(\bar{z}) - \bar{\varepsilon}_2^2(\bar{z})\right] / \bar{\varepsilon}_1(\bar{z})$ at these values of κ is negative. In the vicinity of the frequency $\kappa = \kappa_r = 0.5\left(\sqrt{4\kappa_p^2 + \kappa_h^2} - \kappa_h\right)$ the value of $|R(\kappa)|$ is minimal. $\operatorname{Re}\varepsilon_{\text{eff}}(\bar{z}) = 0$ within this frequency range and at certain specific magnitude of $\bar{z}$; and $\operatorname{Im}\varepsilon_{\text{eff}}(\bar{z})$ takes its minimum (Figure 10b). A more detailed analysis has shown that in this case the absorption coefficient defined as $W = 1 - |R|^2 - |T|^2$ is minimal. At higher values of κ, $\kappa >> \kappa_p$ the real part of $\varepsilon_{\text{eff}}(\bar{z})$ is non-negative and dependence $|R(\kappa)|$ has an oscillating character.

In Figure 11a the values of $|R(\kappa)|$ and $|T(\kappa)|$ for the substrate with $\bar{\varepsilon}_1(\bar{z}) = 12.53\left[1 - \sin(20\pi(\bar{z}+1))\right] + i0.05$ and $\bar{\varepsilon}_2(\bar{z}) = 12.53\cos(20\pi(\bar{z}+1))$ are presented within the frequency interval $0 \le \kappa \le 50$, $\alpha = 0$. Abrupt changes in $|R(\kappa)|$ are caused by the periodic dependence on $\bar{z}$ of the anisotropic layer effective permittivity $\varepsilon_{\text{eff}}(\bar{z}) = \left[\bar{\varepsilon}_1^2(\bar{z}) - \bar{\varepsilon}_2^2(\bar{z})\right] / \bar{\varepsilon}_1(\bar{z})$ (Figure 11b). At frequencies providing $|R(\kappa)|$ with minimum, the electromagnetic field inside the layer is localized in a small neighborhood of its upper boundary $\bar{z} = 0$, (the incident wave arrives onto this plane) (Figure 11c). At the same frequencies $|T(\kappa)| << 1$. It means that there is a resonant absorption of an energy of the incident wave inside the layer. This effect is directly related to the existence of intervals of $\bar{z}$, where $\mathrm{Re}(\varepsilon_{\mathrm{eff}}(\bar{z}))<0$.

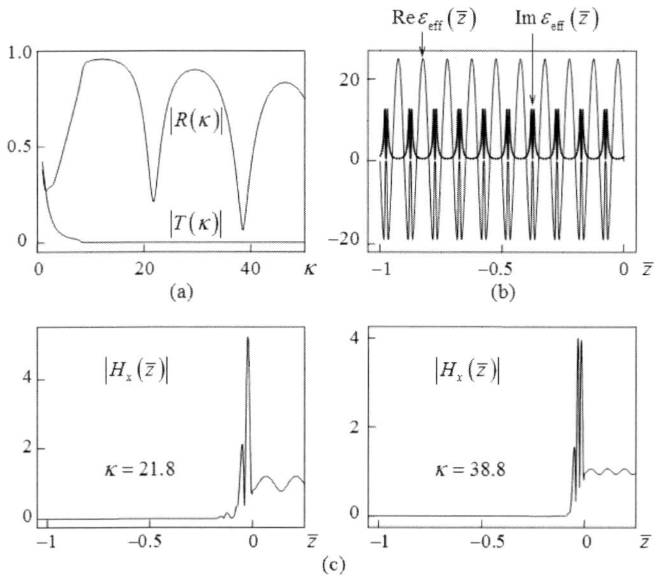

Figure 11. H-polarization. (a) Amplitude-frequency characteristics $|R(\kappa)|$ and $|T(\kappa)|$. (b) The distribution of $\varepsilon_{\mathrm{eff}}(\bar{z})$ along the thickness of the anisotropic layer. (c) The patterns of $|H_x(\bar{z})|$ at $\kappa = 21.8$ and $\kappa = 38.8$.

5. Periodic Stripe Grating on the Boundary of the Transverse Inhomogeneous Stratum

Let us consider an electrodynamic structure consisting of a plane transversely inhomogeneous dielectric layer occupying the part $\mathrm{D}=\{p=\{x,y,z\}:-h\le z<0,-\infty<x,y<\infty\}$ of the space R^3, and infinitely thin perfectly conducting periodic stripe grating located on its upper boundary in the plane $z=0$ (Figure 12). $\varepsilon_1(z)$ is the stratum's permittivity, $\mu_1 = const$ is its permeability. Half spaces $z>0$ and $z<-h$ are filled with a homogeneous medium with constitutive parameters $\varepsilon = \mu = 1$. From the half space $z > 0$ a monochromatic plane E-polarized wave is coming on the structure (1). It is necessary to find out the diffraction field (secondary field), which is defined by the only non-zero component of the electric field $E_x = u(y,z)$, satisfying [16] equations

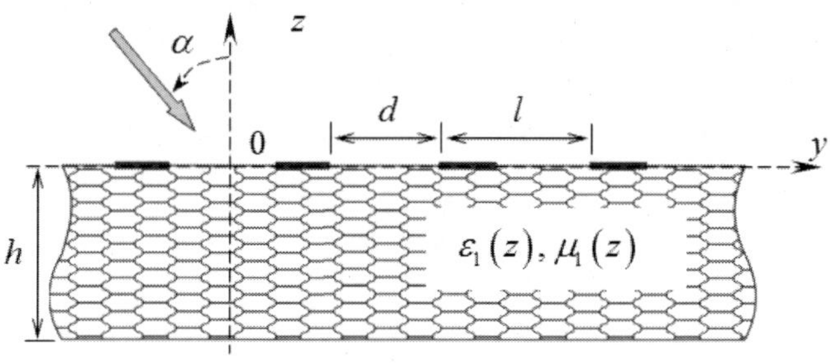

Figure 12. Geometry of the problem.

$$\left[\Delta + k^2\gamma(z)\right]u(y,z) = 0;\quad \gamma(z) = \begin{cases} 1; & z > 0 \text{ and } z < -h \\ \varepsilon_1(z)\mu_1, & -h < z < 0, \end{cases} \tag{49}$$

quasi periodicity conditions

$$u(y+l,z) = \exp(ikl\sin\alpha)u(y,z) \tag{50}$$

(l is the length of grating's period), the Meixner condition near the edges of the grating and the radiation condition at infinity. In addition, the usual requirements of the electromagnetic theory related to the behavior of the tangential components of the total field on the surface of perfect conductors and on the surfaces, separating regions with different constitutive parameters must be fulfilled.

By separating the variables in (49) the general representation for the function $u(y,z)$ is obtained:

$$\bar{u}(\bar{y},\bar{z})=\begin{cases}\sum_{n=-\infty}^{+\infty} R_n \exp\left[i\left(\Gamma_n \bar{z}+\Phi_n \bar{y}\right)\right]; & \bar{z}>0\\ \sum_{n=-\infty}^{+\infty}\left[A_n^+ u_n^+(\bar{z})+A_n^- u_n^-(\bar{z})\right]\exp(i\Phi_n \bar{y}); & -\bar{h}<\bar{z}<0\\ \sum_{n=-\infty}^{+\infty} T_n \exp\left[-i\left(\Gamma_n\left(\bar{z}+\bar{h}\right)-i\Phi_n \bar{y}\right)\right]; & \bar{z}<-\bar{h}.\end{cases} \quad (51)$$

Here, $\bar{u}(\bar{y},\bar{z})=u(\bar{y}l/2\pi,\bar{z}l/2\pi)$; $\bar{y}=2\pi y/l$, $\bar{z}=2\pi z/l$ $\bar{h}=2\pi h/l$; $\Gamma_n=\sqrt{\kappa^2-\Phi_n^2}$ ($\mathrm{Im}\Gamma_n \geq 0$ and $\mathrm{Re}\Gamma_n \geq 0$), $\Phi_n=n+\kappa\sin\alpha$; $\kappa=kl/2\pi$; R_n, T_n and $A_n^{\pm}$ are unknown amplitude coefficients; the functions $u_n^{\pm}(\bar{z})$ are linearly independent solutions of ordinary second-order differential equations

$$\left[\frac{d^2}{d\bar{z}^2}+\left(\kappa^2\bar{\varepsilon}_1(\bar{z})\mu_1-\Phi_n^2\right)\right]u_n^{\pm}(\bar{z})=0;\ -\bar{h}<\bar{z}<0,\ \bar{\varepsilon}_1(\bar{z})=\varepsilon_1(\bar{z}l/2\pi). \quad (52)$$

Suppose that solutions to equations (52) satisfying the following conditions

$$u_n^{\pm}\left(-\bar{h}\right)=1,\ \left.\frac{du_n^{\pm}(\bar{z})}{d\bar{z}}\right|_{\bar{z}=-\bar{h}}=\pm ip_n;\quad p_n=\mu_1\Gamma_n \quad (53)$$

at $\bar{z}=-\bar{h}$ are found. Using representation (51) and satisfying all the conditions imposed on the complete field at the lower and upper boundaries of the layer, we conclude that $A_n^+=0$ for all $n=0,\pm1,\ldots$. To determine the values of A_n^-, it is necessary to invert the duel infinite system of functional equations

$$\begin{cases}\sum_{n=-\infty}^{\infty} x_n \exp(in\bar{y})=0; & \pi d/l<|\bar{y}|\leq\pi\\ \sum_{n=-\infty}^{\infty} x_n\left(\Gamma_n+\frac{i}{\mu_1}\frac{u_n^{-\prime}(0)}{u_n^-(0)}\right)\exp(in\bar{y})=2\kappa\cos\alpha; & |\bar{y}|\leq\pi d/l\end{cases} \quad (54)$$

(d is the width of the slit between the metal stripes of the grating). Here $x_n = A_n^- \overline{u}_n^-(0)$. Find out these values from (54) by the method of the Riemann-Hilbert problem (by the method of analytical regularization) [21]; the values $u_n^{-\prime}(0)$ and $u_n^-(0)$ find out from (52), (53) by the method of the Riccati equation, that was discussed in details in the previous sections.

The amplitudes $\{R_n\}_{n=-\infty}^{\infty}$ and $\{T_n\}_{n=-\infty}^{\infty}$ of diffraction fields in the reflection and transmission zones of the periodic structure are defined in terms of $\{x_n\}_{n=-\infty}^{\infty}$ and $\{u_n^-(0)\}_{n=-\infty}^{\infty}$ according to relations

$$R_n = x_n - \delta_n^0 \quad \text{and} \quad T_n = x_n / u_n^-(0); \quad n = 0, \pm1, \pm2, ...$$

Now consider several numerical results obtained by the method presented above. In Figure 13a the curves $W^T(\kappa) = \sum_n W_n^T(\kappa) = \sum_n |T_n|^2 \operatorname{Re}\Gamma_n / \Gamma_0$ calculated for the homogeneous ($\varepsilon_1(z) = 2.25$ – lines 1 in Figures 13a and 13b) and inhomogeneous ($\varepsilon_1(z) = 2.25 + \sin[4\pi(z/h+1)]$ – lines 2 and $\varepsilon_1(z) = (z/h+1)^3 + 1.25$ – lines 3) layers are presented. All parameters of the problem in the case of a homogeneous layer are the same as in [22] ($d/l = 0.45$, $h/l = 1.1$, $\alpha = 10°$). It was shown in [22] that for this periodic structure in the frequency range $0.597 < \kappa < 0.852$ the number of waves propagating without attenuation in the region D is greater than in the reflection ($z > 0$) and transmission zones ($z < -h$). As a result, [22], it is possible to conclude that the formation of the amplitude-frequency characteristics of the structure is substantially affected by the excitation of the number of high-Q free field oscillations. In the case of the inhomogeneous layers, the same picture is observed. But, for obvious reasons, the values of κ, providing the resonant regime occurrences, in particular, the regimes of complete transmission ($W^T(\kappa) = 1$) and reflection ($W^T(\kappa) = 0$), depends on the specific type of function $\varepsilon_1(z)$.

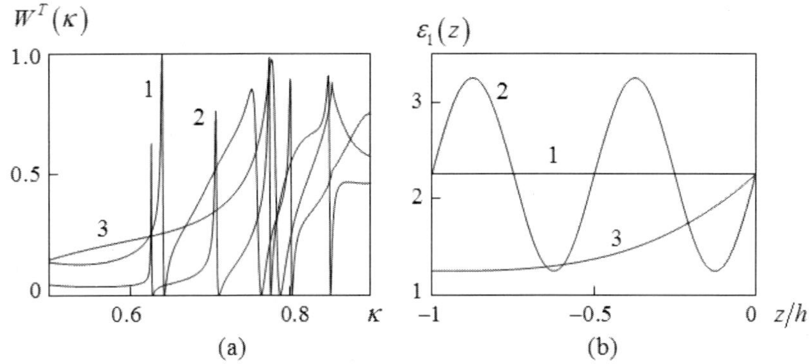

Figure 13. E -polarization. (a) Frequency characteristic of transmitted energy $W^T(\kappa)$. (b) Distribution of $\varepsilon_1(z)$ along the thickness of the anisotropic layer.

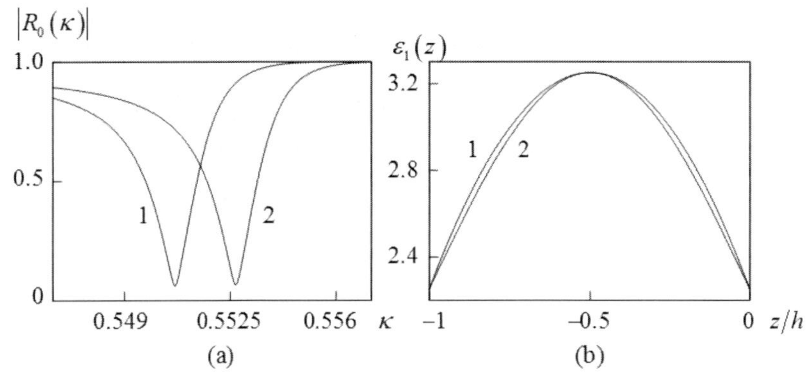

Figure 14. E -polarization. (a) Amplitude-frequency characteristic of $|R_0(\kappa)|$. (b) Distribution of $\varepsilon_1(z)$ along the thickness of the dielectric layer.

Let us dwell briefly on the long-wave resonances (the first ones when moving along the axis κ) and find out how strongly the frequencies corresponding to them depend on the form of the function $\varepsilon_1(z)$. In particular, is it possible to recognize inhomogeneous layers with the help of these resonances if the values of their permittivity $\varepsilon_1(z)$ differ from each other very slightly? For example, consider two problems ($d/l = 0.5$, $h/l = 1.1$, $\alpha = 45°$) with permittivity functions $\varepsilon_1(z)$: $\varepsilon_1(z) = 3.25 - 4(z/h + 0.5)^2$ (lines 1 in Figures

14a and 14b) and $\varepsilon_1(z) = 2.25 + \sin(\pi(z/h+1))$ (lines 2). The difference in the permittivity of the layer, in this case, is insignificant (the maximum deviation in the absolute values of $\varepsilon_1(z)$ within the interval $-h < z < 0$ does not exceed 0.055), but the values of the resonant frequencies (points κ providing $|R_0(\kappa)| \approx 0$) differ significantly. So, the shift of the resonant frequency allows one to record minor changes in the permittivity behavior transverse the layer.

In the framework of the approach presented above, it is possible to consider urgent problems for stratified inhomogeneous plasma-like media, which have recently received close attention (see, for example, [23]), and, in particular, to study the effects of the free (reflectionless or tunneling) propagation of electromagnetic waves through a plasma layer with large-amplitude subwavelength inhomogeneities [23]. As an example, consider a layer, characterized by parameters $h/l = 1.1$, $\varepsilon_1(z) = 1 - \omega_p^2 N(z)/\omega(\omega + i\nu)$, with a stripe grating placed on its upper boundary, $d/l = 0.8$. Here dimensionless function $0 \le N(z) \le 1$ defines the distribution of electrons along the thickness of the layer; ω_p is a plasma frequency corresponding to a maximum value of electron density of the layer; $\nu = 0.01\omega_p$ is an electron collision frequency.

In Figure 15 the curves $|R_0(\kappa)|$ for $N(z) = z/h$ (line 1), $N(z) = (z/h)^2$ (line 2) and $N(z) = (z/h)^3$ (line 3) are presented. The same figure shows the behavior $|R_0(\kappa)|$ for a layer without grating ($N(z) = z/h$, line 4) and for a layer with grating, having an average for the cubic dependence value $N(z) = 1/3$ (line 5). The presence of grating leads to the appearance of almost complete transition resonance of an E-polarized wave through the plasma substrate. It is important to note that the values of frequency parameter κ, providing the resonant decrease of $|R_0(\kappa)|$, are laying within the interval $\kappa < \kappa_p = \omega_p l/2\pi c = 0.6$; and $\mathrm{Re}\,\varepsilon_1(z)$ takes the negative values near the lower boundary of the layer.

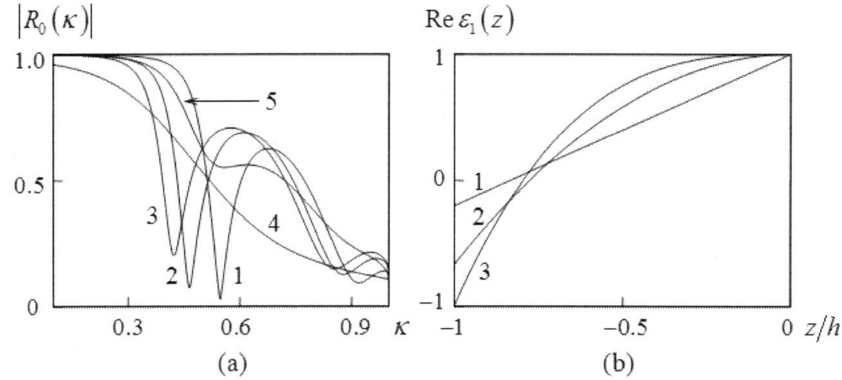

Figure 15. E -polarization, $\alpha = 0$. (a) Amplitude-frequency characteristic of $|R_0(\kappa)|$. (b) Distribution of $\mathrm{Re}\varepsilon_1(z)$ along the thickness of the dielectric layer.

By changing the grating parameters (for example, the filling factor d/l), or plasma parameters (for example, plasma frequency κ_p), it is possible to optimize a tunneling effect (regime of reflectionless) of an electromagnetic wave passing through an inhomogeneous plasma layer. Figure 16 presents the results of calculations allowing the carrying out such optimization.

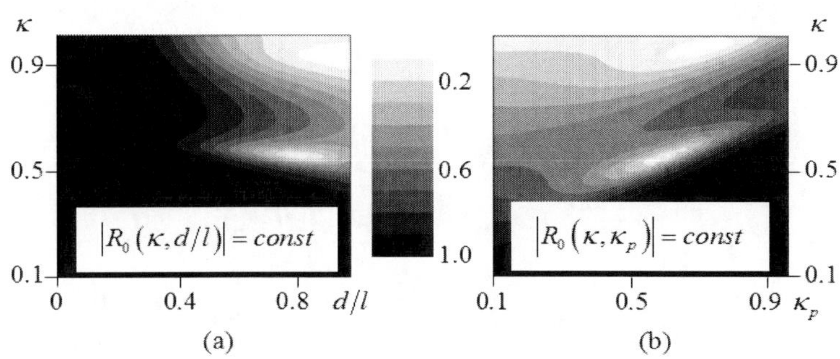

Figure 16. E -polarization, $h/l = 1.1$, $N(z) = z/h$, $\alpha = 0$. Patterns of $|R_0| = const$ (a) in the coordinates κ, d/l at $\kappa_p = 0.6$ and (b) in the coordinates κ, κ_p at $d/l = 0.8$.

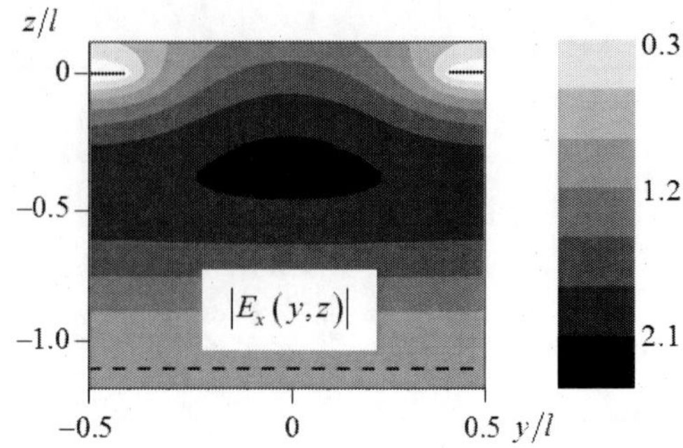

Figure 17. E -polarization, $\kappa = 0.5475$, $h/l = 1.1$, $d/l = 0.8$, $N(z) = z/h$, $\alpha = 0$. Patterns of $|E_x(y,z)|$ along the structure period.

The developed algorithm enables calculation of the electromagnetic field both outside the inhomogeneous plasma layer and inside it. Figure 17 shows the pattern of $E_x(y,z)$ within the period of structure, characterized by $N(z) = z/h$, at the frequency $\kappa = 0.5475$, corresponding to the minimum value of $|R_0(\kappa)| \approx 0.028$. The electromagnetic field is concentrated inside the layer, and the normalized amplitude of the E_x -component of the field in the transition zone ($z < -h$) is close to unity. Such pattern of the field confirms the effect of almost reflectionless tunneling of plane E -polarized electromagnetic wave passing through the plasma layer.

6. Nonlinear Dielectric Layer

The section discusses the issues of mathematical modeling of diffraction and propagation of electromagnetic waves in medium including nonlinear planar dielectric structures [24, 25]. Nonlinearity, generally speaking, leads to multifrequency and to the need to solve non-stationary diffraction problems [25]. However, in the case when the permittivity depends on the electric field intensity averaged over the period, the nonlinear problem can be formulated as a single-frequency problem in spectral region [26]. Previously, for the problems with Kerr nonlinearity, the solution method based on the explicit use of the first integral for the corresponding nonlinear equation has been developed (see, for example, [27]). However, obtaining of such integral in an explicit form is far from always possible. Approaches that are free from any a restriction are needed. Below we present such approach. It is based on the construction of special solutions to the Cauchy problem for the Riccati type equation.

The formulation of the original problem is the same as in Section 2. Only E-polarized wave arrives onto the layer at the angle $\alpha = 0$ (the dependence of the field on the coordinate y disappears), and the dielectric constant of the layer is described by the Kerr law [25] $\varepsilon_1(z) = \tilde{\varepsilon} + \beta |E_x(z)|^2$, where $\tilde{\varepsilon} = const$ is the permittivity in the absence of a field and β is a nonlinearity coefficient. Presume $\mu_1(z) = 1$, and half spaces $z > 0$ and $z < -h$ are filled by media with $\varepsilon = \mu = 1$.

The problem is reduced to finding out the function $u(z) = E_x(z)$ (E_x - diffraction field components), satisfying the equations

$$\left[\frac{d^2}{dz^2} + k^2\right] u(z) = 0; \quad z > 0 \quad \text{and} \quad z < -h$$

$$\left[\frac{d^2}{dz^2} + k^2\left(\tilde{\varepsilon} + \beta |u(z)|^2\right)\right] u(z) = 0; \quad -h < z < 0,$$

continuity conditions for the tangential components of the total field at the media interfaces (transparency conditions), and radiation conditions at infinity.

Introduce a dimensionless variable $\overline{z} = z/h$ and present a general solution to the problem in this form:

$$\overline{u}(\overline{z}) = \begin{cases} R(\kappa)\exp(i\kappa\overline{z}\cos\alpha); & \overline{z} \geq 0 \\ T(\kappa)\exp\left(\int_{-1}^{\overline{z}} V(\tau)d\tau\right); & -1 \leq \overline{z} \leq 0 \\ T(\kappa)\exp\left[-i\kappa(\overline{z}+1)\cos\alpha\right]; & \overline{z} \leq -1. \end{cases}$$

Here, $\bar{u}(\bar{z}) = u(\bar{z}h)$, $\kappa = kh$, $R(\kappa)$ and $T(\kappa)$ are unknown amplitudes, and the unknown function $V(\bar{z})$ should be the solution to the following Cauchy problem for the Riccati equation with the additional nonlinear term $\kappa^2 |T|^2 \exp\left(2\int_{-1}^{\bar{z}} \operatorname{Re} V(x)dx\right)$:

$$\begin{cases} \dfrac{dV(\bar{z})}{d\bar{z}} + V(\bar{z})^2 + \kappa^2 \tilde{\varepsilon} + \kappa^2 |T|^2 \exp\left(2\int_{-1}^{\bar{z}} \operatorname{Re} V(\tau)d\tau\right) = 0; \quad -1 < \bar{z} < 0 \\ V(-1) = -i\kappa. \end{cases} \tag{55}$$

Solve this problem by approximating the function $V(\bar{z})$ with a grid function $V_n \approx V(\bar{z}_n)$ (see also Section 2). Here, $\bar{z}_n = -1 + \delta(n-1)$, $n = 1, 2, \ldots, N$, $\delta = (N-1)^{-1}$, N is a natural number. It follows from (55) that $V_1 = -i\kappa$. It is necessary to find the values V_n, $n = 2, 3, \ldots, N$. Suppose $\operatorname{Im} \tilde{\varepsilon} = 0$, the case of complex valued $\tilde{\varepsilon}$ can be considered similarly.

It follows from (55)

$$\operatorname{Re} \frac{dV(\bar{z})}{d\bar{z}} + (\operatorname{Re} V(\bar{z}))^2 - (\operatorname{Im} V(\bar{z}))^2 + \kappa^2 \tilde{\varepsilon} + \kappa^2 |\bar{T}|^2 \exp\left(2\int_{-1}^{\bar{z}} \operatorname{Re} V(\tau)d\tau\right) = 0, \tag{56}$$

$$\operatorname{Im} \frac{dV(\bar{z})}{d\bar{z}} + 2 \operatorname{Re} V(\bar{z}) \operatorname{Im} V(\bar{z}) = 0. \tag{57}$$

And from (57) and (55) –

$$\operatorname{Im} V(\bar{z}) = -\kappa \exp\left(-2\int_{-1}^{\bar{z}} \operatorname{Re} V(\tau)d\tau\right).$$

Substituting this expression into (56), we obtain

$$\begin{aligned}
&\operatorname{Re}\frac{dV(\bar{z})}{d\bar{z}}+\left(\operatorname{Re}V(\bar{z})\right)^2-\kappa^2\exp\left(-4\int_0^{\bar{z}}\operatorname{Re}V(\tau)d\tau\right)+\\
&+\kappa^2\tilde{\varepsilon}+\kappa^2\left|\overline{T}\right|^2\exp\left(2\int_{-1}^{\bar{z}}\operatorname{Re}V(\tau)d\tau\right)=0.
\end{aligned}\tag{58}$$

Integrate (58) in the range from $\bar{z}_n$ to $\bar{z}_{n+1}$, denote $W_n=\operatorname{Re}V(\bar{z}_n)$ and approximate the integrals in the resulting expression

$$\begin{aligned}
&W_{n+1}-W_n+\int_{\bar{z}_n}^{\bar{z}_{n+1}}W^2(\tau)d\tau-\kappa^2\int_{\bar{z}_n}^{\bar{z}_{n+1}}\exp\left(-4\int_{-1}^{\tau}\operatorname{Re}V(\xi)d\xi\right)d\tau+\\
&+\kappa^2\tilde{\varepsilon}\delta+\kappa^2\left|\overline{T}\right|^2\int_{\bar{z}_n}^{\bar{z}_{n+1}}\exp\left(2\int_{-1}^{\tau}\operatorname{Re}V(\xi)d\xi\right)d\tau=0
\end{aligned}$$

by two-point quadrature trapezoidal formula. As a result, the representations

$$\begin{aligned}
&W_{n+1}=\frac{-2C_n}{1+A_n+\sqrt{(1+A_n)^2-2\delta C_nB_n}},\quad I_{n+1}=I_n\exp\left(\delta\left(W_n+W_{n+1}\right)\right);\\
&n=1,2,\ldots,N-1,\quad W_1=0,\quad I_1=1
\end{aligned}$$

was derived. Here,

$$\begin{aligned}
&A_n=\kappa^2\delta^2I_n^{-2}\left(1-2\delta W_n\right)+0.5\kappa^2\delta^2|T|^2I_n\left(1+\delta W_n\right),\\
&B_n=1-2\kappa^2\delta^2I_n^{-2}+0.5\kappa^2\delta^2|T|^2I_n,\\
&C_n=-W_n\left[1-\kappa^2\delta^2\left(I_n^{-2}+0.5|T|^2I_n\right)\right]+0.5W_n^2\delta B_n+\kappa^2\delta\left(\tilde{\varepsilon}-I_n^{-2}+|T|^2I_n\right).
\end{aligned}$$

These representations allow to consistently calculate all values $\operatorname{Re}V_n=W_n$ and $\operatorname{Im}V_v=-\kappa I_n^{-1}$, only if $|T|$ is known. Error in calculation of $V(\bar{z}_n)$ (error in determining function $V(\bar{z})$, $-1\le\bar{z}\le 0$) is $O\left(\delta^2\right)$. From the continuity conditions at the upper boundary of the layer, we further obtain

$$T = \frac{2i\kappa \exp\left(-\int_{-1}^{0} V(\tau)d\tau\right)}{i\kappa - V(0)} \quad \text{and} \quad R = \frac{i\kappa + V(0)}{i\kappa - V(0)}.$$

These relations are nonlinear equations with respect to normalized reflection and transmission coefficients. The diffraction problem is solved.

The corresponding algorithm found its code implementation and was used to study the effects of the complete transmission of plane waves through nonlinear dielectric layers. One of the results of the computational experiments is presented in Figure 18 (curves $|R(\kappa)|$ and $|T(\kappa)|$ for various β, $\tilde{\varepsilon} = 2.04$). It should be noted here the existence of the frequencies κ at which there are two or more different values of the analyzed characteristics, including values significantly differing in magnitude.

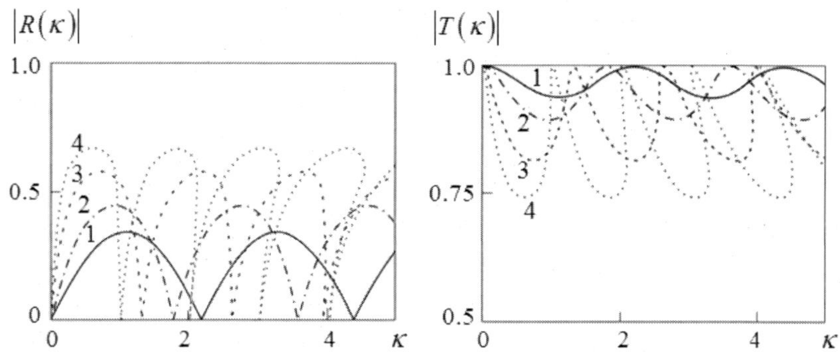

Figure 18. E -polarization. Amplitude-frequency characteristics of $|R(\kappa)|$ and $|T(\kappa)|$ for different values of the nonlinearity parameter β : 1 – $\beta = 0$, 2 – $\beta = 1$, 3 – $\beta = 4$, 4 – $\beta = 9$.

7. Spectral Problems of Electrodynamics of Layered Inhomogeneous Media

Over the past decades, the interest of researchers in the spectral problems of the theory of diffraction of monochromatic waves preserved its actuality [9, 28–

36]. From a formal mathematical point of view, these problems describe the features of the analytic continuation of solutions of diffraction problems to the region of complex, sometimes physically impracticable, parameter values (frequency, energy, etc.) [37–42]. In such approach, the central question is: what conditions on the infinity should replace the classical conditions – the Sommerfeld radiation conditions [43] and the partial radiation conditions [44]. It turns out that, for example, in the 2D case with a compact scatterer, this should be the Reichardt radiation condition [28]. The problem formulated in this way for the real valued frequency parameter k is equivalent to the problem with the Sommerfeld radiation condition. It is "analytically continued" into the corresponding Riemann surface K of complex-valued parameter k and is uniquely solvable everywhere in K outside of some, no more than countable, set $\Omega=\left\{\bar{k}_n\right\}$ with a single accumulation point at infinity (in the case of dispersive inhomogeneities, this statement should be refined [45]). Points $\bar{k}_n$ (the finite multiple resolvent's poles of the diffraction problem operator continued into K [38]) are the eigenvalues of the corresponding homogeneous problem (no sources) or, otherwise, the eigenfrequencies. They correspond to non-trivial solutions of the homogeneous problem that are the eigenfunctions or free oscillations of the field in the object under consideration.

The theory of non-self-adjoint operators and meromorphic operator functions [31, 38, 46, 47] today is the only adequate mathematical apparatus for obtaining all results of the spectral theory of open electrodynamic structures. It was possible to apply the spectral theory methods due to procedures of the analytical regularization of the boundary value problems of the diffraction theory [32–34, 36]. These procedures allow (in several particular cases important for theory and practical applications) to reduce the initial boundary-value problem for complex-valued frequencies k to the equivalent operator equation of the second kind $x+A(k)x=b$, $k\in\mathrm{K}$, $x,b\in\mathrm{H}$. Here, H is Hilbert space, and $A(k)$ is a meromorphic operator function defining for each $k\in\mathrm{K}\setminus\Omega$ compact, or even kernel type operator $A(k):\mathrm{H}\to\mathrm{H}$. For spectral problems, we obtain the homogeneous operator equation $\bar{x}+A(\bar{k})\bar{x}=0$, $\bar{x}\in\mathrm{H}$,

$\bar{k} \in \Omega$. Its non-trivial solutions determine the spectrum Ω of the object under the study and the free oscillations of the field that can exist in this object.

The described approach, based on the application of regularization procedures to solve the spectral problems of electromagnetic theory, was first used in [48, 49]. Within its framework, the key results of the spectral theory of open compact, waveguide and periodic resonators [19, 32, 33] and the theory of resonant scattering of monochromatic and pulsed electromagnetic waves by such structures have been obtained [50, 51]. Below we present the results enabling the efficient application of this approach to solving urgent electromagnetic problems associated with the analysis and synthesis of layered-inhomogeneous dielectric media.

As a model of such a medium, we consider a transversely inhomogeneous layer placed in the space with $\varepsilon = \mu = 1$ (see Section 2), and suppose that in the fields of free oscillations, the dependence on the coordinate y is determined by the factor $\exp(iFy)$, $\operatorname{Im} F = 0$. The consideration here is restricted by the case of an E-polarization ($E_y = E_z = H_x = 0$); the H-case ($H_y = H_z = E_x = 0$) is considered similarly. As before, move on to dimensionless coordinates $\bar{y} = y/h$, $\bar{z} = z/h$ (the problem is independent of x coordinate). The problem is transformed into such form

$$\begin{cases} \left[\dfrac{d}{d\bar{z}}\left(\dfrac{1}{\mu(\bar{z})}\dfrac{d}{d\bar{z}}\right)+\left(\kappa^2\varepsilon(\bar{z})-\dfrac{\Phi^2}{\mu(\bar{z})}\right)\right]u(\bar{z})=0; \quad |\bar{z}|<\infty \\ u(\bar{z})\Big|_{\bar{z}=0,-1}=0, \quad \dfrac{1}{\mu(\bar{z})}\dfrac{du(\bar{z})}{d\bar{z}}\Bigg|_{\bar{z}=0,-1}=0. \end{cases} \tag{59}$$

Here, $u(\bar{z}) = \bar{E}_x(\bar{y},\bar{z})\exp(-i\Phi\bar{y})$, $\bar{E}_x(\bar{y},\bar{z}) = E_x(\bar{y}h,\bar{z}h)$, $\kappa = kh$, $\Phi = Fh$, and

$$\varepsilon(\bar{z}) = \begin{cases} 1; & \bar{z} > 0 \text{ and } \bar{z} < -1 \\ \varepsilon_1(\bar{z}h); & -1 < \bar{z} < 0 \end{cases}, \quad \mu(\bar{z}) = \begin{cases} 1; & \bar{z} > 0 \text{ and } \bar{z} < -1 \\ \mu_1(\bar{z}h); & -1 < \bar{z} < 0 \end{cases}.$$

Let us represent the general solution to problem (59) in the form

$$u(\bar{z})=\begin{cases} R(\kappa)\exp(i\Gamma\bar{z}); & \bar{z}>0 \\ A_{+}(\kappa)u_{+}(\bar{z})+A_{-}(\kappa)u_{-}(\bar{z}); & -1<\bar{z}<0, \\ T(\kappa)\exp(-i\Gamma(\bar{z}+1)); & \bar{z}<-1 \end{cases} \tag{60}$$

$R(\kappa)$, $T(\kappa)$ and $A_{\pm}(\kappa)$ are unknown complex amplitudes; $\Gamma=\sqrt{\kappa^2-\Phi^2}$. Unknown functions $u_{\pm}(\bar{z})$ must satisfy the equations

$$\left[\frac{d}{d\,\bar{z}}\left(\frac{1}{\mu(\bar{z})}\frac{d}{d\,\bar{z}}\right)+\left(\kappa^2\varepsilon(\bar{z})-\frac{\Phi^2}{\mu(\bar{z})}\right)\right]u_{\pm}(\bar{z})=0 \tag{61}$$

on the interval $-1<\bar{z}<0$. Suppose that solutions to the equations (61) are constructed and satisfy the following conditions at $\bar{z}=-1$:

$$u_{\pm}(-1)=1,\quad \left.\frac{du_{\pm}(\bar{z})}{d\,\bar{z}}\right|_{\bar{z}=-1}=\pm iq;\quad q=\mu(-1)\sqrt{\kappa^2-\Phi^2}. \tag{62}$$

Obviously, the functions $u_{\pm}(\bar{z})$ are linearly independent if $\mu(-1)\neq 0$ and $\kappa\neq\pm\Phi$. Assuming these conditions to be fulfilled, substitute (60) into the transparency conditions for $u(\bar{z})$ in the lower boundary of the layer and obtain $A_{+}=0$, $A_{-}=T$. The conditions at the upper boundary of the layer lead to a homogeneous system of equations

$$\begin{cases} R(\kappa)-T(\kappa)u_{-}(0)=0, \\ R(\kappa)\left[i\Gamma\mu(0)\right]-T(\kappa)\left.\dfrac{du_{-}(\bar{z})}{d\,\bar{z}}\right|_{\bar{z}=0}=0. \end{cases} \tag{63}$$

A non-trivial solution of (63) can be obtained only if

$$\left.\frac{du_-(\bar{z})}{d\,\bar{z}}\right|_{\bar{z}=0} - i\Gamma(\kappa)\mu(0)u_-(0) = 0. \tag{64}$$

Find out the eigenfrequencies $\bar{\kappa} = \bar{k}h$ from (64). The corresponding dispersion equation must be considered on the two-sheeted Riemann surface K of function $\Gamma(\kappa) = \sqrt{\kappa^2 - \Phi^2}$ with two algebraic branch points $\kappa_\pm = \pm\Phi$. In K we make the cuts along the lines $(\mathrm{Re}\,\kappa)^2 - (\mathrm{Im}\,\kappa)^2 - \Phi^2 = 0;\ \ \mathrm{Im}\,\kappa \le 0$ [19]. On the first sheet of surface at $0 \le \arg\kappa \le \pi$ $\mathrm{Im}\,\Gamma(\kappa) \ge 0$, and $\mathrm{Re}\,\Gamma(\kappa) \ge 0$ in the interval $0 \le \arg\kappa \le \pi/2$ and $\mathrm{Re}\,\Gamma(\kappa) \le 0$ in $\pi/2 \le \arg\kappa \le \pi$. For points κ: $3\pi/2 \le \arg\kappa \le 2\pi$, located to the right (left) of the cut, $\mathrm{Re}\,\Gamma(\kappa) > 0$ and $\mathrm{Im}\,\Gamma(\kappa) \le 0$ ($\mathrm{Re}\,\Gamma(\kappa) \le 0$ and $\mathrm{Im}\,\Gamma(\kappa) > 0$). For points κ: $\pi \le \arg\kappa \le 3\pi/2$ the situation is mirroring the previous one up to a sign reversal of $\mathrm{Re}\,\Gamma(\kappa)$.

So, according to (60), (63), (64), a solution of a spectral problem for a transversely inhomogeneous dielectric layer will be constructed if a solution to the problem (61), (62) $u_-(\bar{z})$, $-1 \le \bar{z} \le 0$ is found. But to do this thoroughly and correctly, as it has been already demonstrated more than once in this chapter (see, for example, Section 2), it is possible within the framework of a fairly general approach, which in the case analyzed here is based on the transition from (61), (62) to the Cauchy problem for the Riccati equation. Below we briefly describe the relevant procedures.

New unknown function $V(\bar{z})$ is defined by the relation

$$u_-(z) = \exp\left(\int_{-1}^{\bar{z}} \mu(\tau)V(\tau)d\tau\right). \tag{65}$$

Substituting (65) into (61) and taking into account that $du_-(\bar{z})/d\,\bar{z} = \mu(\bar{z})u_-(\bar{z})V(\bar{z})$, reduce the problem (61), (62) to the following Cauchy problem for the Riccati equation:

$$\begin{cases} \dfrac{dV(\bar{z})}{d\bar{z}} + \mu(\bar{z})V^2(\bar{z}) + \kappa^2\varepsilon(\bar{z}) - \dfrac{\Phi^2}{\mu(\bar{z})} = 0; \quad -1 < \bar{z} < 0 \\ V(-1) = -i\sqrt{\kappa^2 - \Phi^2}. \end{cases} \tag{66}$$

In terms of $V(\bar{z})$, the dispersion equation (64) can be written as

$$V(0) - i\sqrt{\kappa^2 - \Phi^2} = 0. \tag{67}$$

Let us approximate $V(\bar{z})$ with the grid function $V_n \approx V(\bar{z}_n)$ ($\bar{z}_n = -1 + \delta(n-1)$, $n = 1, 2, ..., N$, N is a positive integer) and integrate the equation of the problem (66) in the range from $\bar{z}_n$ to $\bar{z}_{n+1}$. It results in

$$\begin{gathered} V_{n+1} - V_n + \int_{\bar{z}_n}^{\bar{z}_{n+1}} \mu(\tau)V^2(\tau)d\tau + \int_{\bar{z}_n}^{\bar{z}_{n+1}} p(\tau)d\tau = 0; \\ n = 1, 2, ...N, \quad p(\tau) = \kappa^2\varepsilon(\tau) - \Phi^2/\mu(\tau). \end{gathered} \tag{68}$$

Choosing a sufficiently large N and approximating the integrals in (68) by the quadrature sum of the two-point trapezoidal formula, we arrive to

$$\begin{gathered} V_{n+1} = \frac{-2F_n}{1 + \sqrt{1 - 2\delta\mu_{n+1}F_n}} \quad \text{and} \quad V_1 = V(-1) = -i\sqrt{\kappa^2 - \Phi^2}; \\ F_n = -V_n + 0.5\delta\left(\mu_n V_n^2 + p_{n+1} + p_n\right), \quad \mu_n = \mu(\bar{z}_n), \quad p_n = p(\bar{z}_n), \\ n = 1, 2, ..., N-1. \end{gathered} \tag{69}$$

The formulas (69), (67), (65), (63) and (60) give a solution to the original problem, with the error estimated as $O\left(\delta^2\right)$.

The same algorithm can be used to solve the problem of eigenwaves and eigen propagation constants of these waves $\bar{\Phi} = h\bar{F}$. For this, it is necessary to fix the real value $\kappa > 0$ in the dispersion equation (67), and to consider the value $\Phi \in \mathrm{F}$ as a spectral parameter. Here F is the two-sheeted Riemann surface of function $\Gamma(\Phi)=\sqrt{\kappa^2-\Phi^2}$ with two algebraic branch points $\Phi_\pm=\pm\kappa$. The cuts on this surface are the same as in [19].

Consider now two spectral problems (problems for free field oscillations in a transversely inhomogeneous dielectric layer), the solution of one of them can be obtained in closed form. This is a problem for a homogeneous magnetodielectric layer ($\mu(\bar{z})=\mu_1\geq 1$, $\varepsilon(\bar{z})=\varepsilon_1\geq 1$, $-1<\bar{z}<0$) at parameter value $\Phi=0$. It is easy to show that in this case the solution to the Cauchy problem (66) is written as

$$V(\bar{z})=\frac{i\Gamma_1}{\mu_1}\cdot\frac{(\Gamma_1-\mu_1\Gamma)\exp\left[i\Gamma_1(\bar{z}+1)\right]-(\Gamma_1+\mu_1\Gamma)\exp\left[-i\Gamma_1(\bar{z}+1)\right]}{(\Gamma_1-\mu_1\Gamma)\exp\left[i\Gamma_1(\bar{z}+1)\right]+(\Gamma_1+\mu_1\Gamma)\exp\left[-i\Gamma_1(\bar{z}+1)\right]};$$
$$\Gamma_1=\sqrt{\kappa^2\varepsilon_1\mu_1-\Phi^2}=\kappa\sqrt{\varepsilon_1\mu_1},$$

and dispersion equation (67) takes the form

$$\exp(i2\Gamma_1)(\Gamma_1-\mu_1\Gamma)^2-(\Gamma_1+\mu_1\Gamma)^2=0. \tag{70}$$

From (70) we find the eigenfrequencies of the layer:

$$\bar{\kappa}_n=\frac{\pi n}{\sqrt{\varepsilon_1\mu_1}}-\frac{i}{\sqrt{\varepsilon_1\mu_1}}\ln\left|\frac{\sqrt{\varepsilon_1}+\sqrt{\mu_1}}{\sqrt{\varepsilon_1}-\sqrt{\mu_1}}\right|;\quad n=0,\pm1,\pm2,... \tag{71}$$

They are the complex numbers with negative imaginary parts, and therefore the corresponding free field oscillations decay exponentially in time. The distance between adjacent eigenfrequencies $\bar{\kappa}_{n+1}$ and $\bar{\kappa}_n$ is independent of the index n and is $\bar{\kappa}_{n+1}-\bar{\kappa}_n=\pi/\sqrt{\varepsilon_1\mu_1}$; that is the set Ω of natural frequencies of a homogeneous magnetodielectric layer is equidistant. In Figure 19, the dependence $|R(\kappa)|$ calculated for the case of a normally incident E-polarized plane wave reflected from the layer with parameters $\varepsilon_1=4$ and $\mu_1=1$ is shown. The oscillations' period in $|R(\kappa)|$ coincides with the distance between adjacent eigenfrequencies of the layer, and the value $|R(\kappa)|$ vanishes at points $\kappa = \mathrm{Re}\,\bar{\kappa}_n$.

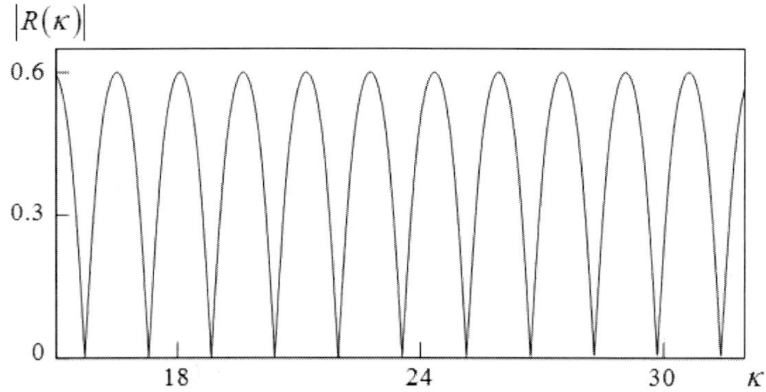

Figure 19. E-polarization. Amplitude-frequency characteristic of $|R(\kappa)|$, normal incidence, $\varepsilon_1 = 4$ and $\mu_1 = 1$.

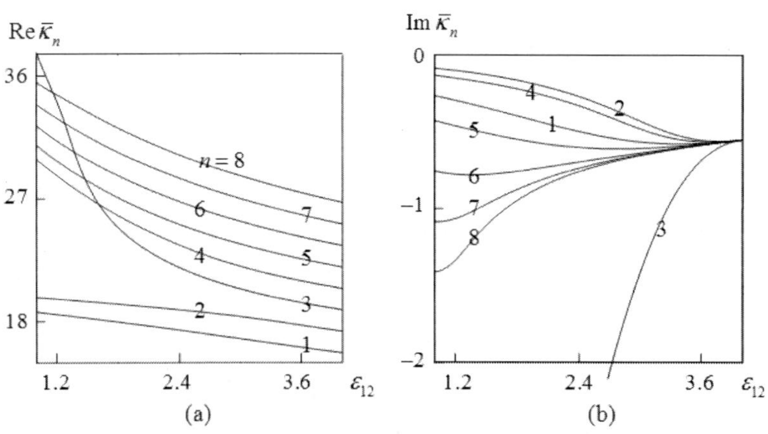

Figure 20. E-polarization. The drift of the eigenfrequencies of the layer (72) with change in ε_{12}.

In the second problem $\Phi = 0$, we consider the layer with $\mu_1 = 1$, and the permittivity is described by a piecewise constant periodic function

$$\varepsilon_1(z) = \begin{cases} \varepsilon_{11}; & -\frac{(p+0.5)h}{P} < z < -\frac{ph}{P} \\ \varepsilon_{12}; & -\frac{(p+1)h}{P} < z < -\frac{(p+0.5)h}{P} \end{cases}; \quad n = 0,1,...,P-1. \tag{72}$$

Obviously, for $\varepsilon_{11} = \varepsilon_{12}$ we have a homogeneous layer of a thickness h with a permittivity $\varepsilon_1 = \varepsilon_{11} = \varepsilon_{12}$, its eigenfrequencies are determined by the formula (71). Figure 20 illustrates the evolution of real and imaginary parts of eigenfrequencies of inhomogeneous layer when permittivity ε_{12} changes from the value $\varepsilon_{12} = \varepsilon_{11} = 4$ (homogeneous layer) to $\varepsilon_{12} = 1$ (inhomogeneous layer, $P = 12$). The characteristic feature of the eigenfrequencies' spectrum of an inhomogeneous layer is the presence of gaps – intervals in the real axis of the frequency parameter κ, where there are no real parts of the eigenfrequencies and outside which the real parts of the eigenfrequencies are distributed almost equidistantly (Figure 20a). The analysis showed that the presence of spectral points $\bar{\kappa}$ with $|\mathrm{Im}\,\bar{\kappa}| >> 1$ in the frequency spectrum Ω is one of the main reasons for the formation of such gaps. Figure 20b illustrates the going to infinity of one of the eigenfrequencies of a homogeneous layer (line 3) when it is continuously converting to a multilayered inhomogeneous medium with a periodic change of permittivity.

Note that the size and location of the gaps are close to the parameters of the so-called forbidden zones formed by the corresponding unbounded periodic medium [52]. A wave propagating in such a medium upon frequency transition to the band gap due to multiple reflections cannot leave a limited spatial region. A peculiar effect is also associated with the specific behavior of $|R(\kappa)|$ (case of normal incidence) in the frequency range $20 \le \kappa \le 29.5$ (Figure 21). Here, $\alpha = 0$, $\varepsilon_{12} = 1$; the frequency range where $|R(\kappa)| \approx 1$, almost exactly coincides with the gap corresponding to this value of ε_{12}.

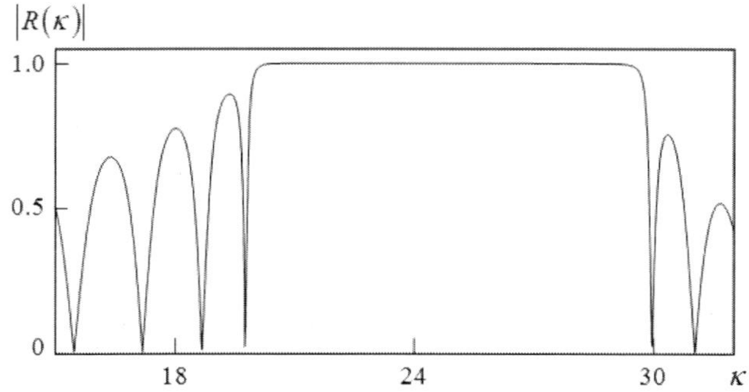

Figure 21. E -polarization. Amplitude-frequency characteristic of $|R(\kappa)|$ for a layer with a permittivity $\varepsilon_1(z)$ periodic along its thickness.

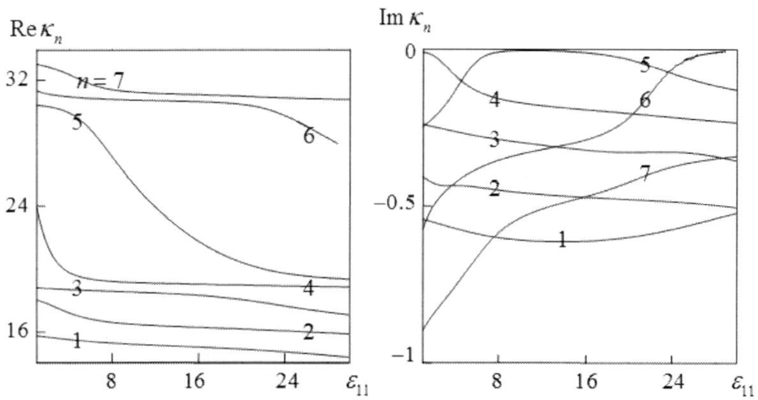

Figure 22. E -polarization. The drift of the eigenfrequencies of the layer upon violation of the periodicity of the function $\varepsilon_1(z)$.

Let us finalize the study with the case when the periodic dependence of the permittivity of the layer along its thickness is violated. Let in (72) $\varepsilon_{11} = 4$ and $\varepsilon_{12} = 1$, but on the fifth period of function $\varepsilon_1(z)$ ($p = 5$), ε_{11} is changing from $\varepsilon_{11} = 1$ (the layer is absent) via $\varepsilon_{11} = 4$ (periodic layer) and till the value $\varepsilon_{11} = 30$. If the periodicity is distorted, the real parts of several eigenfrequencies κ_n "penetrate" into the gap, and their imaginary parts sharply decrease (Figure 22). Violation of the periodicity of the layer affects the behavior of $|R(\kappa)|$

(Figure 23). Within the frequency range where the level of reflection of a normally incident plane E-polarized wave reaches its maximum, the frequencies, at which $|R(\kappa)|$ drops almost to zero, appear. These frequencies coincide with the real parts of the eigenfrequencies penetrating into a gap.

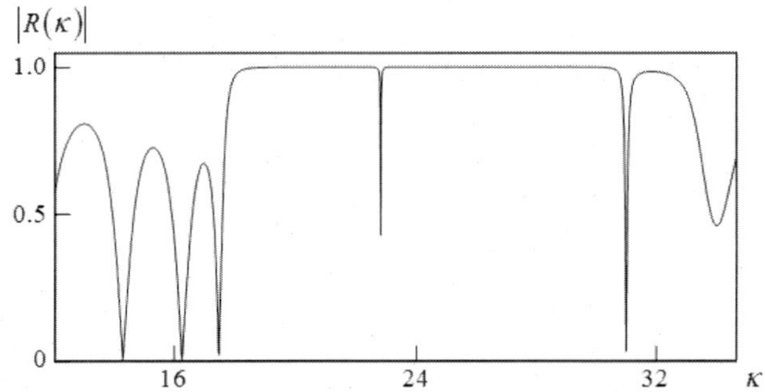

Figure 23. E -polarization. Amplitude-frequency characteristic of $|R(\kappa)|$ for a layer with a disturbed periodicity of function $\varepsilon_1(z)$, $\varepsilon_{11} = 1.3$.

CONCLUSION

Closing the discussion, we would like to give a concise estimation of the computational efficiency and advantages for practical applications of the method for solving the boundary value problems of electromagnetic waves diffraction by layered inhomogeneous media presented in this chapter.

The structure of the method may be split up into two principal steps. At the first step, the original boundary-value diffraction problem is reduced to the Cauchy problem for the Riccati equation. The solution to this problem determines all the main characteristics of the diffraction field both outside and inside the layered inhomogeneous medium. It takes place due to the analytically established connection between the solutions of the Riccati equation and the ordinary linear homogeneous differential equation of the second order. The second, but no less significant, step is the development of the authentic and rather efficient numerical algorithm for solving the Cauchy problem for the Riccati equation. Conventionally, the approaches to solving such problems consist of the discretization of the sought-for function variation interval, followed by the application of the Runge-Kutta method. We avoid the application of this method; instead, we almost completely analytically inversed the 'quadratic nonlinearity' of the Riccati equation and obtained a simple recursive formula for calculating the desired solution at the discretization points. This enabled us to create the algorithm having the unique properties in terms of calculation rate, efficiency, stability, and reliability.

As a result, we arrive to rather obvious and quite logical conclusion. The solution and corresponding algorithm developed and constructed in this paper for solving the direct problems of electromagnetic wave diffraction by layered-inhomogeneous structures will be in demand for solving urgent and practically important inverse problems. Those are the problems of remote atmospheric and subsurface sounding, computer tomography and diagnostics of layered inhomogeneous structures of natural and artificial origin, creating new artificial materials with unusual electrodynamic characteristics. Efficient solving of such problems (and they often need to be solved in the "in situ monitoring" mode) is possible only if the researchers have at their disposal methods comparable in operational characteristics to the method we described in this chapter.

REFERENCES

[1] Epstein, Paul. "Reflection of Waves in an Inhomogeneous Absorbing Medium." *Proceedings of the National Academy of Sciences of USA* 16, no. 10 (1930): 627–637.

[2] Brekhovskykh, Leonid. "Reflection of Plane Waves from Layered Inhomogeneous Media." *Zhurnal Tehnicheskoy Fiziki* 19, no. 10 (1949): 1126–1135.

[3] Wait, James. *Electromagnetic Waves in Stratified Media*. Oxford: Pergamon Press, 1970.

[4] Brekhovskikh, Leonid. *Waves in Layered Media*. Amsterdam: Elsevier Academic Press, 1960.

[5] Brekhovskikh, Leonid and Godin, Oleg. *Acoustics of Layered Media. I. Plane and Quasi-Plane Waves*. Berlin: Springer, 1990.

[6] Kliatskin, Valeriy. *The Invariant Embedding Method in Wave-Propagation Theory*. Moscow: Nauka, 1986.

[7] Schwarzburg, Alexander. "Dispersion of Electromagnetic Waves in Stratified and Non-Stationary Media (Exactly Solvable Models)." *Uspehi Fizicheskih Nauk* 170, no. 12 (2000): 1297–1324.

[8] Pazynin, Leonid, Pazynin, Vadim, and Sliusarenko, Hanna. "Negative Refraction of Plane Electromagnetic Waves in Non-Uniform Double-Negative Media." *Optics Letters* 44, no. 5 (2019): 1125–1128.

[9] Babich, Vasiliy and Buldyrev, Vladimir. *Asymptotic Methods in Short-Wavelength Diffraction Theory*. Oxford: Alpha Science, 2009.

[10] Melezhik, Petr, Brovenko, Andrey, and Poyedinchuk, Anatoliy. "Spectral Problems in the Wave Diffraction by Layered Media." *Telecommunications and Radio Engineering* 72, no. 20 (2013): 1821–1838.

[11] Sveshnikov, Alexey and Tikhonravov, Andrey. "Mathematical Methods for Analysis and Synthesis of Layered Media." *Matematicheskoye Modelirovaniye* 1, no. 7 (1989): 13–38.

[12] Seleznev, Igor, Krivonos, Yuri, and Yakovlev, Vitaly. *Wave Scattering by Local Inhomogeneities in Solid Media*. Kiyiv: Naukova Dumka, 1985.

[13] Brovenko, Andrey, Melezhik, Petr, Panin, Sergey, and Poyedinchuk, Anatoliy. "Numerical-Analytical Method for Solving Problems of Diffraction of Waves on Layered Inhomogeneous Media." *Fizicheskiye Osnovy Priborostroyeniya* 2, no. 1 (2013): 34–47.

[14] Melezhik, Petr, Brovenko, Andrey, and Poyedinchuk, Anatoliy. "Semi-Analytical Method for Solving Problems of Electromagnetics Wave Diffraction by Inhomogeneous Anisotropic Layers." *Telecommunications and Radio Engineering* 74, no. 1 (2015): 9–27.

[15] Brovenko, Andrey, Melezhik, Petr, Panin, Sergey, and Poyedinchuk, Anatoliy. "Semi-Analytical Method for Solving Problems of Electromagnetics Wave Diffraction by Circular Radially Inhomogeneous Magnetodielectric Cylinder." *Telecommunications and Radio Engineering* 76, no. 1 (2017): 1–18.

[16] Brovenko, Andrey, Melezhik, Petr, Poyedinchuk, Anatoliy, and Troshchylo, Aleksandr. "Wave Diffraction on Band Grating, Situated on

the Boundary of Layered-Inhomogeneous Medium: Method of Analytic Regularization." *Radiophysics and Quantum Electronics* 56, no. 4 (2013): 238–248.

[17] Brovenko, Andrey, Melezhik, Petr, Poyedinchuk, Anatoliy, and Troshchylo, Aleksandr. "A Numerical Algorithm of Solving Problems of Electromagnetic Wave Diffraction by a Plane Layer with the Kerr Nonlinearity." *Telecommunications and Radio Engineering* 76, no. 16 (2017): 1405–1415.

[18] Sirenko, Yuriy and Velychko, Lyudmyla, eds. *Electromagnetic Waves in Complex Systems: Selected Theoretical and Applied Problems*. New York: Springer, 2016.

[19] Shestopalov, Victor and Sirenko, Yuriy. *A Dynamical Theory of Gratings.* Kiyiv: Naukova Dumka, 1989.

[20] Krylov, Vladimir, Bobkov, Vladimir, and Monastiniy, Petr. *Computational Methods*. Moscow: Nauka, 1977.

[21] Shestopalov, Victor. *The Method of the Riemann-Hilbert Problem in the Theory of Electromagnetic Wave Diffraction and Propagation*. Kharkiv: Kharkiv State University Press, 1971.

[22] Shestopalov, Victor, Kirilenko, Anatoliy, Masalov, Sergey, and Sirenko, Yuriy. *Resonance Scattering of Waves. Vol. 1. Diffraction gratings*. Kiyiv: Naukova Dumka, 1986.

[23] Merkulov, Evgeniy and Erokhin, Nikolai. "Resonant Tunneling of an Electromagnetic Wave Through a Layer of Inhomogeneous Plasma with the Generation of Wavefield Bursts." *Voprosy Atomnoy Nauki i Tehniki*, no. 7 (2010): 128–130.

[24] Ponath, Hans-Ernst and Steqeman, George, eds. *Modern Problems in Condensed Sciences*. Amsterdam: Elsevier Academic Press, 1991.

[25] Davidovich, Michael. "Diffraction of a Plane Electromagnetic Wave by a Nonlinear Dielectric Layer." *Izvestiya Saratovskogo Universiteta. Seriya Fizika* 7, no. 1 (2007): 32–40.

[26] Bass, Fridrih and Gurevich, Yuriy. *Hot Electrons and Strong Electromagnetic Waves in Semiconductors and Gas Discharge Plasmas*. Moscow: Nauka, 1975.

[27] Shurman, Hans-Ernst. "Reflection and Transmission of a Plane TE-Wave at a Lossless Nonlinear Dielectric Film." *Phisika D* 158, (2001): 197–215.

[28] Reichardt, Hans. "The Radiation Condition for the Wave Equation." *Abhandlungen aus dem Mathematischen Seminar, Universitat Hamburg*, no. 24 (1960): 41–53.

[29] Weinberg, Boris. *Asymptotic Methods for Equations of Mathematical Physics.* Moscow: Moscow University Publ. House, 1982.

[30] Shestopalov, Victor, Litvinenko, Leonid, Masalov, Sergey, and Sologub, Vladimir. *Wave Diffraction by Gratings.* Kharkiv: Kharkiv State University Press, 1973.

[31] Dolph, Charl. "Recent Developments in Some Non-Self-Adjoint Problems of Mathematical Physics." *Bulletin of the American Mathematical Society* 67, no. 1 (1961): 1–69.

[32] Shestopalov, Victor. *A Spectral Theory and Excitation of Open Structures.* Kiyiv: Naukova Dumka, 1987.

[33] Shestopalov, Victor. *Morsean Critical Points of Secular Equations.* Kiyiv: Naukova Dumka, 1992.

[34] Shestopalov, Victor, Tuchkin, Yuriy, Poyedinchuk, Anatoliy, and Sirenko, Yuriy. *New Solution Methods for Direct and Inverse Diffraction Problems. Analytical Regularization Techniques for Boundary Value Problems in Electrodynamics.* Kharkiv: Osnova, 1997.

[35] Voytovich, Nikolay, Katzenelenbaum, Boris, and Sivov, Aleksey. *A Generalized Eigenoscillation Method in Diffraction Theory.* Moscow: Nauka, 1977.

[36] Sirenko, Yuriy and Strom, Staffan, eds. *Modern Theory of Gratings. Resonant Scattering: Analysis Techniques and Phenomena*. New York: Springer, 2010.

[37] Muravey, Leonid. "On the Asymptotic Character at Large Time Parameters of the Green Function of the First Outer Boundary Value Problem for Waveguide Equation with Two Space Variables." *Doklady Akademii Nauk SSSR* 220, no. 6, (1975): 1271–1273.

[38] Waynberg, Boris. "On the Operator Eigenfunctions Corresponding to Poles of its Resolvent Analytically Continued Through the Continuous Spectrum." *Matematicheskiy Sbornik* 87, no. 2 (1972): 293–308.

[39] Muravey, Leonid. "Analytical Extensions with Respect to a Parameter of Green Functions of External Boundary Value Problems for the Two-Dimensional Helmholtz Equation." *Matematicheskiy Sbornik* 105, no. 1 (1978): 63–108.

[40] Arsenyev, Aleksandr. "On the Existence of Resonant Poles and Resonances in Scattering Problems (the Case of Boundary Conditions of the II and III Kind)." *Zhurnal Vychislitel'noy Matematiki i Matematicheskoy Fiziki* 16, no. 3 (1976): 718–724.

[41] Sirenko, Yuriy. "Analytic Extensions of Diffraction Problems and Threshold Effects in Electrodynamics." *Doklady Academii Nauk USSR, Seriya A*, no. 2 (1968): 65–68.

[42] Poyedinchuk, Anatoliy, Tuchkin, Yuriy, and Shestopalov, Victor. "Regularizing the Spectral Problems of Wave Diffraction by Unclosed Shields." *Doklady Akademii Nauk SSSR* 295, no. 6 (1987): 1358–1362.

[43] Sommerfeld, Arnold. "Die Greensche Funktion der Schwingleichung." *Jahresbericht der Deutschen Mathematiker Vereinigung* 21, (1912): 309–353.

[44] Sveshnikov, Alexey. "The Principle of Radiation." *Doklady Akademii Nauk SSSR* 3, no. 5 (1950): 517–520.

[45] Brovenko, Andrey, Melezhik, Petr, Poyedinchuk, Anatoliy, and Yashina, Nataliya. "Surface Resonances of Metal Stripe Grating Put on the Plane Surface of Metamaterial." *Progress in Electromagnetics Research* 63, (2006): 209–222.

[46] Gokhberg, Izrail and Krein, Mark. *Introduction to the Theory of Linear Non-Self-Adjoint Operators.* Moscow: Nauka, 1965.

[47] Keldysh, Mstislav. "On Completeness of Eigenfunction Systems of Some Classes on Non-Self-Adjoint Operators." *Uspehi Matematicheskih Nauk* 26, no. 4 (1971): 15–41.

[48] Shestopalov, Yuriy. "Justifying the Spectral Method of Identifying Proper Waves of Microstrip Lines." *Differential Equations* 16, no. 8 (1980): 1504–1512.

[49] Shestopalov, Yuriy. "Properties of the Spectrum of One Class of Non-Self-Adjoint Boundary Value Problems for a Set of Helmholmtz Equations." *Doklady Akademii Nauk SSSR* 252, no. 5 (1980): 1108–1111.

[50] Sirenko, Yuriy, Strom, Staffan, and Yashina, Nataliya. *Modeling and Analysis of Transient Processes in Open Resonant Structures. New Methods and Techniques*. New York: Springer, 2007.

[51] Sirenko, Kostyantyn, Sirenko, Yuriy, and Yashina, Nataliya. "Modeling and Analysis of Transients in Periodic Gratings. I. Fully Absorbing Boundaries for 2-D Open Problems." *Journal of the Optical Society of America A* 27, no. 3 (2010): 532–543.

[52] Elachi, Charles. "Waves in Active and Passive Periodic Structures." *Proceedings of the IEEE* 64, no. 12 (1976): 1666–1698.

In: A Closer Look at Boundary Value Problems ISBN: 978-1-53617-857-9
Editor: Mustafa Avci

Chapter 5

THE LINEARIZATION METHODS AS A BASIS TO DERIVE THE RELAXATION AND THE SHOOTING METHODS

István Faragó[1,*] and Stefan M. Filipov[2], PhD

[1]Department of Differential Equations, Institute of Mathematics, Budapest University of Technology and Economics, and MTA-ELTE Research Group, Budapest, Hungary

[2]Department of Computer Science, University of Chemical Technology and Metallurgy, Sofia, Bulgaria

ABSTRACT

This chapter investigates numerical solution of nonlinear two-point boundary value problems. It establishes a connection between three important, seemingly unrelated, classes of iterative methods, namely: the linearization methods, the relaxation methods (finite difference methods), and the shooting methods. It has recently been demonstrated that using finite differences to discretize the sequence of linear problems obtained by quasi-linearization, Picard linearization, or constant-slope linearization, leads to the usual iteration formula of the respective relaxation method. Thus, the linearization methods can be used as a basis to derive the relaxation methods. In this work we demonstrate that the shooting methods can

[*] Corresponding Author's Email: faragois@cs.elte.hu ().

be derived from the linearization methods, too. We show that relaxing a shooting trajectory, i.e., an initial value problem solution, is in fact a projection transformation. The obtained function, called projection trajectory, can be used to correct the initial condition. Using the new initial condition, we can find a new shooting trajectory, and so on. The described procedure is called shooting-projection iteration (SPI). We show that using the quasi-linearization equation to relax (project) the shooting trajectory leads to the usual shooting by Newton method, the constant-slope linearization leads to the usual shooting by constant-slope method, while the Picard linearization leads to the recently proposed shooting-projection method. Therefore, the latter method can rightfully be called shooting by Picard method. A possible application of the new theoretical results is suggested and numerical computer experiments are presented. MATLAB codes are provided.

Keywords: nonlinear two-point boundary value problem, linearization, relaxation, shooting-projection

1. INTRODUCTION

This chapter considers two-point boundary value problems (TPBVPs) of the form

$$u''(x) = f\big(x, u(x), u'(x)\big), x \in (a, b), \tag{1}$$

$$u(a) = u_a, u(b) = u_b, \tag{2}$$

where a, b, u_a, and u_b are given constants, $u(x)$ is an unknown real-valued function of a real independent variable $x \in [a, b]$, and f is a given function that specifies the differential equation (1) [1-3]. Although there is a lot of literature on TPBVPs, the subject is still an active area of research [4-12]. If the function f is linear with respect to u and u', then the problem is called linear. Otherwise the problem is called nonlinear. In the sequel we use the notation

$$v(x) = u'(x). \tag{3}$$

The following conditions guarantee that the problem (1)-(2) has a solution and that this solution is unique:

f is continuous on the domain $D = \{(x, u, v) \mid x \in [a, b], u, v \in \mathbb{R}\}$,
$q = \partial_2 f = \frac{\partial f}{\partial u}$ and $p = \partial_3 f = \frac{\partial f}{\partial v}$ exist and are continuous on D,
$q > 0$ on D,
p is bounded on D. (4)

These conditions are sufficient but not necessary. Besides this statement, there are other theorems that guarantee the existence and uniqueness of solution. In the sequel, we assume that the TPBVP (1)-(2) has a unique solution. Often, this solution cannot be defined in a closed form (there is no analytic solution). Therefore, we define a numerical (approximate) solution, which is obtained as the result of some, appropriate for the case, numerical procedure.

This work studies numerical methods for solving nonlinear TPBVPs of the form (1)-(2). Two important classes of iterative methods for the solution of such problems are the relaxation methods [13], also known as finite difference methods (FDMs) [1-3, 13-17], and the shooting methods [1-3, 13, 14, 18, 19]. In the relaxation methods, one starts from an arbitrary function (function 0 in Figure 1), and each next function (functions 1, 2, ... in Figure 1) is brought into a better agreement with the differential equation (1). To achieve this, the differential equation (1) is first discretized using finite differences, and then the resulting nonlinear system, together with the two boundary conditions, is solved by some iterative method for algebraic systems, e.g., Newton method, fixed-point iteration, etc. Each iterative method generates a particular sequence of functions, called *relaxation trajectories*, that satisfy the two boundary conditions but do not satisfy the differential equation. However, the relaxation trajectories satisfy the differential equation (1) *approximately* and, if the particular iterative method is convergent for the given TPBVP, converge to the exact solution (dashed line in Figure 1). As shown in section 4, the relaxation trajectories obtained by the FDM can, in principle, be obtained as continuous functions (solid lines in Figure 1) from the corresponding linearization method. In the shooting methods, the TPBPV (1)-(2) is replaced by an initial value problem (IVP) (Cauchy problem) and a guess is made for the initial condition $v(a)$. The IVP is solved using some numerical technique. The obtained function (function 0 in Figure 2), called *shooting trajectory*, satisfies the differential equation (1) and the first (left) boundary condition but does not satisfy the second (right) boundary condition. Then, using the end-value of the shooting trajectory and some iterative approach for solving algebraic equations, e.g., Newton method, fixed-point iteration, etc., the initial condition is corrected. The new initial condition is used to obtain a new shooting trajectory (function 1 in Figure 2), and so on. If the particular iterative method turns out to be convergent for the given TPBVP, then the sequence of shooting trajectories converges to the exact solution (dashed line in Figure 2).

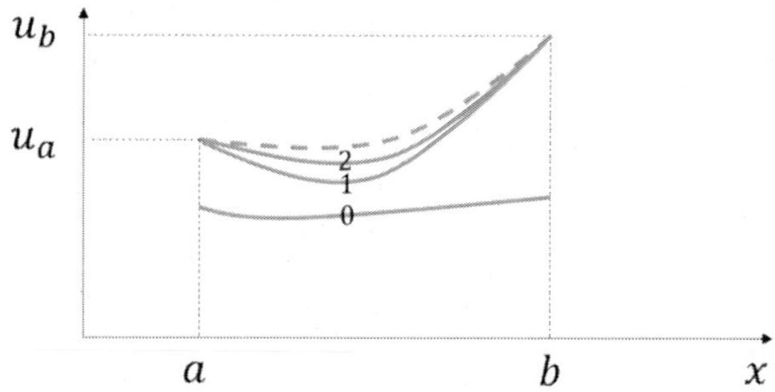

Figure 1. Relaxation method. Three successive relaxation trajectories (functions), marked 0, 1, 2, and the exact solution (dashed line) are shown. The function 0 is arbitrary. The functions 1, 2, ... satisfy the two boundary conditions but do not satisfy the differential equation.

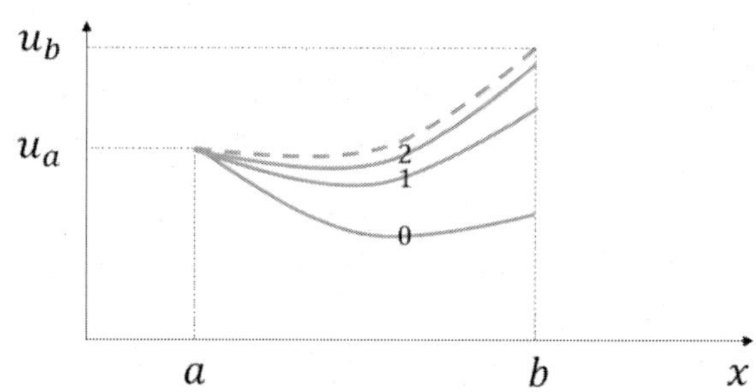

Figure 2. Shooting method. Three successive shooting trajectories (functions), marked 0, 1, 2, and the exact solution (dashed line) are shown. The functions 0, 1, 2, ... satisfy the differential equation and the left boundary condition but do not satisfy the right boundary condition.

Other important iterative methods for solving nonlinear TPBVPs are the linearization methods, e.g., the quasilinearization method [20, 2, 14, 17], the Picard linearization method (Picard successive approximations) [21, 17], etc. In these methods, one starts from an arbitrary function and uses this function to obtain a linear TPBVP that approximates the nonlinear TPBVP (1)-(2). Then, the linear problem is solved by some technique, and the solution is used to obtain a new linear TPBVP which is a better approximation to (1)-(2). Provided the procedure is convergent, the generated sequence of functions converges to the solution of the nonlinear problem (1)-(2).

In this work we establish some important connections between the linearization methods, the relaxation methods, and the shooting methods. The chapter comprises some of our recent research plus some new unpublished results. The new results are presented in sections 5, 6, and 7.

2. Iterative Methods for Algebraic Equations

This section considers three important iterative methods for finding roots of nonlinear algebraic equations, namely, the Newton method, the Picard method (fixed-point iteration), and the constant-slope method. The constant-slope method is a particular case of a fixed-point iteration. As shown in the section, all three methods can be viewed as linearization methods because at each iteration step the nonlinear function in the considered nonlinear algebraic equation is replaced by some linear function. The solution to the obtained in this way linear equation is taken as the next root approximation. Consider the algebraic equation

$$F(x) = 0, \tag{5}$$

where F is some given nonlinear function.

Table 2.1. Three different types of linearization (8) for the nonlinear algebraic equation (7)

	(i) Newton	(ii) Picard	(iii) Constant-slope
$\rho_k =$	$\phi'(x_k)$	0	$\phi'(x_0)$

Using the nonlinear function F, we introduce a new nonlinear function ϕ:

$$\phi(x) = x - \frac{F(x)}{m}, \tag{6}$$

where m is some fixed number different from zero. Then Eqn. (5) is equivalent to the equation

$$x = \phi(x), \tag{7}$$

i.e., the fixed points of ϕ are roots of (5) and vice versa. Let $\tilde{x}$ be a unique root of (5) and let x_k denote a suitable approximation of $\tilde{x}$. The nonlinear function $\phi(x)$ can be approximated by the linear function $\phi(x_k) + \rho_k(x - x_k)$, where ρ_k is some number. We assume that $\rho_k \neq 1$. Thus, instead of equation (7) we write

$$x \approx \phi(x_k) + \rho_k(x - x_k). \tag{8}$$

Clearly, (8) is obtained by linearizing the right-hand side of (7). Depending on the choice of ρ_k, we have three different types of linearization: Newton, Picard, or constant-slope (Table 2.1). In the table, x_0 denotes some initial guess for $\tilde{x}$. Replacing in (8) the sign $\approx$ by exact equality and, accordingly, x by x_{k+1} gives:

$$x_{k+1} = \phi(x_k) + \rho_k(x_{k+1} - x_k). \tag{9}$$

We can take x_{k+1} as the next approximation to $\tilde{x}$. If iteration (9) turns out to be convergent, then the sequence $x_0, x_1, \ldots$ converges to $\tilde{x}$.

Using (6) in Eqn. (9), we get the usual form of the considered iterative methods:

$$x_{k+1} = x_k - \frac{F(x_k)}{m_k}, \tag{10}$$

where $m_k = m(1-\rho_k)$. The three different definitions of ρ_k (Table 2.1) yield three different m_k-s (Table 2.2) corresponding to the three different iterative methods: Newton, Picard, or constant-slope.

Table 2.2. Three iterative methods (10) for soling the nonlinear equation (5)

	(i) Newton	(ii) Picard	(iii) Constant-slope
$m_k =$	$F'(x_k)$	m	$F'(x_0)$

As can be seen from (10) and Table 2.2, the constant-slope method is just the Picard method with $m = F'(x_0)$. Both methods are fixed-point iterations and have linear convergence. This means that, for any $k > k_0$ (k_0 is a certain integer), $|e_{k+1}| \le C|e_k|$, where $e_k = x_k - \tilde{x}$ and $0 < C < 1$. The value of m_k for the Newton method is readjusted at each iteration step. It is equal to the slope of the straight line tangent to $F(x)$ at $x = x_k$. The Newton method has quadratic convergence. This means that, if $k > k_0$, $|e_{k+1}| \le C|e_k|^2$, where $C > 0$. Whether a particular iterative method is convergent or not depends on the nonlinearity F and the initial guess x_0 (and the choice of m for the Picard method) [22].

The specific presentation (9) of the Newton, Picard, and constant-slope iterative methods (Table 2.1) is very convenient since it gives us a way to carry out the linearization of the differential equation (1) in an analogous fashion. Note that the right-hand side of (8) for the Newton method is just the first two terms of the Taylor expansion of $\phi(x)$ around $x = x_k$.

3. Linearization Methods for Nonlinear TPBVPs

Let the function $u_k(x), x \in [a,b]$ be some suitable approximation to the solution of the nonlinear problem (1)-(2). Analogously to (8) we can approximate the right-hand side of (1) by a linear function:

$$u''(x) \approx f_k(x) + q_k(x)\big(u(x) - u_k(x)\big) + p_k(x)\big(v(x) - v_k(x)\big), \quad (11)$$

where $v_k(x) = u_k'(x)$, $f_k(x) = f\big(x, u_k(x), v_k(x)\big)$, and $q_k(x)$, $p_k(x)$ are given in Table 3.1. In the table, $u_0(x)$ is some initial guess for the solution of (1)-(2). Adding the boundary conditions (2), replacing in (11) the sign $\approx$ by exact equality and, accordingly, $u(x)$ by its approximation $u_{k+1}(x)$, we get

$$u_{k+1}''(x) = f_k(x) + q_k(x)\big(u_{k+1}(x) - u_k(x)\big) + p_k(x)\big(u_{k+1}'(x) - u_k'(x)\big), \tag{12}$$

$$u_{k+1}(a) = u_a, u_{k+1}(b) = u_b. \tag{13}$$

If the function $u_k(x)$ is given (fixed), then equations (12)-(13) constitute a linear TPBVP for the unknown function $u_{k+1}(x)$.

Thus, if we start from some initial guess $u_0(x)$, we can solve successively (12)-(13) for $k = 0,1,\ldots$ and obtain $u_1(x), u_2(x), \ldots$. If the sequence of solutions $u_1(x), u_2(x), \ldots$ is convergent, it converges to the solution of the nonlinear problem (1)-(2). The method (12)-(13) for the Newton case (column (i) of Table 3.1) is the well-known quasilinearization method (QLM). Originally, it was introduced by Bellman and Kalaba [20] as a generalization of the Newton-Raphson method. It can be viewed as Newton method on operator level [2]. We note that, under conditions (4) and provided that $u_k(x)$ is continuously differentiable, the linear problem (12)-(13) arising from quasilinearization has a unique solution. The Picard linearization method and the constant-slope linearization method are also given by equations (12)-(13) but with $q_k(x)$, $p_k(x)$ given in columns (ii) and (iii) of Table 3.1.

To solve the linear problem (12)-(13) we can apply different numerical techniques. In the next section we prove that applying the finite difference method leads to the usual iteration formula of the respective relaxation method. Then we show that the shooting-projection iteration, with (12)-(13) as a projection transformation, leads to the respective shooting method. Thus, the linearization methods (12)-(13) can be used as a basis to derive both the relaxation and the shooting methods. A possible application of these theoretical results is suggested in the final sections.

Table 3.1. Three different types of linearization (11) of the differential equation (1)

	(i) Newton (QLM)	(ii) Picard	(iii) Constant-slope
$q_k(x) =$	$\partial_2 f\left(x, u_k(x), u_k'(x)\right)$	0	$\partial_2 f\left(x, u_0(x), u_0'(x)\right)$
$p_k(x) =$	$\partial_3 f\left(x, u_k(x), u_k'(x)\right)$	0	$\partial_3 f\left(x, u_0(x), u_0'(x)\right)$

4. DERIVING THE RELAXATION METHODS

Let us divide the interval $[a, b]$ by N equally separated mesh-points:

$$x_i = a + (i-1)h, h = \frac{b-a}{N-1}, i = 1,2, \dots, N, \tag{14}$$

where $N > 1$. The points x_i (14), where $x_1 = a$ and $x_N = b$, define a uniform mesh on the interval $[a, b]$. We are going to prove that using finite differences to discretize (12)-(13) on the mesh (14) yields the usual iteration formula of the respective relaxation method. Let u_i^k and u_i^{k+1} denote, respectively, approximations of the values of $u_k(x)$ and $u_{k+1}(x)$ at $x = x_i$ (14).

To approximate the first derivatives of $u_k(x)$ and $u_{k+1}(x)$ at the inner mesh-points, we could use forward (a), backward (b), or central (c) difference approximation:

$$\begin{aligned}
&\text{(a) } \mathrm{D}_+ u_i^k = \frac{u_{i+1}^k - u_i^k}{h}, \mathrm{D}_+ u_i^{k+1} = \frac{u_{i+1}^{k+1} - u_i^{k+1}}{h},\\
&\text{(b) } \mathrm{D}_- u_i^k = \frac{u_i^k - u_{i-1}^k}{h}, \mathrm{D}_- u_i^{k+1} = \frac{u_i^{k+1} - u_{i-1}^{k+1}}{h},\\
&\text{(c) } \mathrm{D}_0 u_i^k = \frac{u_{i+1}^k - u_{i-1}^k}{2h}, \mathrm{D}_0 u_i^{k+1} = \frac{u_{i+1}^{k+1} - u_{i-1}^{k+1}}{2h}.
\end{aligned} \tag{15}$$

Discretizing (12) using the central difference approximation to approximate the second derivative of $u_{k+1}(x)$, we get

$$\frac{u_{i-1}^{k+1} - 2u_i^{k+1} + u_{i+1}^{k+1}}{h^2} = f_i^k + q_i^k\left(u_i^{k+1} - u_i^k\right) + p_i^k\left(\mathrm{D}u_i^{k+1} - \mathrm{D}u_i^k\right),$$
$$i = 2,3, \dots, N-1, \tag{16}$$

where $\mathrm{D} \in \{\mathrm{D}_+, \mathrm{D}_-, \mathrm{D}_0\}$, $f_i^k = f(x_i, u_i^k, \mathrm{D}u_i^k)$, and q_i^k, p_i^k are given in Table 4.1. Note that, since h is a constant, $\mathrm{D}u_i^k$ and $\mathrm{D}u_i^{k+1}$ are linear homogenous functions of their arguments. Therefore, according to Euler's theorem on homogenous functions,

$$\mathrm{D}u_i^k = \frac{\partial \mathrm{D}u_i^k}{\partial u_{i-1}^k} u_{i-1}^k + \frac{\partial \mathrm{D}u_i^k}{\partial u_i^k} u_i^k + \frac{\partial \mathrm{D}u_i^k}{\partial u_{i+1}^k} u_{i+1}^k, \tag{17}$$

$$\mathrm{D}u_i^{k+1} = \frac{\partial \mathrm{D}u_i^{k+1}}{\partial u_{i-1}^{k+1}} u_{i-1}^{k+1} + \frac{\partial \mathrm{D}u_i^{k+1}}{\partial u_i^{k+1}} u_i^{k+1} + \frac{\partial \mathrm{D}u_i^{k+1}}{\partial u_{i+1}^{k+1}} u_{i+1}^{k+1}. \tag{18}$$

Table 4.1. Three different relaxation methods (25) for solving the nonlinear TPBVP (1)-(2)

	(i) Newton	(ii) Picard	(iii) Constant-slope
$q_i^k =$	$\partial_2 f(x_i, u_i^k, \mathrm{D}u_i^k)$	0	$\partial_2 f(x_i, u_i^0, \mathrm{D}u_i^0)$
$p_i^k =$	$\partial_3 f(x_i, u_i^k, \mathrm{D}u_i^k)$	0	$\partial_3 f(x_i, u_i^0, \mathrm{D}u_i^0)$

Substituting (17) and (18) into (16) and using the obvious relations

$$\frac{\partial \mathrm{D}u_i^{k+1}}{\partial u_{i-1}^{k+1}} = \frac{\partial \mathrm{D}u_i^k}{\partial u_{i-1}^k}, \frac{\partial \mathrm{D}u_i^{k+1}}{\partial u_i^{k+1}} = \frac{\partial \mathrm{D}u_i^k}{\partial u_i^k}, \frac{\partial \mathrm{D}u_i^{k+1}}{\partial u_{i+1}^{k+1}} = \frac{\partial \mathrm{D}u_i^k}{\partial u_{i+1}^k}, \tag{19}$$

equation (16) is transformed into the form

$$L_{i,i-1}^k u_{i-1}^{k+1} + L_{i,i}^k u_i^{k+1} + L_{i,i+1}^k u_{i+1}^{k+1} = L_{i,i-1}^k u_{i-1}^k + L_{i,i}^k u_i^k + L_{i,i+1}^k u_{i+1}^k - G_i^k, \tag{20}$$

where

$$L_{i,i-1}^k = \left(1 - h^2 p_i^k \frac{\partial \mathrm{D}u_i^k}{\partial u_{i-1}^k}\right),$$
$$L_{i,i}^k = \left(-2 - h^2 q_i^k - h^2 p_i^k \frac{\partial \mathrm{D}u_i^k}{\partial u_i^k}\right),$$
$$L_{i,i+1}^k = \left(1 - h^2 p_i^k \frac{\partial \mathrm{D}u_i^k}{\partial u_{i+1}^k}\right), \tag{21}$$

and

$$G_i^k = u_{i-1}^k - 2u_i^k + u_{i+1}^k - h^2 f_i^k. \tag{22}$$

Equation (20), together with the boundary conditions

$$u_1^{k+1} = u_a, u_N^{k+1} = u_b, \tag{23}$$

can be written in a matrix form

$$\mathbf{L}^k \mathbf{u}^{k+1} = \mathbf{L}^k \mathbf{u}^k - \mathbf{G}^k. \tag{24}$$

The components of the column-vector $\mathbf{G}^k$ are $G_1^k = u_1^k - u_a$, $G_N^k = u_N^k - u_b$, and (22) for $i = 2,3,\dots,N-1$. The nonzero components of the $N \times N$ matrix $\mathbf{L}^k$ are $L_{1,1}^k = 1$, $L_{N,N}^k = 1$, and (21) for $i = 2,3,\dots,N-1$. In (24) $\mathbf{u}^k = \left[u_1^k, u_2^k, \dots, u_N^k\right]^T$ and $\mathbf{u}^{k+1} = \left[u_1^{k+1}, u_2^{k+1}, \dots, u_N^{k+1}\right]^T$. Under conditions (4) and providing $h < 2/P$ when $\mathrm{D} = \mathrm{D}_0$ or $h < 1/P$ when $\mathrm{D} \in \{\mathrm{D}_+, \mathrm{D}_-\}$, where P is the upper bound of $|p|$, the matrix $\mathbf{L}^k$ is strictly diagonally dominant, hence nonsingular. Multiplying (24) from the left by the inverse of $\mathbf{L}^k$, we finally get

$$\mathbf{u}^{k+1} = \mathbf{u}^k - \left(\mathbf{L}^k\right)^{-1} \mathbf{G}^k. \tag{25}$$

Formula (25) is nothing else but the iteration formula of the respective relaxation method. When (12)-(13) is the Newton linearization method (quasilinearization), then (25) is the Newton relaxation method (see columns (i) in Tables 3.1 and 4.1). When (12)-(13) is the Picard linearization method, then (25) is the Picard relaxation method (see columns (ii) in Tables 3.1 and 4.1). When (12)-(13) is the constant-slope linearization method, then (25) is the constant-slope relaxation method (see columns (iii) in Tables 3.1 and 4.1). The usual way to derive formula (25) is first to discretize the nonlinear problem (1)-(2) using the FDM and then to apply one of three considered iterative methods to the obtained nonlinear algebraic system. As we have just proven, the relaxation methods (25) can be derived directly from the linearization methods (12)-(13) by simply discretizing equations (12)-(13) with the FDM. Therefore, given a function u_k, we can obtain u_{k+1} by either applying (25) or solving (12)-(13) by some technique other than the FDM. No matter which way we use, the resulting function u_{k+1}, called *relaxation trajectory*, is, within numerical precision, the same. This equivalence was first proven by the authors in [17]. The result was used to construct a way of replacing the relaxation methods (25) by successive application of the linear shooting method.

5. Deriving the Shooting Methods

This and the following sections of the chapter present some unpublished new results of our recent research work. In this section we show that the linearization methods (12)-(13) can be used a basis to derive the shooting methods. First, we demonstrate that if the function u_k is a shooting trajectory (IVP solution), then equations (12)-(13) define a projection transformation of u_k into u_{k+1}. In this case the relaxation trajectory u_{k+1} is called a projection trajectory. Then, we introduce the shooting-projection iteration. In essence, the shooting-projection iteration is the following: the shooting trajectory is transformed into a projection trajectory, then the first derivative of the projection trajectory at the left boundary is used as a new initial condition and a new shooting trajectory is found, and so on. We show that applying the shooting-projection iteration with the quasi-linearization equation as a projection transformation results in the usual shooting by Newton method. When the Picard linearization is used as a projection, then we get the recently proposed by the authors shooting-projection method [19], while the constant-slope linearization leads to the shooting by constant-slope method.

5.1. Relaxing the Shooting Trajectory – Projection Transformation

Let $u(x; v_a^k), x \in [a, b], v_a^k \in \mathbb{R}$ be a solution to the following initial value problem:

$$u''(x; v_a^k) = f\left(x, u(x; v_a^k), u'(x; v_a^k)\right), \tag{26}$$

$$u(a; v_a^k) = u_a, u'(a; v_a^k) = v_a^k. \tag{27}$$

The function $u(x; v_a^k)$ is called a shooting trajectory. It satisfies the differential equation (1) and the left boundary condition in (2) but, typically, does not satisfy the right boundary condition in (2). The number v_a^k is some approximation for the first derivative at $x = a$ of the solution of the nonlinear problem (1)-(2). Let us relax the shooting trajectory into $u_{k+1}(x)$ using the linearization equations (12)-(13). In other words, we replace the function $u_k(x)$ in (12)-(13) by the shooting-trajectory $u(x; v_a^k)$. Taking into account (26), equations (12)-(13) become

$$u''_{k+1}(x) = u''_k(x) + q_k(x)\big(u_{k+1}(x) - u_k(x)\big) + \\ +p_k(x)\big(u'_{k+1}(x) - u'_k(x)\big), \tag{28}$$

$$u_{k+1}(a) = u_a, u_{k+1}(b) = u_b, \tag{29}$$

where $u_k(x) = u(x; v_a^k)$. Let $\hat{P}$ be an operator defined by the equation

$$\hat{P}u_k = u_{k+1}, \tag{30}$$

where u_k is *any* fixed function (not necessarily a shooting trajectory) for which (28)-(29) has a unique solution and u_{k+1} is this unique solution. Obviously, if u_k satisfies the boundary conditions, then the unique solution to (28)-(29) is just $u_{k+1} = u_k$, i.e., $\hat{P}u_k = u_k$. Since u_{k+1} satisfies the boundary conditions, it follows that

$$\hat{P}u_{k+1} = u_{k+1}. \tag{31}$$

Hence, from (30) and (31) we get

$$\hat{P}^2 u_k = \hat{P}u_k. \tag{32}$$

Since this is true for any u_k, it follows that $\hat{P}^2 = \hat{P}$, i.e., $\hat{P}$ is idempotent. In general, an idempotent mapping of a set (or other mathematical structure) into a subset (or a sub-structure) is a projection [23]. Therefore, the transformation of u_k into u_{k+1} defined by (28)-(29) can be considered a projection transformation. [1]

5.2. Shooting-Projection Iterative Procedure

As shown in the previous section, the linear problem (28)-(29) defines a projection transformation of u_k into u_{k+1}. Depending on what values of $q_k(x)$, $p_k(x)$ we choose (Table 3.1), we have three types of projection: Newton, Picard, or constant-slope. The function u_{k+1}, i.e., the solution to (28)-(29), is called a projection trajectory. The projection trajectory satisfies the boundary conditions (2) exactly and the differential equation (1) approximately. Therefore, it is an approximate solution to the nonlinear TPBVP (1)-(2). Hence, we could use the first derivative of the projection trajectory at the first (left) boundary as a new initial condition:

$$v_a^{k+1} = u'_{k+1}(a). \tag{33}$$

Then, we could use the new initial condition v_a^{k+1} to find a new shooting trajectory, and so on. This is the shooting-projection iterative procedure or *shooting-projection iteration* (SPI). In summary, the algorithm of the SPI is:

1. Use v_a^k and find the shooting trajectory $u(x; v_a^k)$ that satisfies (26)-(27).
2. Solve (28)-(29) with $u_k(x) = u(x; v_a^k)$ to find the projection trajectory u_{k+1}.
3. Use (33) to find the new initial condition v_a^{k+1} and repeat 1)-3) with v_a^{k+1}.

If the iteration is convergent, then $u(x; \tilde{v}_a)$, where $\tilde{v}_a = \lim v_a^k$ as $k \to \infty$, is the sought solution to the nonlinear TPBVP (1)-(2).

[1] Strictly speaking, in linear algebra and functional analysis, an operator must be linear and idempotent to be a projection. The operator $\hat{P}$ is linear if the boundary conditions are homogenous. However, all the results and conclusions reached in this chapter hold for the non-homogenous BCs (2) as well.

According to the results of section 4, to find the projection trajectory, i.e., to solve (28)-(29), we can apply (25) in which $\mathbf{u}^k$ represents the shooting trajectory $u(x; v_a^k)$. We note that, since the shooting trajectory satisfies the first boundary condition, i.e., $u_1^k = u_a$, and the nonlinear differential equation (1), i.e., in a discretized form $u_{i-1}^k - 2u_i^k + u_{i+1}^k = h^2 f_i^k$, as $\mathbf{G}^k$ in (25) we can take

$$\mathbf{G}^k = [0,0,\dots,0,u(b; v_a^k) - u_b]^T. \quad (34)$$

The option of calculating the projection trajectory via (25) will be used later on in the chapter when we consider possible applications of the theoretical results.

It turns out that, for all three considered SPI methods, namely Newton, Picard, and constant-slope, it is possible to find a relation between the k-th and the improved $(k+1)$-th guess for the initial derivative values, namely v_a^{k+1} as a function of v_a^k. These relations are derived in the next three sections.

Table 5.2.1. Three different iteration formulas (51) corresponding to the three different projections (28)-(29)

	(i) Newton	(ii) Picard	(iii) Constant-slope
$l_k =$	$\dfrac{\partial u(b; v_a^k)}{\partial v_a^k}$	$b - a$	$\dfrac{\partial u(b; v_a^0)}{\partial v_a^0}$

5.3. Deriving the Shooting by Newton Method

Let

$$y(x) = u_{k+1}(x) - u_k(x), \quad (35)$$

where $u_k(x) = u(x; v_a^k)$ is a shooting trajectory and $u_{k+1}(x)$ is the corresponding Newton projection trajectory satisfying (28)-(29). Using (35), equations (28)-(29) can be written as

$$y''(x) = q_k(x)y(x) + p_k(x)y'(x), \quad (36)$$

$$y(a) = 0, y(b) = u_b - u\left(b; v_a^k\right). \tag{37}$$

where $q_k(x)$, $p_k(x)$ are given in Table 3.1 column (i). Taking the derivative of (35) and using (33) and the second initial condition in (27), we replace the boundary conditions (37) with the initial conditions

$$y(a) = 0, y'(a) = v_a^{k+1} - v_a^k. \tag{38}$$

Obviously, the initial value problem (36), (38) is equivalent to the two-point boundary value problem (36), (37). Now, introducing the function $z(x)$ such that

$$y(x) = \left(v_a^{k+1} - v_a^k\right)z(x), \tag{39}$$

the initial value problem (36), (38) becomes

$$z''(x) = q_k(x)z(x) + p_k(x)z'(x), \tag{40}$$

$$z(a) = 0, z'(a) = 1. \tag{41}$$

Differentiating (26)-(27) with respect to v_a^k (as in [18]), comparing the result to (40)-(41), and taking into account that (40)-(41) has a unique solution, we get

$$z(x) = \frac{\partial u\left(x; v_a^k\right)}{\partial v_a^k}. \tag{42}$$

At $x = b$, using (39) and the second boundary condition in (37), equation (42) yields

$$v_a^{k+1} = v_a^k - \frac{u\left(b; v_a^k\right) - u_b}{\frac{\partial u\left(b; v_a^k\right)}{\partial v_a^k}}. \tag{43}$$

Equation (43) is the well-known iteration formula of the shooting by Newton method. Therefore, the shooting-projection iteration with projection transformation (28)-(29), where $q_k(x), p_k(x)$ are given in Table 3.1 column (i), is equivalent to the shooting by Newton method.

5.4. Deriving the Shooting by Picard Method

The Picard projection transformation (28)-(29) with $q_k(x)$, $p_k(x)$ given in Table 3.1 column (ii) is

$$u''_{k+1}(x) = u''_k(x), \tag{44}$$

together with the boundary conditions (29), where $u_k(x) = u(x; v_a^k)$ is a shooting trajectory and $u_{k+1}(x)$ is the corresponding projection trajectory. Integrating (44) on $[a, x]$, and then integrating the result on $[a, b]$, we get

$$\begin{aligned} u_{k+1}(b) - u_{k+1}(a) - u'_{k+1}(a)(b-a) = \\ = u_k(b) - u_k(a) - u'_k(a)(b-a). \end{aligned} \tag{45}$$

Taking into account that the shooting trajectory $u_k(x) = u(x; v_a^k)$ satisfies the initial conditions (27) and the projection trajectory $u_{k+1}(x)$ satisfies the boundary conditions (29), we get

$$v_a^{k+1} = v_a^k - \frac{u(b; v_a^k) - u_b}{b-a}. \tag{46}$$

Formula (46) is the iteration formula of the recently proposed shooting-projection method [19]. Since the projection transformation (44), (29) stems from the Picard linearization method (12)-(13) Table 3.1 column (ii), the shooting-projection method (46) can rightfully be called *shooting by Picard method.*

5.5. Deriving the Shooting by Constant-Slope Method

Using again the functions $y(x)$ (35) and $z(x)$ defined by (39), the initial value problem (40)-(41) for the constant-slope case is derived:

$$z''(x) = q_0(x)z(x) + p_0(x)z'(x), \tag{47}$$

$$z(a) = 0, z'(a) = 1, \tag{48}$$

where $q_0(x)$, $p_0(x)$ are given in Table 3.1 column (iii). Differentiating (26)-(27) for $k = 0$ with respect to v_a^0, comparing the result to (47)-(48), and taking into account that the solution to (47)-(48) is unique, we get

$$z(x) = \frac{\partial u(x; v_a^0)}{\partial v_a^0}. \tag{49}$$

Finally, using (39) and the second boundary condition in (37), equation (49), at $x = b$, yields

$$v_a^{k+1} = v_a^k - \frac{u(b; v_a^k) - u_b}{\frac{\partial u(b; v_a^0)}{\partial v_a^0}}. \tag{50}$$

Equation (50) is the iteration formula of the shooting by constant-slope method. Therefore, the shooting-projection iteration with projection transformation (28)-(29), where $q_k(x)$, $p_k(x)$ are given in Table 3.1 column (iii), is equivalent to the shooting by constant-slope method.

5.5.1. General Iteration Formula of the SPI

Formulas (43), (46), and (50) can be unified in one single formula:

$$v_a^{k+1} = v_a^k - \frac{u(b; v_a^k) - u_b}{l_k}, \tag{51}$$

where l_k depends on the choice of projection: Newton, Picard, or constant-slope (Table 5.2.1). This is the shooting-projection iteration formula for correcting the initial condition. It allows us to avoid (skip) the explicit calculation of the projection trajectory. As discussed, iteration (51), with l_k given in Table 5.2.1, is the same as the iteration formulas of known shooting methods, namely: (i) shooting by Newton method, (ii) shooting-projection method [19], and (iii) shooting by constant-slope method. The theoretical importance of the result is that it shows that the SPI with the usual linearization methods as projections leads to the respective shooting methods. The practical importance of the result is that it gives us an alternative of calculating v_a^{k+1} not through formula (51) but via the projection trajectory. This is discussed in detail in the next sections.

6. Application of the Theoretical Results

Let $u(x; v_a^k)$ be the shooting trajectory satisfying the initial value problem (26)-(27). Using the iteration formula of the shooting by Newton method (43) we can obtain a new initial condition v_a^{k+1}. According to the result in section 5.3, the value of v_a^{k+1} is exactly equal to the first derivative at $x = a$ of the corresponding projection trajectory $u_{k+1}(x)$ obtained from the projection transformation (28)-(29) with $q_k(x)$, $p_k(x)$ given in Table 3.1 column (i). As discussed in section 5.1, this projection transformation is just the quasilineartization (12)-(13) Table 3.1 column (i) applied on the shooting trajectory. This result gives us an alternative to finding the new initial condition v_a^{k+1}, namely, instead of using (43), we could use (33). In order to use (33) we need to find the projection trajectory $u_{k+1}(x)$ that satisfies (28)-(29) . According to the result in section 4, the projection trajectory can be obtained in a discretized form, i.e., $\mathbf{u}^{k+1}$, by applying the Newton relaxation equation (25), where q_i^k, p_i^k are given in Table 4.1 column (i) and $\mathbf{G}^k$ is given by (34). Then, we can use

$$\frac{u_2^{k+1} - u_1^{k+1}}{h} \tag{52}$$

as an approximation to v_a^{k+1}.

This section discusses a possible situation in which the option of calculating v_a^{k+1} via (52) could be very useful. Suppose we need to solve a TPBVP of the form

$$u''(x) = f(u(x)), \tag{53}$$

$$u(a) = u_a, u(b) = u_b, \tag{54}$$

but for some reason the derivative of f with respect to u is not available. For example, if u is a position and x is time, then (53) is the equation of motion for a particle (small object) of unit mass travelling under the influence of the force f. Launching the particle with initial velocity v_a^k and tracing its motion is equivalent to finding the shooting trajectory $(x; v_a^k)$, i.e., to solving the equation of motion (53) with initial conditions (27). If the particle does not arrive at the prescribed position u_b at time b, which is typically the case, we are faced with the need to correct (change) the initial condition v_a^k to a new one v_a^{k+1}. However, if we want to use the shooting by Newton method (43) we need to solve the IVP (40)-(41) and obtain $\partial u(b; v_a^k)/\partial v_a^k$ needed for formula (43) from $z(x)$ at $x = b$ (see Eqn. (42)). Since $q = \partial f/\partial u$ is not available, this is impossible (at least not in a direct way). Another (approximate) method would be to launch a second trial with initial velocity $v_a^k + \Delta v_a^k$, where Δv_a^k is a small change in the initial velocity, and then calculate $\Delta u(b; v_a^k)/\Delta v_a^k$. However, if launching a second trial is impossible or undesirable, then the described in this paragraph new approach may come in very handy. Taking the derivative of (53) we get

$$u'''(x) = \partial_u f(u(x))u'(x), \tag{55}$$

i.e., using v (3) and q,

$$v''(x) = q(x)v(x). \tag{56}$$

For any trajectory $u(x)$ that satisfies the differential equation (53), equation (56) must hold. Discretizing equation (56) on the mesh (14) using the central difference approximation for $v''(x)$, we get

$$\frac{v_{i-1} - 2v_i + v_{i+1}}{h^2} = q_i v_i, i = 2,3, \dots, N-1, \tag{57}$$

where $q_i = \partial_u f(u_i)$. Let $\mathbf{u}^k = [u_1^k, u_2^k, \dots, u_N^k]^T$ represent the shooting trajectory $u(x; v_a^k)$ in a discretized form, and let $\mathbf{v}^k = [v_1^k, v_2^k, \dots, v_N^k]^T$ represent its first derivative $v(x; v_a^k) = u'(x; v_a^k)$. Since the shooting trajectory satisfies the differential equation (53), it follows that v_i^k satisfy (57). Hence, after substituting v_i^k into (57) and rearranging, we get

$$-2 - h^2 q_i^k = -\frac{v_{i-1}^k + v_{i+1}^k}{v_i^k}, i = 2,3, \dots, N-1, \tag{58}$$

where $q_i^k = \partial_u f(u_i^k)$. For the TPBVP (53)-(54), since $p = \partial_v f = 0$, the nonzero elements of the Jacobian matrix $\mathbf{L}^k$ for the Newton relaxation method are

$$L_{1,1}^k = 1, L_{N,N}^k = 1,$$
$$L_{i,i-1}^k = 1, L_{i,i}^k = -2 - h^2 q_i^k, L_{i,i+1}^k = 1, i = 2,3, \dots, N-1. \tag{59}$$

Hence, using (58), the only unknown elements of the matrix $\mathbf{L}^k$ can be found from the shooting trajectory itself:

$$L_{i,i}^k = -\frac{v_{i-1}^k + v_{i+1}^k}{v_i^k}, i = 2,3, \dots, N-1. \tag{60}$$

Of course, equation (60) can be used when $v_i^k \neq 0$. If $v_i^k = 0$ for some particular i, we could simply extrapolate the value of $L_{i,i}^k$ from the values at the neighboring mesh-points. In the example provided in the next section, there are no such complications. Having calculated the matrix elements (60), we could use the Newton relaxation (25) with (34) for $\mathbf{G}^k$ to transform the shooting trajectory $\mathbf{u}^k$ into the projection trajectory $\mathbf{u}^{k+1}$. Then, (52) gives the next initial condition v_a^{k+1} for the shooting by Newton method.

7. Computer Experiments and Matlab Codes

Consider the TPBVP

$$u''(x) = \left(u(x)\right)^3, \tag{61}$$

$$u(0) = 1/2, u(1) = 1. \tag{62}$$

Table 7.1. Solving the TPBVP (61)-(62) by the traditional shooting by Newton method (43)

k	v_a^k – initial condition	E_k – deviation from 2nd BC
0	0	-0.433349035739307
1	0.379948530223661	0.026009489270876
2	0.359783026933729	0.000100006717963

Table 7.2. Solving the TPBVP (61)-(62) by the proposed new approach (52)

k	v_a^k – initial condition	E_k – deviation from 2nd BC
0	0	-0.433349035739307
1	0.379942276669709	0.026001423439514
2	0.359786457564626	0.000104397665347

The problem is solved by the shooting by Newton method, first by using the traditional approach to correct the initial condition, i.e., Eqn. (43), and then by the new approach described in the previous section, namely (52) to approximate v_a^{k+1}. For starting initial condition we choose $v_a^0 = 0$. To find the shooting trajectory, at each iteration step, the explicit Euler method EE_ [17] is used. The discretization step is $h = 0.001$. The deviation from the second boundary condition is denoted by

$$E_k = u\left(b; v_a^k\right) - u_b. \tag{63}$$

The results are shown in Table 7.1 for the traditional shooting by Newton method and in Table 7.2 for the new approach.

Table 7.3. MATLAB codes for the solution of the TPBVP (61)-(62)

Traditional shooting by Newton (43)

```
function main
 N=1001;
 a=0; b=1;
 ua=0.5; ub=1;
 va=0;
 h=(b-a)/(N-1);
 x=zeros(N,1); u=zeros(N,1);
 v=zeros(N,1);
 z=zeros(N,1); w=zeros(N,1);
 for i=1:N
 x(i)=a+h*(i-1);
 end
 u(1)=ua; v(1)=va;
 z(1)=0; w(1)=1;
 hold on; format long;
 E=1;
 while(abs(E)>0.001)
 for i=2:N
 u(i)=u(i-1)+h*v(i-1);
 v(i)=v(i-1)+h*f(x(i),u(i),v(i-
1));
 z(i)=z(i-1)+h*w(i-1);
 qi=q(x(i),u(i),v(i-1));
 pi=p(x(i),u(i),v(i-1));
 w(i)=w(i-1)+h*(qi*z(i)+pi*w(i-
1));
 end
 plot(x,u);
 E=u(N)-ub
 v(1)=v(1)-E/z(N);
 end
end
function f_=f(x,u,v)
 f_=u*u*u;
end

function p_=p(x,u,v)
 p_=0;
end
function q_=q(x,u,v)
 q_=3*u*u;
end
```

The proposed new approach (52)

```
function main
 N=1001;
 a=0; b=1;
 ua=0.5; ub=1;
 va=0;
 h=(b-a)/(N-1);
 x=zeros(N,1); u=zeros(N,1);
 v=zeros(N,1);
 L=zeros(N,N); G=zeros(N,1);
 for i=1:N
 x(i)=a+h*(i-1);
 end
 L(1,1)=1; L(N,N)=1;
 for i=2:N-1
 L(i,i-1)=1; L(i,i+1)=1;
 end
 u(1)=ua; v(1)=va;
 hold on; format long;
 E=1;
 while(abs(E)>0.001)
 for i=2:N
 u(i)=u(i-1)+h*v(i-1);
 v(i)=v(i-1)+h*f(x(i),u(i),v(i-
1));
 end
 for i=2:N-1
 L(i,i)=-(v(i-1)+v(i+1))/v(i);
 end
 plot(x,u,'-g');
 E=u(N)-ub
 G(N)=E;
 uProj=u-L\G;
 v(1)=(uProj(2)-uProj(1))/h;
 end
end

function f_=f(x,u,v)
 f_=u*u*u;
end
```

MATLAB codes for the solution of (61)-(62) by the traditional shooting by Newton method and by the proposed new approach are presented in Table 7.3. In the first code we have used the notation $w = z'$. Note that, for the proposed new approach, the derivatives $q = \partial_u f$ and $p = \partial_v f$ are not defined at all.

Conclusion

This chapter revealed important connections between the linearization, the relaxation, and the shooting methods for nonlinear TPBVPs. It was demonstrated that discretizing the linearization equations by using the FDM yields the respective relaxation method. It was also shown that the shooting-projection iteration with the linearization equations as projections leads to the usual shooting methods. An application of the theoretical results was proposed whereby the shooting by Newton method is carried out not through the traditional iteration formula but via explicit calculation of the projection trajectory. The approach is useful because for certain TPBVPs it requires knowledge only of the nonlinear function f but not of its partial derivatives.

Acknowledgments

This research was partially carried out in the ELTE Institutional Excellence Program (1783-3/2018/FEKUTSRAT) supported by the Hungarian Ministry of Human Capacities, and it was supported by the Hungarian Scientific Research Fund SNN125119. The research was also supported by the program for young scientists and postdocs in Bulgaria.

References

[1] Keller, H. B. (1968). *Numerical Methods for Two-Point Boundary-Value Problems.* Blaisdell Publishing Co. Ginn and Co., Waltham, Mass.-Toronto, Ont.-London.

[2] Ascher, U. M., Mattheij, R. M. and Russell, R. D. (1995). Numerical Solution of Boundary Value Problems for Ordinary Differential

Equations. In: *Classics in Applied Mathematics*, vol. 13, SIAM, Philadelphia.

[3] Faragó, I. (2014). *Numerical Methods for Ordinary Differential Equations.* TypoTech, Budapest http://faragois.web.elte.hu//ODE_angol.pdf.

[4] Pandey, P. K. (2018). Solution of two point boundary value problems, a numerical approach: parametric difference method. *Applied Mathematics and Nonlinear Sciences*, 3(2): 649-658. https://doi.org/10.2478/AMNS.2018.2.00049.

[5] Han, S. H. and McEneaney, W. M. (2018). Fundamental solutions for two-point boundary value problems in orbital mechanics. *Applied Mathematics & Optimization*, 77(1): 129-172. https://doi.org/10.1007/s00245-016-9369-x.

[6] Jena, M. K. and Sahu, K. S. (2018). Operational Matrices from a Frame and their Applications in Solving Boundary Value Problems with Mixed Boundary Conditions. *International Journal of Applied and Computational Mathematics*, 4(5): 127. https://doi.org/10.1007/s40819-018-0566-y.

[7] Justine, H., Chew, J. V. L. and Sulaiman, J. (2017). Quartic non-polynomial spline solution for solving two-point boundary value problems by using Conjugate Gradient iterative method. *Journal of Applied Mathematics and Computational Mechanics*, 16(1): 41-50 doi: 10.17512/jamcm.2017.1.04.

[8] Kanwal, G., Ghaffar, A., Hafeezullah, M. M., Manan, S. A., Rizwan, M. and Rahman, G. (2019). Numerical Solution of 2-Point Boundary Value Problem by Subdivision Scheme. *Communications in Mathematics and Applications*, 10(1): 19–29 doi: 10.26713/cma.v10i1.980.

[9] Filipov, S. M. and Faragó, I. (2018). Implicit Euler time discretization and FDM with Newton method in nonlinear heat transfer modeling, *Mathematical Modeling*, 2(3): 94-98 arXiv preprint arXiv:1811.06337 https://arxiv.org/ftp/arxiv/papers/1811/1811.06337.pdf.

[10] Bello, N., Alkali, A. J. and Roko, A. (2018). A fixed point iterative method for the solution of two-point boundary value problems for a second order differential equations. *Alexandria Engineering Journal*, 57(4): 2515-2520 https://doi.org/10.1016/j.aej.2017.09.010.

[11] Baccouch, M. (2019). Analysis of optimal superconvergence of a local discontinuous Galerkin method for nonlinear second-order two-point boundary-value problems. *Applied Numerical Mathematics*, 145: 361-383 https://doi.org/10.1016/j.apnum.2019.05.003.

[12] Makarov, V. L. and Dragunov, D. V. (2019). An efficient approach for solving stiff nonlinear boundary value problems. *Journal of Computational and Applied Mathematics*, 345: 452-470. https://doi.org/ 10.1016/j.cam.2018.06.025.

[13] Press, W. H., Teukolsky, S. A., Vetterling, W. T. and Flannery, B. P. (2002). *Numerical Recipes in C: The Art of Scientific Computing*. Cambridge University Press.

[14] Cuomo, S. and Marasco, A. (2008). A numerical approach to nonlinear two-point boundary value problems for ODEs. *Computers & Mathematics with Applications,* 55(11): 2476-2489 https://doi.org/10.1016/ j.camwa.2007.10.002.

[15] Lambers, J. (2010). Finite Difference Methods for Two-Point Boundary Value Problems. MAT 461/561, Lecture 26 Notes, Department of Mathematics, University of Southern Mississippi http://www.math.usm.edu/lambers/mat461/spr10/lecture26.pdf.

[16] Lakoba, T. (2012). Finite-difference methods for BVPs. Numerical Diff Equations MATH 337, Lecture Notes 8, Department of Mathematics and Statistics, University of Vermont https://www.studocu.com/en-us/s/ 1333883.

[17] Filipov, S. M., Gospodinov, I. D. and Faragó, I. (2019). Replacing the finite difference methods for nonlinear two-point boundary value problems by successive application of the linear shooting method. *Journal of Computational and Applied Mathematics*, 358: 46-60. https://doi.org/10.1016/j.cam.2019.03.004.

[18] Ha, S. N. (2001). A nonlinear shooting method for two-point boundary value problems. *Computers & Mathematics with Applications*, 42(10-11): 1411-1420. https://doi.org/10.1016/S0898-1221(01)00250-4.

[19] Filipov, S. M., Gospodinov, I. D. and Faragó, I. (2017). Shooting-projection method for two-point boundary value problems. *Applied Mathematics Letters*, 72: 10-15 https://doi.org/10.1016/j.aml.2017.04.002.

[20] Bellman, R. E. and Kalaba, R. E. (1965). Quasilinearization and nonlinear boundary-value problems. In: *Modern Analytic and Computational Methods in Science and Mathematics*, vol. 3, American Elsevier Publishing Co., Inc., New York.

[21] Picard E. (1893). Sur l'application des méthodes d'approximations successives à l'étude de certaines équations différentielles ordinaires. *Journal de mathématiques pures et appliquées 4e série*, 9: 217-272 [Picard E. (1893). On the application of successive approximation methods to the study of some ordinary differential equations. *Journal of Pure and Applied Mathematics 4th Series*, 9: 217-272] http://sites.mathdoc.fr/JMPA/PDF/JMPA_1893_4_9_A4_0.pdf.

[22] Quarteroni, A., Sacco, R. and Saleri, F. (2010). Numerical mathematics (Vol. 37). Springer Science & Business Media.

[23] https://en.wikipedia.org/wiki/Projection_(mathematics).

In: A Closer Look at Boundary Value Problems ISBN: 978-1-53617-857-9
Editor: Mustafa Avci

Chapter 6

THE EXISTENCE OF BOUNDARY VALUE PROBLEMS OF FIFTH PAINLEVÈ EQUATION IN A COMPLEX DOMAIN

Rabha W. Ibrahim[1,2,*]
[1]Informetrics Research Group, Ton Duc Thang University, Ho Chi Minh City, Vietnam
[2]Faculty of Mathematics & Statistics, Ton Duc Thang University, Ho Chi Minh City, Vietnam

Abstract

We consider the class of second order differential equations

$$\psi''(z) = ((\psi'(z))^2 - 1)\cot(\psi(z)) + \frac{1}{z}(1 - \psi'(z)) + cz,$$

where $z \in U := \{z \in \mathbb{C} : |z| < 1\}$ which subjects to the boundary condition

$$\psi \in \Lambda := \{g : g(z) = z + \sum_{n=2}^{\infty} a_n z^n, \; z \in U\},$$

where $c \in \mathbb{R}$. These types of equations can be reduced to the fifth Painlevè equation under a Möbius transform. We shall deliver a set of results describing the analytic solution in view of the geometric function theory.

*Corresponding Author's Email: rabhaibrahim@tdtu.edu.vn.

Particularly, based on the concept of second order, subordination, we discuss the upper and lower bound of these solutions. This concept provides us the univalent solution and connection problem, by estimate the coefficients bound. In addition, the idea of the subordination brings the extreme solution of the BVP and the upper bound of the asymptotic behavior of solutions in a complex domain. Moreover, we are concerned about the convexity of the solution under some conditions by suggesting some special cases of the class. We present the asymptotic behavior by integral formulas involving the solution. We illustrate some examples and numerical cases of the BVP.

Keywords: boundary value problem, Painlevè equation, unit disk, univalent function, analytic function, subordination

AMS Subject Classification: 30C45, 30E25

1. Introduction

The Painlevè equations have not only enticed helpfulness because of their analytic possessions, but also for their arrival in numerous applications, mainly in mathematical physics (developmental the Darboux transformations), approximation theory, symmetric studies, special functions (modified the Bessel functions and Hankel functions), Heisenberg-Weyl algebra, gauge theory, fractional calculus (arbitrary order calculus) and differential geometry of surfaces [1].

In this chapter, we investigate the analytic univalent solution for the second order differential equation in the open unit disk U

$$\psi''(z) = ((\psi'(z))^2 - 1)\cot(\psi(z)) + \frac{1}{z}(1 - \psi'(z)) + cz, \quad z \in U; \tag{1.1}$$

subject to the boundary condition

$$\psi \in \Lambda = \{g : g(z) = z + \sum_{n=2}^{\infty} a_n z^n,\ z \in U\},$$

where $c \in \mathbb{R}$.

Inserting the change of variable

$$\omega(\xi) = \exp(-2i\psi(z)), \quad \xi = \frac{z}{2},$$

in Eq.(1.1), we obtain the special fifth Painlevè equation of a complex variable

$$\omega''(\xi)=\Big(\frac{1}{2\omega(\xi)}+\frac{1}{\omega(\xi)-1}\Big)(\omega'(\xi))^2-\frac{1}{\xi}\omega'(\xi)-4i\frac{\omega(\xi)}{\xi}+8\frac{\omega(\xi)(\omega(\xi)+1)}{\omega(\xi)-1}+2c\xi \quad (1.2)$$

$$\Big(z,\xi\in U,\ c\in\mathbb{R}\Big).$$

For the real case with $c=0$, Eq.(1.2) is studied in [2]. Now, let $\xi=i\tau$ and $\rho(\tau):=\dfrac{\sqrt{\omega}+1}{\sqrt{\omega}-1}$, we have the special third Painlevè equation of a complex variable

$$\rho''(\tau)=\frac{1}{\rho(\tau)}(\rho'(\tau))^2-\frac{1}{\tau}\rho'(\tau)-\frac{1}{\tau}(\rho^2(\tau)-1)+\rho^3(\tau)-\frac{1}{\rho(\tau)}+2ci\tau, \quad (1.3)$$

$$\Big(\tau,\xi\in U,\ c\in\mathbb{R}\Big),$$

where the real case with $c=0$, Eq.(1.3) is suggested in [3, 4]. Again, by using the translation $\varpi(\tau)=-\rho(\tau)$, we arrive at another type of the special third Painlevè equation

$$\varpi''(\tau)=\frac{1}{\varpi(\tau)}(\varpi'(\tau))^2-\frac{1}{\tau}\varpi'(\tau)+\frac{1}{\tau}(\varpi^2(\tau)-1)+\varpi^3(\tau)-\frac{1}{\varpi(\tau)}+2ci\tau, \quad (1.4)$$

which is established for real type with $c=0$ in [2]. By using the following transform

$$\hbar(\ell)=\frac{\varpi'(\tau)-(\varpi(\tau))^2-1}{\varpi'(\tau)-(\varpi(\tau))^2+1},\qquad \ell:=\frac{\tau^2}{2},$$

which is presented in [5] to obtain the special Painlevè equation

$$\hbar''(\ell)=\Big(\frac{1}{2\hbar(\ell)}+\frac{1}{\hbar(\ell)-1}\Big)(\hbar'(\ell))^2-\frac{1}{\ell}\hbar'(\ell)-\frac{1}{8}\frac{(\hbar(\ell)-1)^2}{\ell^2\hbar(\ell)}-\frac{\hbar(\ell)}{\ell}+2ci\sqrt{2\ell}, \quad (1.5)$$

where the real formula with $c=0$ is studied in [5].The existence and uniqueness of analytic solution of (1.5) are established in [2] when $c=0$ by using the technique of uniform asymptotic behaviors given in [6]. In the proof of the existence, they delivered the following solution

$$\hbar(\ell)=1+\frac{2\sin^2(\frac{\psi(x)}{2})}{\psi'(x)-1},\quad \ell:=-\frac{x^2}{8}.$$

The chapter is organized as follows: Section 2 delivers the method of study, including some studies of BVP (1.1); Section 3 deals with the technique to solve (1.1) and Section 4 includes some special cases.

2. THE METHOD OF STUDY

Let $\mathbb{Q}(U)[a,n] = \{\phi : \phi(z) = a + a_n z^n + ...,\}$ be the class of all analytic functions in U, where $\mathbb{Q}(U)[0,1] = \Lambda$. We demand the following requisite concepts throughout this paper (see [7]). A function $\phi \in \Lambda$ is said to be univalent in U if it never occupies the same value twice; i.e., f $\xi_1 \neq \xi_2$ in U then $\phi(\xi_1) \neq \phi(\xi_2)$ or equivalently, if $\phi(\xi_1) = \phi(\xi_2)$ then $\xi_1 = \xi_2$. We denote by $\mathcal{S}$ the class of functions $\phi \in \Lambda$ that are univalent in U. A function $\phi \in \mathcal{S}$ is called starlike with respect to origin in U if the linear segment joining the origin to every other point of $\phi(\xi : |\xi| = r < 1)$ lies entirely in $\phi(\xi : |\xi| = r < 1)$. In more scenic view, the requirement is that every point of $\phi(\xi : |\xi| = r < 1)$ be manifest from the origin.

A function $\phi \in \mathcal{S}$ is called convex in U if the linear segment joining any two points of $\phi(\xi : |\xi| = r < 1)$ lies entirely in $\phi(\xi : |\xi| = r < 1)$. More information, a function $\phi \in \mathcal{S}$ is called convex in U if it is starlike with respect to each and every of its points. This class of functions $\phi \in \mathcal{S}$ that are starlike with respect to the origin is denoted by $\mathcal{S}^*$ and convex in U by $\mathcal{C}$ (note that $\mathcal{C} \subset \mathcal{S}^*$ [8],p10). Obviously, $\phi \in \mathcal{S}^*$ if and only if $\xi\phi'(\xi)/\phi(\xi) \in \mathcal{P}$ and $\phi \in \mathcal{C}$ if and only if $1 + \xi\phi''(\xi)/\phi'(\xi) \in \mathcal{P}$, where $\mathcal{P}$ is the class of all functions ϕ analytic in U and owning positive real part in U with $\phi(0) = 1$.

For functions ϕ and ψ in Λ, ϕ is subordinate to ψ, symbolized by $\phi \prec \psi$, if there occurs a Schwarz function $\omega \in U$ with $\omega(0) = 0$, $|\omega(\xi)| < 1$ and $\xi \in U$ so that $\phi(\xi) = \psi(\omega(\xi))$ for all $\xi \in U$ (see [7] or [8]). Distinctly $\phi(\xi) \prec \psi(\xi)$ is equivalent to $\phi(0) = \psi(0)$ and $\phi(U) \subset \psi(U)$.

If $\Gamma \neq \mathbb{C}$, and θ is analytic in U satisfying $\theta(0) = a$ then there occurs a conformal mapping $\vartheta : U \to \Gamma$ such that $\vartheta(0) = a$, we have the definition of second order subordination

$$\{\Theta\big(\theta(\xi), \xi\theta'(\xi), \xi^2\theta''(\xi); \xi\big) | \xi \in U\} \subset \Gamma \Rightarrow \theta(\xi) \prec \vartheta(\xi),$$

where $\Theta\big(\theta(\xi), \xi\theta'(\xi), \xi^2\theta''(\xi); \xi\big) : \mathbb{C}^3 \times U \to \mathbb{C}$ is analytic in U. Equivalently, there occurs a conformal mapping $\hbar : U \to \Gamma$ such that $\hbar(0) = \Theta(0)$ achieving the inequality

$$\Theta\big(\theta(\xi), \xi\theta'(\xi), \xi^2\theta''(\xi); \xi\big) \prec \hbar(\xi) \Rightarrow \theta(\xi) \prec \vartheta(\xi). \tag{2.1}$$

If $\hbar(\xi)$ is univalent in U then θ is a solution of the differential subordination (2.1). Moreover, if ϑ is univalent in U then it is called a dominant (upper bound of a solution $\theta(\xi)$) and if it is true for all solutions of (2.1) then it is called the best dominant which is unique (see [8]). This class of analytic function is called the class of admissible functions, which is denoted by $\Psi[\hbar, \vartheta]$.

In our discussion, we shall use the above inequality. We shall process BVP (1.1) in term of the second order differential subordination and prove that the solution is of the form of normalized functions (the boundary condition) in the open unit disk. Based on this idea, one can study more properties of the solution by using the geometric representation of univalent functions in U such as star-likeness and convexity. Moreover, we can find the upper bound of the connection problem (coefficient estimate) of the solution. Note that this problem is studied in [2] for a real case when $c = 0$ utilizing the method of uniform asymptotic.

For this purpose, we request the following preliminaries:

Lemma 2.1. *[8] Let $\Theta \in \Psi[\hbar, \vartheta]$, ϑ be analytic function in U, $\hbar$ be conformal mapping of U onto Γ with $\hbar(0) = a$ and $\theta \in \mathbb{Q}(U)[a, n]$ achieving one of the following relations*

(i) $\Theta\big(\theta(\xi), \xi\theta'(\xi), \xi^2\theta''(\xi); \xi\big) \in \Gamma$;

(ii) $\Theta\big(\theta(\xi), \xi\theta'(\xi), \xi^2\theta''(\xi); \xi\big) \prec \hbar(\xi)$, *where* $\Gamma = \hbar(U)$;

(iii) the differential equation $\Theta\big(\theta(\xi), \xi\theta'(\xi), \xi^2\theta''(\xi); \xi\big) = \hbar(\xi)$ *has a univalent solution* $\theta(\xi)$, *provide* $\hbar(\xi)$ *is univalent;*

(iv) there exists an analytic function υ *in* U *satisfying* $\upsilon(U) \subset \Gamma$ *and the differential equation* $\Theta\big(\theta(\xi), \xi\theta'(\xi), \xi^2\theta''(\xi); \xi\big) = \upsilon(\xi)$ *has a solution;*

(v) the differential equation $\Theta\big(\theta(\xi), \xi\theta'(\xi), \xi^2\theta''(\xi)\big) = \hbar(\xi)$ *has a univalent solution* $\theta(\xi)$;

Then $\theta \prec \vartheta$ *and* ϑ *is the best dominant. Moreover, if* $\upsilon(\xi) = Ke^{i\tau}$ *then*

$$|\Theta\big(\theta(\xi), \xi\theta'(\xi), \xi^2\theta''(\xi); \xi\big)| < K \Rightarrow |\theta(\xi)| < K, \quad \xi \in \partial U,$$

and

$$\Re\Big(\Theta\big(\theta(\xi),\xi\theta'(\xi),\xi^2\theta''(\xi);\xi\big)\Big) > 0 \Rightarrow \Re(\theta(\xi)) > 0, \quad K > 0.$$

Lemma 2.2. *[9] If for analytic functions $\sigma(\xi)$ and $\Sigma(\xi)$, with the inequality*

$$\sigma(\xi) = \sum_{n=1}^{\infty} \sigma_n \xi^n \prec \Sigma(\xi) = \sum_{n=2}^{\infty} e_n \xi^n,$$

where Σ is convex univalent in U then

$$|\sigma_n| \le |e_1|,\ n \ge 1.$$

2.1. Processing of BVP (1.1)

Consider the BVP (1.1),

$$\psi''(z) = ((\psi'(z))^2 - 1)\cot(\psi(z)) + \frac{1}{z}(1 - \psi'(z)) + cz, \quad z \in U;$$

Multiplying by z^2, we have

$$z^2\psi''(z) = z^2((\psi'(z))^2 - 1)\cot(\psi(z)) + z(1 - \psi'(z)) + cz^3.$$

Define the functional

$$\Psi\left(\psi(z), z\psi'(z), z^2\psi''(z); z\right) = z^2\psi''(z) - z^2((\psi'(z))^2 - 1)\cot(\psi(z)) - z(1-\psi'(z)) - cz^3. \tag{2.2}$$

It is clear that $\Psi\big(0,0,0;0\big) = 0$ if and only if $\psi(0) = 0$; which means zero is a solution of Eq. (2.2). In addition, based on a special case of the fifth Painlevè equation [2], Eq. (2.2) have a nontrivial solution in the open unit disk. Our aim is to show that this solution satisfies the boundary condition

$$\psi \in \Lambda := \{g : g(z) = z + \sum_{n=2}^{\infty} a_n z^n,\ z \in U\},$$

where $c \in \mathbb{R}$, which means that $\psi(0) = 0$ and $\psi'(0) = 1$ (the class of normalized functions). Therefore, it is enough to show that $\psi(z) \prec f(z)$, where $f \in \Lambda$.

2.2. Approximate Case of BVP 1.1

In this section we shall deal with the special case, when $\cot(\psi(z)) \approx \dfrac{1}{\psi(z)}$ providing $\psi(z) \neq 0$, $z \in U \setminus \{0\}$. Thus, we have the equation

$$\psi''(z) = \frac{(\psi'(z))^2 - 1}{\psi(z)} + \frac{1}{z}(1 - \psi'(z)) + cz, \quad z \in U; \tag{2.3}$$

subject to the boundary condition $\psi \in \Lambda = \{g : g(z) = z + \sum_{n=2}^{\infty} a_n z^n,\ z \in U\}$, where $c \in \mathbb{R}$. Dividing (2.3) by $\psi'(z)$ then a calculation implies

$$\begin{aligned}
\frac{\psi''(z)}{\psi'(z)} &= \frac{z[(\psi'(z))^2 - 1]}{z\psi(z)\psi'(z)} + \frac{(1 - \psi'(z))}{z\psi'(z)} + \frac{cz^2}{z\psi'(z)}, \quad z \in U \\
&= \frac{z[(\psi'(z))^2 - 1] + \psi(z)(1 - \psi'(z)) + cz^2\psi(z)}{z\psi(z)\psi'(z)} \\
&= \frac{z(\psi'(z))^2 - \psi(z)\psi'(z) + \psi(z)(cz^2 + 1) - z}{z\psi(z)\psi'(z)} \\
&:= F_c(z).
\end{aligned} \tag{2.4}$$

We have the following propositions:

Proposition 2.3. *Eq.* (2.3) *has a nontrivial analytic solution in* $\psi \in U$ *achieving the boundary condition*$\psi \in \Lambda$.

Proof. Since the fifth Painlevè equation has a solution ($c = 0$), then Eq. (2.3) has a nontrivial analytic solution $\psi(z)$ in U for all c. To show that it satisfies the boundary condition $\psi \in \Lambda$, we utilize Eq.(2.4). Integrating Eq.(2.4) with respect to z, we have

$$\begin{aligned}
\psi'(z) &= \exp\Big(\int_0^z F_c(\xi)d\xi\Big), \quad z, \xi \in U \\
&= \exp\left(\int_0^1 F_c(\omega(z))\omega'(z)dz\right)
\end{aligned} \tag{2.5}$$

for some analytic function $\omega(z) = \xi$, with $\omega(0) = 0$ and $|\omega(z)| < 1$. This leads to the subordination

$$\psi'(z) \prec e^z, \quad z \in U.$$

Obviously, $\psi'(0) = e^0 = 1$, and since e^z is univalent for all $|z| < 1$, (see [8], p90); thus, we have $\psi(z) \in \Lambda$. □

Proposition 2.4. *If the solution $\psi(z)$ of Eq.* (2.3) *satisfies the subordination*

$$z\psi'(z) \prec \hbar(z),$$

where $\hbar(z)$ is star-like, then the solution is convex.

Proof. By Proposition 2.3, we have $\psi(z) \in \Lambda$. From (2.5), we obtain

$$z\psi'(z) = z\exp\Big(\int_0^z F_c(\xi)d\xi\Big) := \hbar(z).$$

By logarithmic derivative, we obtain

$$1 + \frac{z\psi''(z)}{\psi'(z)} = \frac{z\hbar'(z)}{\hbar(z)},$$

but

$$\Re(\frac{z\hbar'(z)}{\hbar(z)}) > 0 \Rightarrow \Re(1 + \frac{z\psi''(z)}{\psi'(z)}) > 0.$$

Hence, $\psi(z)$ is convex in U. □

3. SOLVABILITY OF BVP (1.1)

In this section, we deliver the existence outcomes satisfying the boundary condition of Eq. (1.1).

Theorem 3.1. *Let $\Psi\big(\psi(z), z\psi'(z), z^2\psi''(z); z\big)$ be analytic function in U, $\hbar$ be conformal mapping of U onto Γ with $\hbar(0) = 0$. If Ψ lies in the conic domain $\Gamma = \hbar(U)$ then for all analytic solution ψ of Eq.* (2.2) *satisfies that there is an analytic function $\vartheta \in \mathbb{Q}(U)[0,1]$ such that $\psi(z) \prec \vartheta(z)$ and ϑ is the best dominant.*

Proof. Let $c \neq 0$, $\Psi\big(\psi(z), z\psi'(z), z^2\psi''(z); z\big)$ be an analytic function in U, and $\hbar$ be a conformal mapping of U onto Γ with $\hbar(0) = 0$. Since Ψ lies in the conic domain $\Gamma = \hbar(U)$ then $\Psi\big(\psi(z), z\psi'(z), z^2\psi''(z); z\big)$ satisfies one of the relations (i) -(ii) of Lemma 2.1, i.e.,

$$\Psi\big(\psi(z), z\psi'(z), z^2\psi''(z); z\big) \in \hbar(U).$$

Now suppose that $\psi(z) \not\prec \vartheta(z)$, which means there are two points $z_0 \in \partial U \setminus \mathfrak{E}(\vartheta)$, $z \in U$, and $\mu \geq 1$ such that

$$\Re\Big(1+\frac{z^2\psi''(z)}{\psi'(z)}\Big) \geq \mu\Re\Big(1+\frac{z_0^2\vartheta''(z_0)}{\vartheta'(z_0)}\Big), \; z\psi'(z) = \mu z_0\vartheta'(z_0), \; \psi(z) = \vartheta(z_0),$$

$$\Big(\mu \geq 1, z_0 \in \partial U \setminus \mathfrak{E}(\vartheta), z \in U, \mathfrak{E}(\vartheta) \subset \mathbb{C}\Big).$$

Utilizing these facts, we get the definition of admissible function (see [8], p27), i.e.,

$$\Psi\big(\psi(z_0), z_0\psi'(z_0), z_0^2\psi''(z_0); z_0\big) \notin \hbar(U),$$

which contradicts the assumption of the theorem. Thus, in view of Lemma 2.1, we obtain $\psi(z) \prec \vartheta(z)$ and ϑ is the best dominant. Moreover, since the functions Ψ and $\hbar$ have the fact

$$\Psi\big(0, 0, 0; 0\big) = \hbar(0) = 0,$$

then we conclude that

$$\Psi\big(\psi(z), z\psi'(z), z^2\psi''(z); z\big) \prec z \Rightarrow \lim_{z\to +1} \Re\Big(\Psi\big(\psi(z), z\psi'(z), z^2\psi''(z); z\big)\Big) > 0,$$

this implies that $\Re(\psi) > 0$ (by Lemma 2.1). In addition, by the relation

$$\begin{aligned} 0 < k &= \Re\Big(\Psi\big(\psi(z_0), z_0\psi'(z_0), z_0^2\psi''(z_0); z_0\big)\Big) \\ &\leq \Big|\Psi\big(\psi(z_0), z_0\psi'(z_0), z_0^2\psi''(z_0); z_0\big)\Big| < K, \end{aligned}$$

where $k \leq K$; therefore, we obtain $|\psi(z_0)| < K, \forall\, z_0 \in \partial U$. Also, the analytic function $\Psi\big(\psi(z), z\psi'(z), z^2\psi''(z); z\big)$ takes its maximization in the domain $\hbar(U)$, i.e., there is a point $\xi \in U$ such that

$$\Psi\big(\psi(\xi), z\psi'(\xi), z^2\psi''(\xi); \xi\big) = \hbar(\xi)$$

has a solution $\psi(\xi)$ in U. If this solution is univalent (this can be occurred when $\hbar(\xi)$ is univalent) then by Lemma 2.1-(iii) the subordination inequality $\psi(z) \prec \vartheta(z)$ and ϑ is the best dominant. If $\psi(\xi)$ is not univalent (but analytic) with $\psi(U) \subset \hbar(U)$ then the differential equation

$$\Psi\big(\psi(\xi), z\psi'(\xi), z^2\psi''(\xi); \xi\big) = \psi(\xi)$$

has a solution, consequently, by letting $\psi(\xi) = \upsilon(\xi)$ in Lemma 2.1-(iv), we arrive at $\psi(z) \prec \vartheta(z)$ and ϑ is the best dominant.
Let $c = 0$. Since $\hbar \in \Lambda$, then the differential equation

$$\Psi\big(\psi(z), z\psi'(z), z^2\psi''(z)\big) = \hbar(z)$$

has a univalent solution in U, which yields by Lemma 2.1-(v), $\psi(z) \prec \vartheta(z)$ and ϑ is the best dominant. □

Theorem 3.2. *Let the assumption of Theorem 3.1 hold. If* $\vartheta(z) = z + \sum_{n=2}^{\infty} b_n z^n$ *is univalent in* U *then for all solution of the boundary problem* (1.1), $\psi(z) = z + \sum_{n=2}^{\infty} a_n z^n$ *the connection problem satisfies*

$$|a_n| \le |b_m|, \quad \forall\, n \ge m \ge 1.$$

Proof. Let ϑ be univalent in U. Then by Theorem 3.1, the subordination inequality

$$\psi(z) = z + \sum_{n=2}^{\infty} a_n z^n \prec \vartheta(z) = z + \sum_{n=2}^{\infty} b_n z^n, \quad z \in U.$$

If $\vartheta(z)$ is univalent convex in U, then in view of Lemma 2.2, this implies that

$$|a_n| \le |b_1|, \quad n \ge 1 = m.$$

Moreover, equality holds for $|a_1| = |b_1| = 1$. If $\vartheta(z)$ is univalent but not convex in U, then we have (Bieberbach Theorem)

$$|a_n| \le |b_n| \le n, \ \forall n = m \ge 2.$$

This completes the proof. □

Remark 3.3. *Theorem 3.2 (in the case c=0) is a generalization of the upper bound of the connection problem of real case that given in [2] Theorem 1, where* $|a_2| < \dfrac{1}{\pi} \simeq 0.318 < 1$ *for convex case and* $|a_2| < 0.318 < 2$ *for univalent case.*

Remark 3.4. *In this remark, we illustrate some special cases of the behaviour of solution of* (1.1) *to be starlike.*

- *We recall that according to Noshiro-Warschawski Theorem (Duren [7], p. 47) if a solution* ψ *is analytic in the simply connected complex domain* $\mathbb{U}$ *and* $\Re\{\psi'(z)\} > 0$ *in* $\mathbb{U}$ *then* ψ *is univalent in* $\mathbb{U}$ *and according to Kaplan's Theorem (Duren [7], p. 48) such function is of bounded boundary rotation, that is, none of the curves* $\psi(z : |z| = r1)$ *makes a reverse hairpin turn. This class is called the class of bounded turning. In this place, we have the following result:*

  If
  $$\Big|\frac{z\psi'(z)}{\psi(z)} - 1\Big| < \mathbb{M}, \quad 0 < \mathbb{M} \le 1,$$
  then $\psi(z)$ *is of bounded turning function (see [8],Theorem 5.5e);*

- *If*
  $$\Re(z\psi''(z)) > -0.5, \quad z \in U$$
  then $\psi \in \mathcal{S}^*$ *(see [8], Corollary 5.2c.1);*

- *If*
  $$\Big|\frac{z\psi''(z)}{\psi'(z)}\Big| \le \nu\Big|\frac{z\psi'(z)}{\psi(z)}\Big|, \quad \nu \in (0, 2], \frac{\psi(z)}{z} \ne 0,$$
  then the solution $\psi \in \mathcal{S}^*(\beta)$, $\beta = 2/(2 + \nu)$ *(see [8], Theorem 5.3a);*

- *If*
  $$\Big|\frac{z\psi''(z)}{\psi'(z)}\Big| \le \sqrt{2}\Big|\frac{z\psi'(z)}{\psi(z)} + 1\Big|, \frac{\psi(z)}{z} \ne 0,$$
  then the solution $\psi \in \mathcal{S}^*$ *(see [8], Theorem 5.3d);*

- *If*
  $$\Big|\psi'(z) - 1\Big| < \frac{2}{\sqrt{5}},$$
  then the solution $\psi \in \mathcal{S}^*$ *(see [8], Theorem 5.5a);*

- *If*
  $$\Big|\psi''(z)\Big| < \frac{2}{\sqrt{5}},$$
  then the solution $\psi \in \mathcal{S}^*$ *(see [8], Corollary 5.5a.1)*

- *If*

$$\Re \frac{z\psi''(z)}{\psi'(z)} > -\frac{3}{2},$$

then the solution $\psi \in \mathcal{S}^*(1/2)$ (see [8], Theorem 2.6i) and this is sharp with the extreme solution taking the form

$$\psi(z) = \frac{z}{\sqrt{1-z^2}}.$$

- *If*

$$\frac{z\psi''(z)}{\psi'(z)} + 1 \prec \frac{1+z}{1-z} + \frac{2z}{1-z^2},$$

then the solution satisfies $\psi \in \mathcal{S}^*$ (see [8], Theorem 2.6d).

3.1. Examples

In this section, we illustrate some applications, using Mathematica-WolframAlpha (11.2).

Example 3.5. *(see Fig. 3.1) We have the following data*

$$\psi(z) = \frac{z}{1-z}, \ \hbar(z) = -z, \ c = 1,$$

then, we get

$$\begin{aligned}
&lim_{z\to -1}((z^2 d^2/(dz^2)z/(1-z) - z^2((d/(dz)z/(1-z))^2 - 1) \\
&* \cot(z/(1-z)) - z(1 - d/(dz)z/(1-z))) - z^3) \\
&= 2 - 15/16 \cot(1/2) = 0.284
\end{aligned}$$

thus, $\Psi\big(\psi(z), z\psi'(z), z^2\psi''(z); z\big)$ lies in the domain of $\hbar(z)$. Consequently, by Theorem 3.1, we indicate that $\psi(z)$ is a solution having the form

$$\psi(z) = z + z^2 + z^3 + ... \prec -z.$$

Moreover, the connection problem satisfies Theorem 3.2, where $|a_n| = |b_1| = 1$.

Example 3.6. *(see Fig. 3.2) We have the following data*

$$\psi(z) = \frac{z}{1-z}, \ \hbar(z) = \frac{z}{(1-z)^8}, \ c = 2,$$

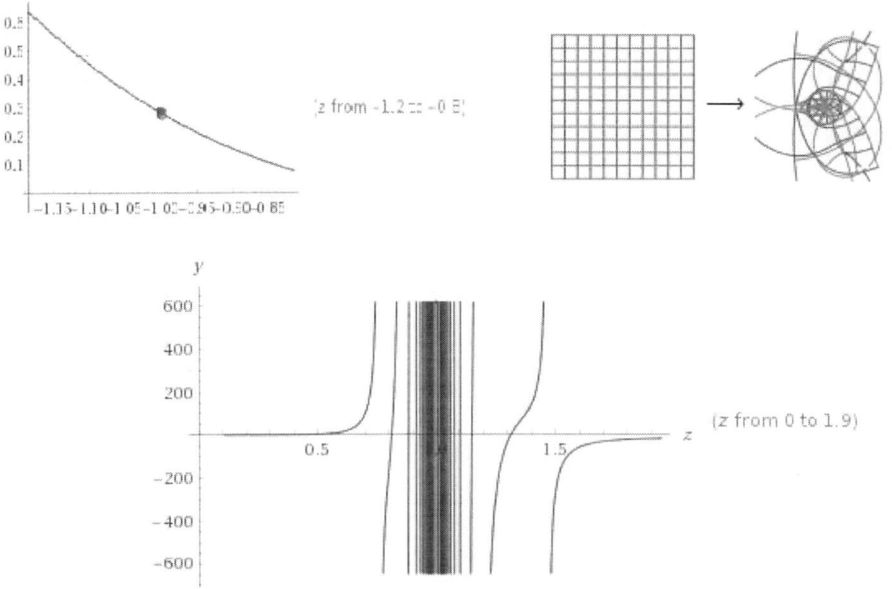

Figure 3.1. $\Psi = z^2(z/(1-z))'' - z^2(((z/(1-z))')^2 - 1) * \cot(z/(1-z)) - z * (1 - (z/(1-z))')) - z^3$.

then, BVP (1.1) *has the series*

$$\psi''(z) - ((\psi'(z))^2 - 1)\cot(\psi(z)) - \frac{1}{z}(1 - \psi'(z)) - 2z$$
$$= z + (22z^2)/3 + (44z^3)/3 + (1189z^4)/45 + (2002z^5)/45 + (67397z^6)/945 + O(z^7),$$

with the limit

$$lim_{z\to -1}((d^2/(dz^2)z/(1-z) - ((d/(dz)z/(1-z))^2 - 1)$$
$$* \cot(z/(1-z)) - (1 - d/(dz)z/(1-z))/z) - 2z)$$
$$= 3 - 15/16\cot(1/2) \simeq 1.$$

Since

$$\frac{z}{(1-z)^8} = z + 8z^2 + 36z^3 + 120z^4 + 330z^5 + 792z^6 + O(z^7)$$

thus, BVP (1.1) *lies in the domain of* $\hbar(z)$. *This implies that* $\psi(z)$ *is a solution (Theorem 3.1) and the connection problem satisfies* $|a_n| \leq |b_n$ *(Theorem 3.2). If we consider* $\hbar(z) = \dfrac{z}{(1-z)^7} = z + 7z^2 + 28z^3 + 84z^4 + 210z^5 + 462z^6 + O(z^7)$, *which achieves* $|a_n| \nleq |b_n$.

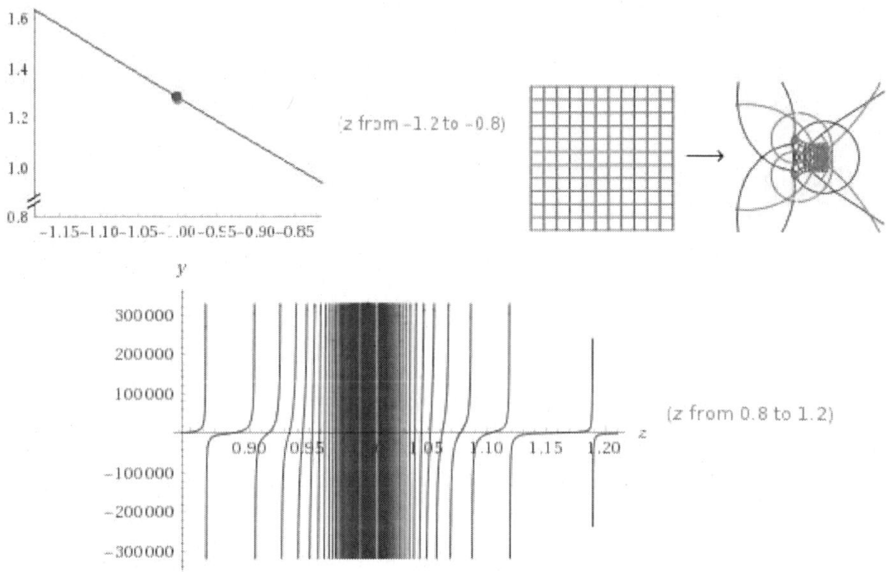

Figure 3.2. $\Psi = z^2(z/(1-z))'' - z^2(((z/(1-z))')^2 - 1) * \cot(z/(1-z)) - z * (1 - (z/(1-z))')) - 2z^3$

3.2. Special Case (c=0)

In this section, we investigate the analytic univalent solution for the second order differential equation in the open unit disk U

$$\phi''(z) = ((\phi'(z))^2 - 1)\cot(\phi(z)) + \frac{1}{z}(1 - \phi'(z)), \quad z \in U; \tag{3.1}$$

subject to the boundary condition

$$\phi \in \Lambda^* = \{\rho : \rho(z) = z + a_3 z^3 + ..., \; z \in U\}.$$

The real case of (3.1) has been studied by Zeng et al. [2], when the boundary condition is given by

$$\phi \in \Omega = \{\rho : \rho(x) = x - a_2 x^2 + O(x^3), x \to 0\}.$$

To check the solution of (3.1), we request the following preliminaries, which can be found in [8]:

Lemma 3.7.

i *If* $g \in \Lambda^*$ *such that* $\Re(\frac{zg''(z)}{g'(z)}) > \frac{-3}{2}$ *then* $g \in \mathcal{S}^*(1/2)$. *The result is sharp and the extreme function is*

$$g(z) = \frac{z}{\sqrt{1-z^2}}, \quad z \in U.$$

ii *If* $g \in \Lambda^* \cap \mathcal{S}^*$, *then* $\Re(\frac{g(z)}{z}) > \frac{1}{2}$. *This inequality is sharp and the extreme function is*

$$g(z) = \frac{z}{1-z^2}, \quad z \in U.$$

iii *Let* f *and* g *be two analytic functions in* U *with* $f(0) = g(0) = 0$. *If* $g \in \mathcal{S}^*$ *then*

$$\Re(\frac{f'(z)}{g'(z)}) > \ell \Rightarrow \Re(\frac{f(z)}{g(z)}) > \ell, \quad \ell \in \mathbb{R}.$$

iv *Let* $g \in \mathcal{S}^*$ *with* $g(0) = 0$. *If* $h \in \Lambda^*$ *satisfies* $zh'(z) \prec g(z)$, *then*

$$h(z) \prec \hbar(z) = \frac{1}{2}\int_0^z \frac{g(\eta)}{\eta} d\eta.$$

Moreover, $\hbar(z)$ *is convex and is the best dominant.*

v *Let* $g \in \mathcal{S}^*$ *with* $g(0) = 0$. *If* $h \in \mathbb{H}[a, n] = \{h : h(z) = a + a_{n+1}z^{n+1} + ..., z \in U\}$ *satisfies* $\frac{zh'(z)}{h(z)} \prec g(z)$, *then*

$$h(z) \prec \hbar(z) = a \exp\Big(\frac{1}{2}\int_0^z \frac{g(\eta)}{\eta} d\eta\Big),$$

such that $\hbar(z)$ *is the best dominant.*

vi Let ρ be convex in U with $\rho(0) = 1$ and $\Re(\rho) > 0$. If

$$\varrho \in \mathbb{H}[1, n] = \{\varrho : \varrho(z) = 1 + a_{n+1}z^{n+1} + ..., \ z \in U\}$$

satisfies $\varrho^2(z) + 2\varrho(z).z\varrho'(z) \prec \rho(z)$, then

$$\varrho(z) \prec \wp(z) = \sqrt{\frac{1}{nz^{1/n}} \int_0^z \rho(\eta).\eta^{(1/n)-1} d\eta},$$

such that $\wp(z)$ is the best dominant.

3.3. Existence of BVP

In this section, we deliver the existence outcomes satisfying the boundary condition of Eq. (3.1).

Theorem 3.8. *The boundary value problem* (3.1) *has an extreme solution $\phi^{**} \in \Lambda^*$ if it satisfies*

$$\Re\Big(\frac{z(\phi^{**}(z))''}{(\phi^{**}(z))'}\Big) > -\frac{3}{2}, \quad z \in U. \tag{3.2}$$

Furthermore, the extreme solution satisfies that for all solution $\phi^*(z) \in \Lambda^*$, there exists a constant $\ell \in (0, 1]$ such that $|\phi^*(z)| \leq \ell|\phi^{**}(z)|$.

Proof. The boundary value problem (3.1) has a solution for the real case (see [2], Theorem 1, where $a_2 = 0$). Hence, it has a solution in a complex domain U. Now, let $\phi^* \in \Lambda^*$, then the condition (3.2) implies that $\phi^* \in \mathcal{S}^*(1/2)$ with extreme formula (see Lemma 3.7 i),

$$\phi^*(z) = \frac{z}{\sqrt{1 - z^2}} = z + (z^3)/2 + (3z^5)/8 + (5z^7)/16 + O(z^9), \quad z \in U.$$

This concludes that $\phi^* \in \Lambda^* \cap \mathcal{S}^*$, which leads to (see Lemma 3.7 ii) to another extreme solution given by

$$\phi^{**}(z) = \frac{z}{1 - z^2} = z + z^3 + z^5 + O(z^7), \quad z \in U.$$

By utilizing Lemma 3.7 iii, we check the dominant solution. Since

$$\min \Re \left\{ \frac{\Big(\frac{z}{1 - z^2}\Big)'}{\Big(\frac{z}{\sqrt{1 - z^2}}\Big)'} \right\} = \max \Re \left\{ \frac{\Big(\frac{z}{\sqrt{1 - z^2}}\Big)'}{\Big(\frac{z}{1 - z^2}\Big)'} \right\} = 1,$$

we have

$$\min \Re \left\{ \frac{\left(\dfrac{z}{1-z^2}\right)}{\left(\dfrac{z}{\sqrt{1-z^2}}\right)} \right\} = \max \Re \left\{ \frac{\left(\dfrac{z}{\sqrt{1-z^2}}\right)}{\left(\dfrac{z}{1-z^2}\right)} \right\} = 1.$$

Now by recognizing the coefficients bound, we arrive at the extreme solution $\phi^{**}(z)$. Therefore, we obtain the following inequality

$$1 = \min \Re \left\{ \frac{\left(\dfrac{z}{1-z^2}\right)}{\left(\dfrac{z}{\sqrt{1-z^2}}\right)} \right\} \leq \left| \frac{\phi^{**}(z)}{\phi^*(z)} \right|,$$

which yields that there is a constant $\ell \in (0,1]$ such that

$$|\phi^*(z)| \leq \ell |\phi^{**}(z)|, \quad z \in U.$$

This completes the proof.

□

Remark 3.9. *The extreme solution*

$$\phi^{**}(z) = \frac{z}{1-z^2} = z + z^3 + z^5 + O(z^7), \quad z \in U,$$

satisfies the coefficient bounds $|a_3| = |a_5|, ... = 1$. *We have the following example:*

Example 3.10. *(see Fig 3.3) Let* $\phi(z) = ze^{(3/2)z^2}$, $z \in U$. *Then, we have the following data*

$$\phi(z) = z + (3z^3)/2 + (9z^5)/8 + (9z^7)/16 + O(z^9) \in \Lambda;$$

$$\Re\left(\frac{z(\phi(z))''}{(\phi(z))'}\right) = \frac{9}{2} > -\frac{3}{2};$$

such that all its roots are in the open unit disk (see Fig 3.3) and

$$\Re \left\{ \frac{\left(ze^{(3/2)z^2}\right)'}{\left(\dfrac{z}{1-z^2}\right)'} \right\} = 1 \Rightarrow \Re \left\{ \frac{\left(ze^{(3/2)z^2}\right)}{\left(\dfrac{z}{1-z^2}\right)} \right\} = 1, \quad z \to 0.$$

But the function $\phi(z)$ *is not starlike ([8],p55). Therefore, it is not a solution for the boundary value problem* (3.1). *Moreover, we have* $|a_n| \geq 1$ *for some* n.

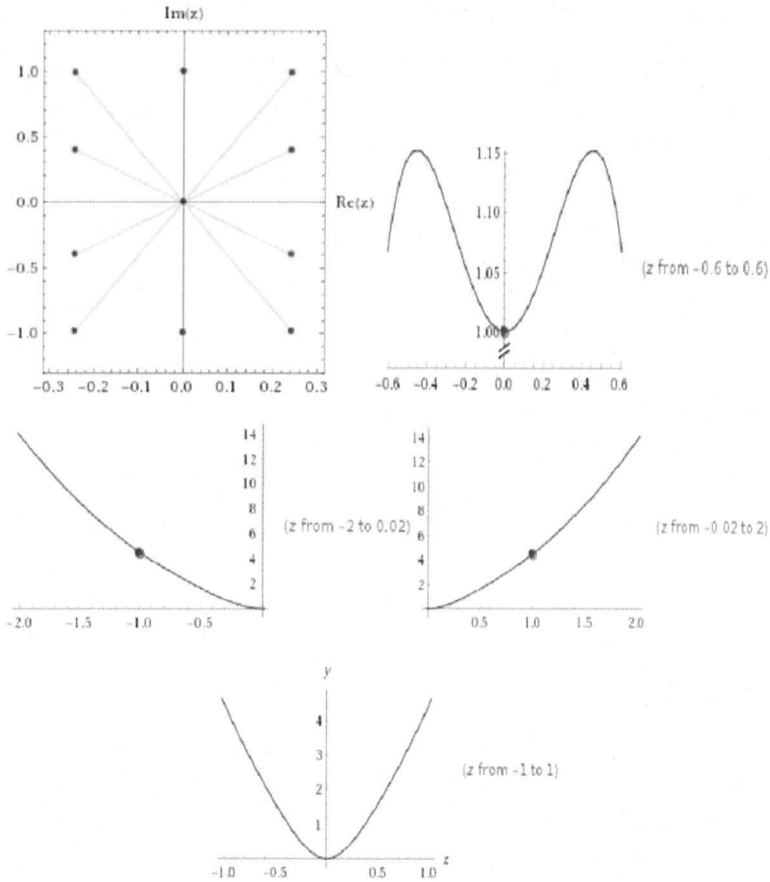

Figure 3.3. Data of $\phi(z) = ze^{(3/2)z^2}$; from the top-left, the graph represents to all roots of the expression $\Phi := \Re\Big(\dfrac{z(\phi(z))''}{(\phi(z))'}\Big)$ the second graph refers to the $\lim_{z\to 0} \Re((d/(dz)(z\exp((3z^2)/2)))/(d/(dz)z/(1-z^2))) = 1$, second row is the $\lim_{z\to -1}\Phi = \lim_{z\to +1}\Phi = \dfrac{9}{2}$ and the last graph is Φ.

Example 3.11. *(see Fig 3.4) Let* $\phi(z) = ze^{z^2}$, $z \in U$. *Then, we have the following data*

$$\phi(z) = z + z^3 + z^5/2 + z^7/6 + O(z^9) \in \Lambda; |a_n| \le 1,$$

$$\Re\Big(\frac{z(\phi(z))''}{(\phi(z))'}\Big) = \frac{10}{3} > -\frac{3}{2};$$

such that not all its roots are in the open unit disk (see Fig 3.4) and

$$\Re\left\{\frac{\left(ze^{z^2}\right)'}{\left(\frac{z}{1-z^2}\right)'}\right\} = 1 \Rightarrow \Re\left\{\frac{\left(ze^{z^2}\right)}{\left(\frac{z}{1-z^2}\right)}\right\} = 1, \quad z \to 0.$$

In addition, the function $\phi(z)$ is starlike, because

$$\Re\Big(\frac{z(d/(dz)(z\exp(z^2)))}{z\exp(z^2)}\Big) = 2\Re(z^2) + 1, \quad -0.5 \leq \Re(z) \leq 0.5,$$

which implies that $2\Re(z^2) + 1 > 0$. Therefore, it is a solution for the boundary value problem (3.1). *The asymptotic behavior [2],*

$$\phi(z) = ze^{-0.25z^2} = z - 0.225z^3 + 0.0253125z^5 - 0.00189844z^7 + O(z^9)$$

achieves $|a_n| \leq 1$, but

$$\Re\Big(\frac{z(\phi(z))''}{(\phi(z))'}\Big) = -2.5 < -\frac{3}{2};$$

Therefore, the condition of Theorem 3.8 is sufficient but not necessary.

Next, we shall describe the integral formula of solutions under some conditions, utilizing the subordination method. This method is provided sufficient conditions for the existence of the BVP (3.1). Moreover, this method is useful in the geometric function theory, because it represented inequality theory in a complex domain. The coming result shows that the asymptotic behavior of the solution is dominated by integral formula including the solution of the BVP (3.1).

Theorem 3.12. *Consider the analytic functions $h(z) = ze^{\alpha z^2} + \epsilon$, $\epsilon \in [0,1]$, $\alpha \in [-1,1]$ and $\omega_\alpha(z) = 2\alpha z^2 + 1$, $z \in U$. Let $\phi(z)$ be a solution of* (3.1) *satisfying the condition of Theorem 3.8.*

i If $\epsilon = 0$, and $\omega_\alpha(z)h(z) \prec \phi(z)$ then

$$h(z) \prec \hbar(z) := \frac{1}{2}\int_0^z \frac{\phi(\eta)}{\eta} d\eta.$$

where $\hbar(z)$ is convex and is the best dominant;

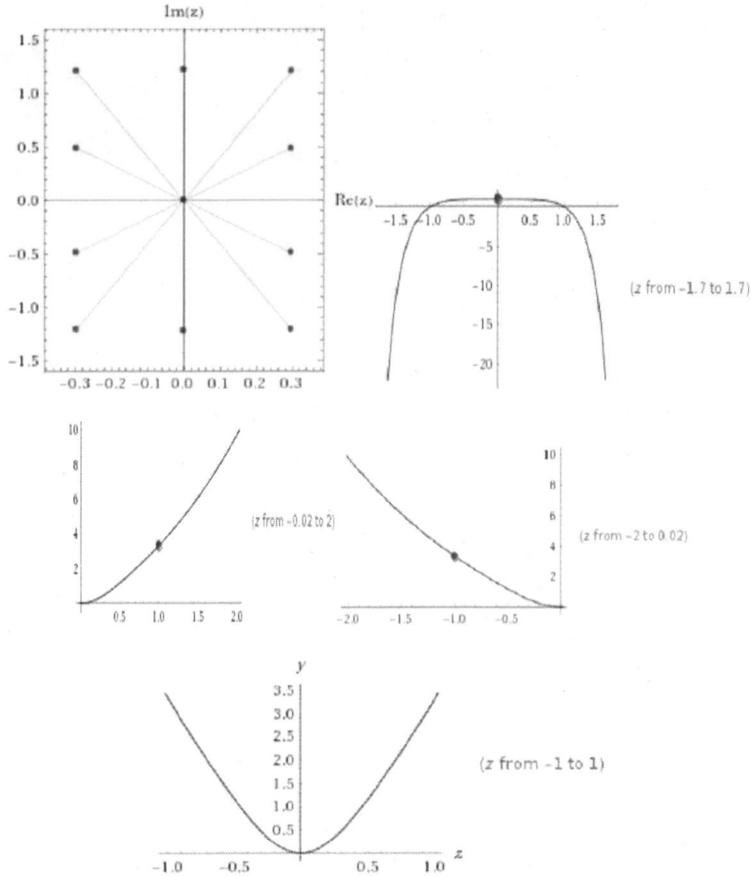

Figure 3.4. Data of $\phi(z) = ze^{z^2}$; from the top-left, the graph represents to all roots of the expression $\Phi := \Re\Big(\frac{z(\phi(z))''}{(\phi(z))'}\Big)$ the second graph refers to the $\lim_{z\to 0}\Re((d/(dz)(z\exp((z^2))))/(d/(dz)z/(1-z^2))) = 1$, second row is the $\lim_{z\to -1}\Phi = \lim_{z\to +1}\Phi = \frac{10}{3}$ and the last graph is Φ.

ii If $\epsilon \in (0,1)$, *and* $\omega_\alpha(z)\left(\frac{h(z)-\epsilon}{h(z)}\right) \prec \phi(z)$ *then*

$$h(z) \prec \hbar(z) := \epsilon \exp\Big(\frac{1}{2}\int_0^z \frac{\phi(\eta)}{\eta}d\eta\Big), \quad \epsilon \neq 0,$$

where $\hbar(z)$ is the best dominant;

iii *If $\epsilon = 1$, and $h^2(z)(1+2\omega_\alpha(z)) - 2h(z)\omega_\alpha(z) \prec \phi'(z)$, $\Re(\phi'(z)) > 0$ then*

$$h(z) \prec \hbar(z) := \sqrt{\frac{1}{2z^{1/2}} \int_0^z \phi'(\eta).\eta^{(1/2)-1} d\eta}.$$

where $\hbar(z)$ is the best dominant.

Proof. By the condition in ϕ, we have $\phi \in \mathcal{S}^*$ with $\phi(0) = 0$ and $\phi'(0) = 1$. It is clear that $h(z) \in \Lambda$, where $\alpha \in [-1, 1]$ satisfying

$$\begin{aligned} zh'(z) &= ze^{\alpha z^2}(2\alpha z^2) + ze^{\alpha z^2} \\ &= \omega_\alpha(z) z e^{\alpha z^2} = \omega_\alpha(z) h(z) \\ &\prec \phi(z). \end{aligned}$$

A direct application of Lemma 3.7- iv, we have the integral formula

$$ze^{\alpha z^2} \prec \frac{1}{2} \int_0^z \frac{\phi(\eta)}{\eta} d\eta.$$

with the best dominant.
Now we let $\epsilon \neq 0$, then $h(z) \in \mathbb{H}[\epsilon, 2]$ achieving $h(z) = \epsilon + a_3 z^3 + \ldots$, $z \in U$ and the subordination conclusion

$$\begin{aligned} \frac{zh'(z)}{h(z)} &= \frac{ze^{\alpha z^2}\omega_\alpha(z) + \epsilon\omega_\alpha(z) - \epsilon\omega_\alpha(z)}{ze^{\alpha z^2} + \epsilon} \\ &= \omega_\alpha(z) \left(\frac{h(z) - \epsilon}{h(z)} \right) \\ &\prec \phi(z). \end{aligned}$$

Thus, in view of Lemma 3.7-v, we have the integral inequality

$$ze^{\alpha z^2} + \epsilon \prec \epsilon \exp\left(\frac{1}{2} \int_0^z \frac{\phi(\eta)}{\eta} d\eta \right), \quad \epsilon \neq 0,$$

as a best dominant.

Now, we let $\epsilon = 1$, we have $h(z) \in\in \mathbb{H}[1,2]$ achieving $h(z) = 1 + a_3z^3 + \ldots$, $z \in U$ and the subordination relation

$$\begin{aligned} h^2(z) + 2h(z)(zh'(z)) &= h^2(z) + 2h(z)\omega_\alpha(z)(h(z) - \epsilon) \\ &= h^2(z) + 2h(z)\omega_\alpha(z)(h(z) - 1) \\ &= h^2(z)(1 + 2\omega_\alpha(z)) - 2h(z)\omega_\alpha(z) \\ &\prec \phi'(z). \end{aligned}$$

Hence, in virtue of Lemma 3.7-vi, we conclude that

$$ze^{\alpha z^2} + 1 \prec \sqrt{\frac{1}{2z^{1/2}} \int_0^z \phi'(\eta).\eta^{(1/2)-1} d\eta}.$$

with the best dominant. This completes the proof. □

Acknowledgment

The author would like to thank the reviewer Dr. Suzan J. Obaiys, School of Mathematical & Computer Sciences, Heriot-Watt University, Malaysia for the extensive information to improve the effort.

References

[1] Gromak, V. I., Laine, I. and Shimomura S.(2008). *Painlevè Differential Equations in the Complex Plane*. Vol. 28. Walter de Gruyter.

[2] Zeng, Zhao-Yun, and Lin Hu., (2019). On the connection problem for nonlinear differential equation, *Boundary Value Problems* 2019(73):1–17. https://doi.org/10.1186/s13661-019-1189-x.

[3] Bobenko, A., Eitner, U., (2000) Painlevè Equations in the Differential Geometry of Surfaces. *Lect. Notes Maths.* Springer, Berlin.

[4] Bobenko, A., Eitner, U., Kitaev, A. (1997), Surfaces with harmonic inverse mean curvature and Painlevè equations. *Geom. Dedic.* (68): 187–227.

[5] Suleimanov, B.I. (1986), On asymptotics of regular solutions for a special kind of Painlevè V equation. *Lect. Notes Math.* (1193): 230–260.

[6] Bassom, A.P., Clarkson, P.A., Law, C.K., McLeod, J.B. (1998), Application of uniform asymptotics to the second Painlevè transcendent. *Arch. Ration. Mech. Anal.* (143): 241–271.

[7] Duren, P. Univalent Functions (1983), *Basic Teaching of Mathematical, Sciences*; **259** Springer-Verlag New York Inc. ISBN 0-387-90795-5. MR0708494.

[8] Miller, S.S. Mocanu, P. T. (2000), Differential subordinations: theory and applications. *Monographs and Textbooks in Pure and Applied Mathematics*, 225. Marcel Dekker, Inc., New York 2000. ISBN: 0-8247-0029-5. MR1760285

[9] Rogosinski, W. (1943), On the coefficients of subordinate functions. *Proc. Lond. Math. Soc.* (48): 48–82.

In: A Closer Look at Boundary Value Problems ISBN: 978-1-53617-857-9
Editor: Mustafa Avci

Chapter 7

EXISTENCE OF SOLUTIONS FOR A STEKLOV PROBLEM WITH VARIABLE EXPONENT

Zehra Yücedağ
Dicle University, Vocational School of Social Sciences,
Diyarbakir, Turkey

ABSTRACT

In this study, we establish the existence of a nontrivial weak solution of Steklov problem under suitable condition on the variable exponent Lebesgue- Sobolev spaces. Our approach is based on variational methods and Ekeland's variational principle.

Keywords: $p(x)-$Laplace operator, variational methods, Steklov Problem, Ekeland's variational principle, variable exponent Lebesgue- Sobolev spaces

1. Introduction

The aim of the present paper is to study for the following Steklov problem involving the $p(x)-$Laplacian

$$\begin{cases} -div\left(a(x)|\nabla u|^{p(x)-2}|\nabla u|\right)+|u|^{p(x)-2}u=f(x,u), x\in\Omega \\ a(x)|\nabla u|^{p(x)-2}\dfrac{\partial u}{\partial \upsilon}=\lambda|u|^{q(x)-2}u, x\in\partial\Omega \end{cases} \qquad (P_\lambda)$$

where $\Omega\subset R^N$ $(N\geq 2)$ is a bounded with smooth boundary, $p(x)\in C(\overline{\Omega})$, $q(x)\in C(\partial\Omega)$, $p^- := \inf_{x\in\overline{\Omega}} p(x)>1$, $q^- := \inf_{x\in\partial\Omega} q(x)>1$ and $p(x)\neq q(y)$ for any $x\in\overline{\Omega}, y\in\partial\Omega$, $\Delta_{p(x)}u := div\left(|\nabla u|^{p(x)-2}\nabla u\right)$ is the $p(x)-$Laplace operator, λ is a positive parameter, $f:\Omega x R\to R$ is a Carathéodory function, $a(x)$ is a function which satisfies the conditions $0<a_1\leq a(x)\leq a_2$ where a_1 and a_2 are positive constants and υ is the outer unit normal to $\partial\Omega$.

The study of differential and partial differential equations with variable exponent growth has attracted considerable attention in recent years.These type of problems appear especially in many applications such as electrorheological fluids and stationary thermo-rheological viscous flows of non-Newtonian fluids, image processing [1, 2, 3, 4, 5], elastic mechanics [6] and the mathematical description of the processes filtration of an idea barotropic gas through a porous medium [7, 8]. Many results have been obtained on these kind of problems, for instance, we here cite [9, 10, 11, 12, 13, 14, 15, 16, 17, 18].

Problems of type (P_λ) have been intensively studied by many authors [19, 20, 21, 22, 23, 24, 25, 26, 27]. In [28], the authors investigated the existence and multiple results by using a variation of the Mountain Pass the following $p(x)-$Laplacian with nonlinear boundary condition in bounded domain

$$\Omega\begin{cases} -div\left(a(x)|\nabla u|^{p(x)-2}\nabla u\right)+b(x)|u|^{p(x)-2}u=\lambda f(x,u), x\in\Omega \\ a(x)|\nabla u|^{p(x)-2}\dfrac{\partial u}{\partial \upsilon}=c(x)|u|^{q(x)-2}u+\eta g(x,u), x\in\partial\Omega \end{cases} \quad (P^*)$$

where f and g functions are satisfy the Ambrosetti-Rabinowitz type condition.

Throughout this paper, we assume the following conditions:

(H1) $F:\Omega xR\to R$ *is a* C^1 *function such that*
$F(x,tu)=t^{s(x)}F(x,u)(t>0)$, $\forall(x,t)\in\Omega xR,$

where $s(x)\in C(\overline{\Omega})$. Put $F(x,\eta):=\int_0^\eta f(x,t)dt$

(H2) There exists $\Omega_0 \subset\subset \Omega$ with $meas(\Omega_0)>0$
such that $F(x,t)>0,\ \forall x\in\Omega_0, t\in R$.
(H3) $1<q^+<p^-, p^-(\partial\Omega)=\inf_{x\in\partial\Omega} p(x)<p^{\partial}(x)$
and $1<s(x)<p(x)$

The main results of this paper are as follows.

Theorem 1.1. If *(H1)*- *(H3)* hold, then for all $\lambda>0$, problem (P_λ) has at least one nontrivial weak solution with negative energy.

Theorem 1.2. If *(H1)*- *(H3)* hold, there exists $\lambda_*>0$ such that, for all $\lambda\in(0,\lambda_*)$, (P_λ) has a nontrivial weak solution.

Inspired by the above problem (P^*), we study problem (P_λ). Firstly, we show under appropriates conditions, problem (P_λ) has a nontrivial weak solution with negative energy. Next, we obtain the existence of continuous spectrum by Ekeland's variational principle. This article is organized as follows. In section 2, we recall the definition of variable exponent Lebesgue-Sobolev spaces. In section 3, we give the proofs of our main results.

2. Preliminaries

In this section, we recall some definitions and basic properties of variable exponent Lebesgue-Sobolev spaces. For details, we refer to [14, 16, 18, 29].

Set $C_+(\overline{\Omega}) = \{p : p(x) \in C(\overline{\Omega}), \inf p(x) > 1, \text{for all } x \in \overline{\Omega}\}$.

For any $p(x) \in C_+(\overline{\Omega})$, we denote

$$1 < p^- := \inf_{x\in\overline{\Omega}} p(x) \leq p^+ := \sup_{x\in\overline{\Omega}} p(x) < \infty$$

Similarly, we can denote $q(x) \in C_+(\partial\Omega)$. Define the variable exponent Lebesgue space by

$$L^{p(x)}(\Omega) = \{u | u : \Omega \to R \text{ is measurable}, \text{such that } \int_\Omega |u(x)|^{p(x)} dx < \infty \}.$$

We define a norm , the so-called Luxemburg norm, on this space by the formula

$$|u|_{p(x)} := \inf\left\{\lambda > 0 : \int_\Omega \left|\frac{u(x)}{\lambda}\right|^{p(x)} dx \leq 1\right\},$$

and $\left(L^{p(x)}(\Omega), |u|_{p(x)}\right)$ becomes a Banach space.

The variable exponent Sobolev space $W^{1,p(x)}(\Omega)$ is denined by

$$W^{1,p(x)}(\Omega) = \left\{u \in L^{p(x)}(\Omega) : |\nabla u| \in L^{p(x)}(\Omega)\right\}$$

and equipped with the norm,

$$\|u\| = |u|_{p(x)} + |\nabla u|_{p(x)}.$$

For $u \in W^{1,p(x)}(\Omega)$, we denote

$$\|u\|_a := \inf\left\{\theta > 0 : \int_\Omega \left(a(x)\left|\frac{\nabla u(x)}{\theta}\right|^{p(x)} + \left|\frac{u(x)}{\theta}\right|^{p(x)} \right) dx \le 1\right\}$$

In view of assumptions of $a(x)$, it is easy to see that $\|u\|_a$ represents a norm on the space $X = W^{1,p(x)}(\Omega)$. Moreover, $\|u\|_a$ and $\|u\|$ are equivalent. In what follows, we will use $\|u\|$ instead of $\|u\|_a$ on $X = W^{1,p(x)}(\Omega)$.

The space $W_0^{1,p(x)}(\Omega)$ is denoted as the closure of $C_0^\infty(\Omega)$ in $W^{1,p(x)}(\Omega)$ with respect to the norm $\|u\|$.

Let $a : \partial\Omega \to R$ be measurable. Define the weighted variable exponent Lebesgue space

$$L_{a(x)}^{p(x)}(\partial\Omega) = \left\{u \mid u : \partial\Omega \to R \textit{ is measurable and} \int_{\partial\Omega} |a(x)||u|^{p(x)} dx < +\infty\right\},$$

with the norm

$$|u|_{(p(x),a(x))} = \inf\left\{\kappa > 0; \int_{\partial\Omega} |a(x)|\left|\frac{u}{\kappa}\right|^{p(x)} d\sigma \le 1\right\}$$

where $d\sigma$ is the measure on the boundary. Then $L_{a(x)}^{p(x)}(\partial\Omega)$ is a Banach space. In particular, $a \in L^\infty(\partial\Omega)$, $L_{a(x)}^{p(x)}(\partial\Omega) = L^{p(x)}(\partial\Omega)$.

Proposition 2.1. [17, 30] The conjugate space of $L^{p(x)}(\Omega)$ is $L^{p'(x)}(\Omega)$ where $\frac{1}{p(x)}+\frac{1}{p'(x)}=1$. For any $u\in L^{p(x)}(\Omega)$ and $v\in L^{p'(x)}(\Omega)$, we have

$$\left|\int_\Omega uvdx\right|\le\left(\frac{1}{p^-}+\frac{1}{p^+}\right)|u|_{p(x)}|v|_{p'(x)}.$$

The modular of the $L^{p(x)}(\Omega)$ space, which is the mapping $\rho_{p(x)}:L^{p(x)}(\Omega)\to R$ defined by

$$\rho_{p(x)}(u)=\int_\Omega |u|^{p(x)}dx,\quad \forall u\in L^{p(x)}(\Omega).$$

Proposition 2.2. [13, 16] For all $u,v\in L^{p(x)}(\Omega)$, we have

$$|u|_{p(x)}<1(=1,>1)\Leftrightarrow \rho_{p(x)}(u)<1(=1,>1),$$

$$\min\left(|u|_{p(x)}^{p^-},|u|_{p(x)}^{p^+}\right)\le\rho_{p(x)}(u)\le\max\left(|u|_{p(x)}^{p^-},|u|_{p(x)}^{p^+}\right),$$

$$\rho_{p(x)}(u-v)\to 0\ (\to\infty)\Leftrightarrow |u-v|_{p(x)}\to 0\ (\to\infty).$$

Proposition 2.3. [12, 22,] Denote $\rho_{L^{p(x)}(\partial\Omega)}(u)=\int_{\partial\Omega}|u|^{p(x)}d\sigma$, $\forall u\in L^{p(x)}(\partial\Omega)$. Then

$$|u|_{L^{p(x)}(\partial\Omega)}\ge 1\Leftrightarrow |u|_{L^{p(x)}(\partial\Omega)}^{p^-}\le\rho_{L^{p(x)}(\partial\Omega)}(u)\le|u|_{L^{p(x)}(\partial\Omega)}^{p^+},$$

$$|u|_{L^{p(x)}(\partial\Omega)}<1\Leftrightarrow |u|_{L^{p(x)}(\partial\Omega)}^{p^+}\le\rho_{L^{p(x)}(\partial\Omega)}(u)\le|u|_{L^{p(x)}(\partial\Omega)}^{p^-}.$$

Proposition 2.4. [16, 28]

(i) If $1 < p^- \leq p^+ < \infty$ then the spaces $L^{p(x)}(\Omega)$, $W^{1,p(x)}(\Omega)$ and $W_0^{1,p(x)}(\Omega)$ are separable, reflexive and uniformly convex Banach spaces,

(ii) If $p_1(x), p_2(x) \in C_+(\overline{\Omega})$ and $p_1(x) < p_2(x)$ for all $x \in \overline{\Omega}$ then the embedding $L^{p_2(x)}(\Omega) \to L^{p_1(x)}(\Omega)$ is continuous,

(iii) If $q(x) \in C_+(\overline{\Omega})$ and $q(x) < p*(x)$ for all $x \in \overline{\Omega}$ then the embedding $W^{1,p(x)}(\Omega) \to L^{q(x)}(\Omega)$ is compact and continuous, where

$$p*(x) = \begin{cases} \dfrac{Np(x)}{N - p(x)}, & \text{if} \quad p(x) < N, \\ \infty, & \text{if} \quad p(x) \geq N, \end{cases}$$

(iv) If $q(x) \in C_+(\partial\Omega)$ and $q(x) < p^{\partial}(x)$ for all $x \in \partial\Omega$ then the trace embedding $W^{1,p(x)}(\Omega) \to L^{q(x)}(\partial\Omega)$ is compact and continuous, where

$$p^{\partial}(x) = \begin{cases} \dfrac{(N-1)\,p(x)}{N - p(x)}, & \text{if} N > p(x), \\ \infty, & N \leq p(x). \end{cases}$$

(v) (Poincaré inequality) There is a positive constant $C > 0$ such that

$$|u|_{p(x)} \leq C|\nabla u|_{p(x)}, \text{ for all } u \in W_0^{1,p(x)}(\Omega).$$

Proposition 2.5. [28] *Let* $\Gamma(u) = \int_\Omega \left(a(x)|\nabla u|^{p(x)} + |u|^{p(x)}\right)dx$ *for* $\forall u \in W^{1,p(x)}(\Omega)$, we have

(i) $\Gamma(u)\geq 1 \Rightarrow \xi_1 \|u\|^{p^-} \leq \Gamma(u) \leq \xi_2 \|u\|^{p^+}$

(ii) $\Gamma(u)\leq 1 \Rightarrow \xi_3 \|u\|^{p^+} \leq \Gamma(u) \leq \xi_4 \|u\|^{p^-}$

where ξ_1, ξ_2, ξ_3 and ξ_4 are positive constants independent of u.

Theorem 2.6. [31] Let f be a continuous function defined on all of R^N If f is coercive, then f has o global minimizer. Furthermore, if the first partial derivatives of f exists on all R^N, then any global minimızers of f can be found among the critical points of f.

PROOF OF MAIN RESULTS

Definition 3.1. We say that $u \in X/\{0\}$ is a weak solution of (P_λ) if

$$\int_\Omega a(x)|\nabla u|^{p(x)-2}\nabla u \nabla v dx + \int_\Omega |u|^{p(x)-2} uv dx - \lambda \int_{\partial\Omega} |u|^{q(x)-2} uv d\sigma$$
$$-\int_\Omega f(x,u) v dx = 0$$

for all $v \in X$.

The energy functional corresponding to problem (P_λ) is defined as $J_\lambda : X \to R$

$$J_\lambda(u) = \int_\Omega \frac{a(x)|\nabla u|^{p(x)} + |u|^{p(x)}}{p(x)} - \lambda \int_{\partial\Omega} \frac{|u|^{q(x)}}{q(x)} d\sigma - \int_\Omega F(x,u) dx$$

where $F(x,u) = \int_0^u f(x,s) ds$.

Proposition 3.2. [28] If one denotes

$$\varphi(u)=\int_{\Omega}\frac{a(x)|\nabla u|^{p(x)}+|u|^{p(x)}}{p(x)},\quad \forall u\in X,$$

then $\varphi\in C^{1}(X,R)$ and the derivative operator of φ, denoted by φ', is

$$\langle\varphi'(u),v\rangle=\int_{\Omega}a(x)|\nabla u|^{p(x)-2}\nabla u\nabla v dx+\int_{\Omega}|u|^{p(x)-2}uvdx,\quad \forall u,v\in X,$$

and one has:

(i) $\varphi':X\to X^{*}$ is a continuous, bounded and strictly monotone operator,
(ii) φ' is a mapping of (S^{+}) type, that is, if $u_n\to u$ (weak convergent) in X and $\limsup_{n\to\infty}\langle\varphi'(u_n)-\varphi'(u),u_n-u\rangle\leq 0$, then $u_n\to u$ (strong convergent) in X,
(iii) $\varphi':X\to X^{*}$ is a homeomorphism.

Proposition 3.3. [20 ,28] If one denotes

$$\psi(u)=\int_{\partial\Omega}\frac{|u|^{q(x)}}{q(x)}d\sigma,\quad \forall u\in X,$$

where $q(x)\in C_{+}(\partial\Omega)$ and $q(x)<p*(x)$ for any $x\in\partial\Omega$, then $\psi\in C^{1}(X,R)$ and the derivative operator of ψ, denoted by ψ', is

$$\langle\psi'(u),v\rangle=\int_{\partial\Omega}|u|^{q(x)-2}uvd\sigma,\quad \forall u,v\in X,$$

and one has $\psi : X \to R$ and $\psi' : X \to X^*$ are sequentially weakly - strongly continuous, bounded, namely, $u_n \to u$ (weakly contiuously) implies $\psi'(u_n) \to \psi(u)$ (strongly contiuous) .

Therefore, Proposition 2.4, Proposition 3.2 and Proposition 3.3, it is easy to see that $J_\lambda(u) \in C^1(X,R)$ and the critical points of J_λ are weak solutions of (**P**). Moreover, the derivate of I_λ is the mapping $J_\lambda : X \to X^*$

$$
\begin{aligned}
\left(J'_\lambda(u), v\right) &= \int_\Omega \left(a(x)|\nabla u|^{p(x)-2} \nabla u \nabla v + |u|^{p(x)-2} uv\right) dx \\
&- \lambda \int_{\partial\Omega} |u|^{q(x)-2} uv d\sigma - \int_\Omega f(x,u) v dx,
\end{aligned}
$$

for any $u, v \in X$ [19].

Lemma 3.4. Assume that (H1)- (H3) hold. Then, there for all $\rho \in (0,1)$, exists two positive real numbers ρ and r and $\lambda_* > 0$ such that for any $\lambda \in (0, \lambda_*)$, we have $J_\lambda(u) \geq r > 0$ with $\|u\| = \rho$.

Proof. For $\|u\| < 1$, By (H1), we have $K > 0$ and $M > 0$ such that

$$
F(x,u) \leq K|u|^{s(x)} \text{ and } f(x,u) \leq M|u|^{s(x)-1} \text{ for all } x \in \Omega. \tag{3.1}
$$

Then, using (3.1) and Proposition 2.5 we have

$$
\begin{aligned}
J_\lambda(u) &= \int_\Omega \frac{a(x)|\nabla u|^{p(x)} + |u|^{p(x)}}{p(x)} - \lambda \int_{\partial\Omega} \frac{|u|^{q(x)}}{q(x)} d\sigma - \int_\Omega F(x,u) dx \\
&\geq \frac{\xi_3}{p^+} \|u\|^{p^+} - \frac{\lambda}{q^-} \int_{\partial\Omega} |u|^{q(x)} d\sigma - K \int_\Omega |u|^{s(x)} dx
\end{aligned} \tag{3.2}
$$

Moreover, using *(H3)*, we have $q(x) < p^{\partial}(x)$, $\forall x \in \overline{\Omega}$. So, we have the continuous embeddings $X \to L^{q(x)}(\partial\Omega)$ and $X \to L^{s(x)}(\Omega)$ (Proposition 2.4). Thus, there exist constants $c_1, c_2 > 0$ such that

$$|u|_{q(x),\partial\Omega} \leq c_1 \|u\| \text{ and } |u|_{s(x),\Omega} \leq c_2 \|u\| \text{ for all } u \in X. \tag{3.3}$$

By (3.3), we write

$$J_\lambda(u) \geq \frac{\xi_3}{p^+} \|u\|^{p^+} - \frac{\lambda}{q^-} c_1^{q^-} \|u\|^{q^-} - K c_2^{s^-} \|u\|^{s^-}.$$

If we choose $0 < \frac{\lambda c_1^{q^-}}{q^-} < \lambda K c_2^{s^-}$ and using *(H3)* for any $u \in X$ with $\|u\| = \rho$, we obtain

$$J_\lambda(u) \geq \frac{\xi_3}{p^+} \rho^{p^+} - 2\lambda K c_2^{s^-} \left(\rho^{s^-} + \rho^{q^-}\right).$$

By the above inequality, we obtain that if we choose λ^* as

$$\lambda^* = \frac{\xi_3 \rho^{p^+}}{2p^+ K c_2^{s^-} \left(\rho^{s^-} + \rho^{q^-}\right)} \tag{3.4}$$

then for any $\lambda \in (0, \lambda^*)$ and any $u \in X$ with $\|u\| = \rho$, there exists such that $J_\lambda(u) \geq r > 0$. The proof is completed.

Lemma 3.5. If (H1)- (H3) hold, there exists $u_1 \in X$ such that $u_1 \geq 0, u_1 \neq 0$ and $J_\lambda(tu_1) < 0$ for $t > 0$ small enough.

Proof. From assumption (H3), we know that $s(x) < p(x)$ for all $x \in \overline{\Omega}_0$. In the sequel, denote by $s_0^- = \inf_{x\in\Omega_0} s(x)$ and $p_0^- = \inf_{x\in\Omega_0} p(x)$Let $\varepsilon_0 > 0$ such that $s_0^- + \varepsilon_0 < p_0^-$.

Since $s \in C(\overline{\Omega_0})$, there is exist an open set Ω_1 *such that* $\Omega_1 \subset\subset \Omega_0 \subset\subset \partial\Omega$ and $\left|s(x) - s_0^-\right| < \varepsilon_0$ for all $x \in \Omega_1$. Hence, we conclude that $s(x) \le s_0^- + \varepsilon_0 < p_0^- < q^+$ for all $x \in \Omega_1$.

Let $\phi \in C_0^\infty(\Omega)$ such that $\operatorname{supp}(\phi) \subset \Omega_1 \subset \Omega_0$, $\phi = 1$ in a subset $\Omega_1^{'} \subset \operatorname{supp}(\phi)$ and $0 \le \phi \le 1$ in Ω_1. So, from *(H2),* we obtain

$$
\begin{aligned}
J_\lambda(t\varphi) &= \int_\Omega \frac{a(x)|\nabla t\varphi|^{p(x)} + |t\varphi|^{p(x)}}{p(x)}dx - \int_{\partial\Omega} \frac{1}{q(x)}|t\varphi|^{q(x)}d\sigma - \lambda\int_\Omega F(x,t\varphi)dx \\
&\le \frac{t^{p_0^-}}{p_0^-}\int_{\Omega_0}\left(a(x)|\nabla\varphi|^{p(x)} + |\varphi|^{p(x)}\right)dx - \frac{1}{q^-}t^{q^+}\int_{\partial\Omega}|\varphi|^{q(x)}d\sigma - \lambda t^{s_0^- + \varepsilon_0}\int_{\Omega_1} F(x,\varphi)dx \\
&\le \frac{t^{p_0^-}}{p_0^-}\int_{\Omega_0}\left(a(x)|\nabla\varphi|^{p(x)} + |\varphi|^{p(x)}\right)dx - \lambda t^{s_0^- + \varepsilon_0}\int_{\Omega_1} F(x,\varphi)dx.
\end{aligned}
$$

Thus,

$$
J_\lambda(t\phi) \le 0,
$$

for $t < \delta^{1/\left(p_0^- - s_0^- - \varepsilon_0\right)}$ with

$$
0 < \delta < \min\left\{1, \frac{\lambda p_0^- \int_{\Omega_1} F(x,\phi)dx}{\int_{\Omega_0}\left(a(x)|\nabla\phi|^{p(x)} + |\phi|^{p(x)}\right)dx}\right\}.
$$

Finally, we point out that $\int_{\Omega_0}\left(a(x)|\nabla\phi|^{p(x)}+|\phi|^{p(x)}\right)dx>0$. In fact if $\int_{\Omega_0}\left(a(x)|\nabla\phi|^{p(x)}+|\phi|^{p(x)}\right)dx=0$, then $\|\phi\|=0$ and consequently $\phi=0$ in Ω which is not true because $\phi=1$ in the subset Ω_1' of Ω. The prof of Lemma 3.5 is complete.

Proof of Theorem 3.1. Let $\lambda\in(0,\lambda_*)$, where λ_* is given by (3.4). From Lemma 3.4, we know that there exists $\rho=(0,1)$ such that the ball centered at the origin and of radius ρ in X, denoted by $B_\rho(0)$, we have

$$\overline{B_\rho(0)}\subset X \text{ and } \inf_{\partial B_\rho(0)} J_\lambda>0.$$

On the other hand, from Lemma 3.5, there exists $\phi\in X$ such that $J_\lambda(t\phi)<0$ for all $t>0$ small enough. Moreover, by Lemma 3.5, we deduce that

$$-\infty<\underline{c}:=\inf_{B_\rho(0)} J_\lambda<0.$$

Let choose $\varepsilon>0$. Then, we have

$$0<\varepsilon<\inf_{\partial B_\rho(0)} J_\lambda-\inf_{B_\rho(0)} J_\lambda.$$

Using the above information, the functional $J_\lambda:\overline{B_\rho(0)}\to R$ is lower bounded on $\overline{B_\rho(0)}$ and $J_\lambda\in C^1(B_\rho(0),R)$. So, applying Ekeland's variational principle [32] there exists $u_\varepsilon\in\overline{B_\rho(0)}$ such that

$$\begin{cases} \underline{c} \le J_\lambda(u_\varepsilon) \le \underline{c} + \varepsilon, \\ J_\lambda(u_\varepsilon) < J_\lambda(u) + \varepsilon.\|u - u_\varepsilon\|, \ \ u_\varepsilon \ne u \end{cases}$$

Since,

$$J_\lambda(u_\varepsilon) \le \inf_{B_\rho(0)} J_\lambda + \varepsilon \le \inf_{B_\rho(0)} J_\lambda + \varepsilon \le \inf_{\partial B_\rho(0)} J_\lambda ,$$

We show that $u_\varepsilon \in B_\rho(0)$. Now, we define $I_\lambda : \overline{B_\rho(0)} \to R$ by

$$I_\lambda(u_\varepsilon) = J_\lambda(u) + \varepsilon.\|u - u_\varepsilon\|.$$

It is clear that u_ε is a minimum point of I_λ and

$$\frac{I_\lambda(u_\varepsilon + tv) - I_\lambda(u_\varepsilon)}{t} \ge 0 ,$$

for $t > 0$ small enough and any $v \in B_1(0)$. The above relation yields

$$\frac{J_\lambda(u_\varepsilon + tv) - J_\lambda(u_\varepsilon)}{t} + \varepsilon.\|v\| \ge 0 .$$

Letting $t \to 0$, we have

$$\langle J'_\lambda(u_\varepsilon), v\rangle + \varepsilon.\|v\| \ge 0 .$$

and this implies that $\|J'_\lambda(u_\varepsilon)\| \le \varepsilon$. Hence, we deduce that there exists a sequence $\{u_n\} \subset B_\rho(0)$ such that

$$J_\lambda(u_n) \to \underline{c} \text{ and } J'_\lambda(u_n) \to 0 \text{ as } n \to \infty \tag{3.5}$$

Since $\{u_n\} \subset B_\rho(0)$ then $\|u_n\| < \rho$, $\{u_n\}$ is bounded in X. Next, we show the strong converges to u_n in X . Since $\{u_n\}$ is bounded in X, there exists u in X such that, up to subsequence, u_n converges weakly to u in X . Taking into account (3.5), we have $\langle I'(u_u), u_n - u\rangle \to 0$. So, we have

$$\begin{aligned}
&\langle I'(u_n), u_n - u\rangle \\
&= \int_\Omega \left(a(x)|\nabla u_n|^{p(x)-2}\nabla u_n(\nabla u_n - \nabla u) + |u_n|^{p(x)-2}u_n(u_n - u)\right)dx \\
&- \lambda\int_{\partial\Omega}|u_n|^{q(x)-2}u_n(u_n - u)d\sigma - \int_\Omega f(x,u_n)(u_n - u)dx.
\end{aligned}$$

Using *(H1)* and Proposition 2.1, it follows

$$\begin{aligned}
&\left|\int_\Omega f(x,u_n)(u_n - u)dx\right| \le \left|\int_\Omega \left(M|u_n|^{s(x)-1}\right)(u_n - u)dx\right| \\
&\le M\left\||u_n|^{s(x)-1}\right|_{L^{s'(x)}(\Omega)}|u_n - u|_{L^{s(x)}(\Omega)}
\end{aligned}$$

where $c_4 > 0$ is constant. Because $s(x) < p^*(x)$ (Proposition 2.6 (iii)), there exists u such that u_n converges weakly to u in X. Thanks to the compact embedding $X \to L^{s(x)}(\Omega)$, we get

$$u_n \to u(strongly) \text{ in } L^{s(x)}$$

$$u_n \to u \; a.e. x \in \Omega$$

So,

$$\int_\Omega f(x,u_n)(u_n - u)dx \to 0.$$

Similarly, by Proposition 2.1, Proposition 2.6 (iv) and Proposition 3.3, we have

$$\int_{\partial\Omega} |u_n|^{q(x)-2} u_n (u_n - u) d\sigma \to 0.$$

Thus,

$$\int_{\Omega} \left(a(x) |\nabla u_n|^{p(x)-2} \nabla u_n (\nabla u_n - \nabla u) + |u_n|^{p(x)-2} u_n (u_n - u) \right) dx \to 0.$$

Finally, from Proposition 3.2, we deduce that u_n converges strongly to u in X Moreover, $J_\lambda \in C^1(X, R)$, we conclude

$$J'_\lambda(u_n) \to J'_\lambda(u) \text{ as } n \to \infty. \tag{3.6}$$

Relations (3.5) and (3.6) show that $J'_\lambda(u) = 0$ and thus u is a weak solution for problem (P_λ). Moreover, by relation (3.5), it follows that $J_\lambda(u) < 0$ and thus, u is a nontrivial weak solution for problem (P_λ).

Lemma 3.6. Under assumptions (H1)- (H3), the functional J_λ is coercive on X.

Proof. Since $s^+ < p^-$, we have the continuous embeddings $X \to L^{s^\pm}(\Omega)$ and $X \to L^{q(x)}(\partial\Omega)$. Then there exist constants $c_3, c_4, c_5 > 0$ such that

$$|u|_{q(x),\partial\Omega} \le c_3 \|u\|, \int_\Omega |u|^{s^+} dx \le c_4 \|u\|^{s^+} \text{ and } \int_\Omega |u|^{s^-} dx \le c_5 \|u\|^{s^-} \tag{3.7}$$

for all $u \in X$.

According to the fact that $|t|^{s(x)} \leq |t|^{s^+} + |t|^{s^-}$ for all $x \in \overline{\Omega}$. with $\|u\| > 1$ Using Proposition 2.4, Proposition 2.5 and inequality (3.7) we obtain

$$J_\lambda(u) = \int_\Omega \frac{a(x)|\nabla u|^{p(x)} + |u|^{p(x)}}{p(x)} - \lambda \int_{\partial\Omega} \frac{|u|^{q(x)}}{q(x)} d\sigma - \int_\Omega F(x,u)dx$$
$$\geq \frac{\xi_1}{p^+}\|u\|^{p^-} - \frac{\lambda}{q^-} c_3 \|u\|^{q^+} - Kc_6\|u\|^{s^+} - Kc_7\|u\|^{s^-}$$

where c_6 and c_7 are positive constants.

As $s^+, s^-, q^+ < p^-$, we infer that $J_\lambda(u) \to \infty$ as $\|u\| \to \infty$. Then, J_λ is coercive on X.

Proof of Theorem 1.2. J_λ has a global minimizer from Lemma 3.6. By using Lemma 3.6 and Theorem 2.6, we conclude that this minimizer is a critical point of J_λ and a solution for problem (P_λ). Moreover, Lemma 3.5 ensures that this minimizer is nontrvial solution for problem (P_λ).

REFERENCES

[1] Diening, Lars, "Theoretical and numerical results for electrorheological fluids." Ph.D. thesis, University of frieburg, Germany, 2002. Zbl 1022.76001.

[2] Halsey, Thomas C, "Electrorheological fluids", *Science* 258, (1992): 761-766.

[3] Mihăilescu, Mihai and Radulescu Vicentiu D, "A multiplicity result for a nonlinear degenerate problem arising in the theory of electrorheological fluids." *Proceedings of the Royal Society A*. 462, (2006): 2625-2641.

[4] Ruzicka, Michael, "Electrorheological fluids: modelling and mathematical theory. Lecture notes in mathematics."1784, Berlin: Springer-Verlag; 2000.

[5] Ruzicka, Michael and Diening Lars, "Non-Newtonian fluids and function spaces." *Institute of Mathematics of the Academy of Sciences of the Czech Republic (Praha)* 8, (2007):95-143.

[6] Zhikov, Vasilii Vasil'evich, "Averaging of functionals of the calculus of variations and elasticity theory." *Math. USSR. Izv.* 29, (1987): 33-66.

[7] Antontsev, Stanislav and Shmarev Sergey, "A model porous medium equation with variable exponent of nonlinearity: existence, uniqueness and localization properties of solutions." *Nonlinear Anal.* 60, no.3 (2005): 515—545.

[8] Chen, Yunmei, Levine Stacey Elizabeth and Rao Murali, "Variable exponent, linear growth functionals in image restoration", *SIAM J. Appl. Math.* 66, no.4 (2006):1383–1406.

[9] Acerbi, Emilio and Mingione, Giuseppe. "Regulary results for a class of functionals with nonstandard growth." *Arch. Rational Mech. Anal.* 156, no.2 (2001):121–140.

[10] Avci, Mustafa, "A topological result for a class of anisotropic difference equations." *Annals of the University of Craiova* 46. no.2 (2013):328-343.

[11] Avci, Mustafa, "Positive ground state solutions to a nonlocal singular elliptic problem." *Canad. J. Appl. Math.* 1, no.1 (2019), 1–14.

[12] Deng, Shao-Gao. "Eigenvalues of the *p(x)*-Laplacian Steklov problem", *J. Math. Anal. Appl.*339, (2008): 925–937.

[13] Edmunds, David and Rakosnik Jiří, "Sobolev embeddings with variable exponent". *Studia Math.* 143, (2000): 267–293.

[14] Fan, Xian-Ling and Han Xiaoyou, "Existence and multiplicity of solutions for p(x)-Laplacian equations in RN." *Nonlinear Anal.* 59 (2004):173–188.

[15] Fan, Xianling, Zhang Qihu and Zhao Dun, "Eigenvalues of p(x)-Laplacian Dirichlet problem", *J. Math. Anal. Appl.* 302, (2015): 306–317.

[16] Kováčik, O., and Răkosnik J., "On spaces $L^{p(x)}(\Omega)$ and $W^{k,p(x)}(\Omega)$", *Czechoslovak Math. J.* 41, no.116, (1991): 592-618.

[17] Mashiyev, Rabil Ayazoglu, Çekiç Bilal, Avci Mustafa and Yücedag Zehra, "The Nehari manifold approach for Dirichlet problem involving the p(x)-Laplacian equation." *Journal of the Korean Mathematical Society* 47, no.4 (2010): 845-860.

[18] Yücedag, Zehra, "Existence of weak solutions for a nonlocal problem involving the p(x)-Laplace Operator." *Journal of Advances in Mathematical Analysis and Applications* 1, no.1 (2016): 8-17.

[19] Allaoui, Mostafa, El Amrouss Abdel Rachid and Ourraoui Anass, "Existence and multiplicity of solutions for a Steklov problem involving the p(x)-Laplace operator." *Electron. J. Differential Equations* 2012, no.132 (2012):1–12.

[20] Ayoujil, Abdesslem, "On the superlinear Steklov problem involving the p(x)-Laplacian", *EJQTDE.*38, (2014):1–13.

[21] Ben, Ali Khaled, Ghanmi Abdeljabbar and Kefi Khaled, "On the Steklov problem involving the p(x)-Laplacian with indefinite weight." *Opuscula Math.* 37, no. 6 (2017): 779–794.

[22] Ben, Ali Khaled., "Existence results for Steklov problem involving the p(x)-Laplacian", *Complex Var. and Elliptic Equ.* 63, (2017):1675-1686.

[23] Ekincioğlu, Ismail and Ayazoğlu Rabil (Mashiyev), "On Steklov Boundary Value Problems for p(x)- Laplacian equations." *Electronic Journal of Math. Analy. And Appl.*5 no. 2 (2017):289-297.

[24] Afrouzi, Ghasem A, Hadjian Armin and Heidarkhani Shapour, "Steklov problem involving the p(x)-Laplacian." 2014, no. 134, (2014): 1–11.

[25] Hsini, Mounir, Irzi Nawal and Kefir Khaled, "Nonhomogeneous p(x)-Laplacian Steklov problem with weights", *Complex Variables and Elliptic Equations* 65, no.3 (2019): 440-454.

[26] Mavinga, N, Nkashama MN., "Steklov spectrum and nonresonance for elliptic equations with nonlinear boundary conditions", *Electron. J. Differ. Equ. Conf.,* Vol.19, pp.197–205, 2010.

[27] Yücedag, Zehra, "Existence results for steklov problem with nonlinear boundary condition." *Middle East Journal of Science* 5, no.2 (2019): 146 – 154.

[28] Wei, Zhiqiang and Chen Zigao, "Existence results for the $p(x)$-Laplacian with nonlinear boundary condition." *Applied Math.,* Article ID 727398, doi:10.5402/2012/727398, (2012).

[29] Mashiyev, Rabil Ayazoglu, Oğraş Sezai and Yücedag Zehra, Avci Mustafa, "Existence and multiplicity of weak solutions for nonuniformly

elliptic equations with nonstandard growth condition," *Complex Variables and Elliptic Equations* 57, no. 5 (2012) : 579-595.

[30] Fan, Xian-Ling and Zhang Qihu, "Existence of solutions for *p(x)*-Laplacian Dirichlet problem". *Nonlinear Anal.* 52, (2003):1843–1852.

[31] Willem, Michel, "Minimax Theorems." Birkhauser, Basel, (1996).

[32] Ekeland, Ivar, "On the variational principle." *J. Math. Anal. Appl.* 47 (1974:324–353.

In: A Closer Look at Boundary Value Problems
Editor: Mustafa Avci
ISBN: 978-1-53617-857-9

Chapter 8

ON A CLASS OF KIRCHHOFF TYPE PROBLEMS INVOLVING CRITICAL EXPONENTS AND CAFFARELLI-KOHN-NIRENBERG INEQUALITIES

Nguyen Thanh Chung*
Department of Mathematics, Quang Binh University,
Dong Hoi, Quang Binh, Vietnam

Abstract

We are interested in the existence of solutions for a class of degenerate nonlocal problems with critical exponents and singular potentials involving Caffarelli-Kohn-Nirenberg inequalities. Under some suitable conditions on the nonlinearities, we establish the existence of infinitely many solutions by using the genus theory, introduced by Krasnoselskii combined with Caffarelli-Kohn-Nirenberg inequalities and a variant of the mountain pass theorem in critical point theory. The main result of this work is to complement and generalize the previous ones on Kirchhoff type problems, presented in the literature.

Keywords: Kirchhoff type problems, Caffarelli-Kohn-Nirenberg inequalities, critical exponents, variational methods

*Corresponding Author's Email: ntchung82@yahoo.com.

AMS Subject Classification: 35J60, 35J92, 58E05

1. INTRODUCTION AND PRELIMINARIES

In this chapter, we are interested in the existence of solutions for a class of Kirchhoff type problems of the form

$$\begin{cases} -M\left(\int_\Omega |x|^{-ap}|\nabla u|^p\,dx\right)\operatorname{div}\left(|x|^{-ap}|\nabla u|^{p-2}\nabla u\right) = \lambda|x|^{-cp^*_{a,c}}|u|^{p^*_{a,c}-2}u \\ \qquad\qquad\qquad\qquad\qquad\qquad + |x|^{-bp}f(x,u), \quad x\in\Omega, \\ u=0, \quad x\in\partial\Omega, \end{cases} \tag{1}$$

where $\Omega\subset\mathbb{R}^N$ is a smooth bounded domain, $0\in\Omega$, $0\le a<\frac{N-p}{p}$, $1<p<N$, $a\le b\le c<a+1$, $p^*_{a,c}=\frac{Np}{N-(1+a-c)p}$, λ is a positive parameter, $f\in C(\overline{\Omega}\times\mathbb{R},\mathbb{R})$ and $M\in C(\mathbb{R}^+_0,\mathbb{R})$ is increasing in $\mathbb{R}^+_0:=[0,+\infty)$. Throughout this work, we assume that the following conditions hold:

$(\mathcal{M}_1)$ There exists $m_0>0$ such that $m(t)\ge m_0$, for all $t\in\mathbb{R}^+_0$.

$(\mathcal{M}_2)$ There exists $\frac{p}{p^*_{a,c}}<\sigma\le 1$ such that $\widehat{M}(t)\ge\sigma M(t)t$ for all $t\in\mathbb{R}^+_0$, where $\widehat{M}(t)=\int_0^t M(\tau)\,d\tau$.

Note that, if $M(t)\equiv 1$ then (1) becomes the elliptic equations with singular potentials of Caffarelli-Kohn-Nirenberg type

$$\begin{cases} -\operatorname{div}\left(|x|^{-ap}|\nabla u|^{p-2}\nabla u\right)=\lambda|x|^{-cp^*_{a,c}}|u|^{p^*_{a,c}-2}u+|x|^{-bp}f(x,u), \quad x\in\Omega, \\ u=0, \quad x\in\partial\Omega. \end{cases} \tag{2}$$

The study of the existence of solutions for problem (2) is a very interesting topic, due to the presence of singular potentials and it has been considered by many authors, we refer to some interesting papers [1, 2, 3, 4, 5, 6, 7]. In [6, 7], Xuan firstly studied the solvability for (2) in the special case $f(x,u)=\lambda|u|^{-(a+1)p+c}|u|^{p-2}u$, $0\le a<\frac{N-p}{p}$ and $c>0$ and obtained the existence of a nontrivial solution by using a variant of the concentration compactness principle due to Lions [8, 9], see [7, Theorem 4.1]. In [1, 2, 4], the authors considered problem (2) in the case $\lambda=1$ and f is related to the first eigenvalue of the operator $-\operatorname{div}\left(|x|^{-ap}|\nabla u|^{p-2}\nabla u\right)$ with zero Dirichlet boundary condition on Ω. In [3], Bouchekif et al. considered (2) in the case when $f(x,u)$ involves a concave term and obtained a multiplicity result by using variational methods. In [5], Liang et al. studied the existence of small infinitely many solutions for problem (2) by using the symmetric mountain pass theorem due to Kajikiya [10]. For further information on problem (2), we refer to [11, 12] in which Miyagaki et al. established a maximum principle and obtained some existence results for elliptic systems involving Caffarelli-Kohn-Nirenberg inequalities by using sub- and supersolutions arguments.

In the case when $a = b = c = 0$, problem (1) becomes a class of p-Kirchhoff type problems with critical exponents

$$\begin{cases} -M\left(\int_\Omega |\nabla u|^p\,dx\right)\Delta_p u = \lambda |u|^{p^*-2}u + f(x,u), & x \in \Omega, \\ u = 0, \quad x \in \partial\Omega, \end{cases} \tag{3}$$

which was considered in many paper, see for examples [13, 14, 15, 16, 17, 18, 19]. We point out the fact that there are many functions $M(t)$ satisfying the conditions $(\mathcal{M}_1)$ and $(\mathcal{M}_2)$, for example $M(t) = a + bt$, $a, b > 0$. Then we call this problem a Kirchhoff type problem. Problems of this type were proposed by Kirchhoff in 1883 as a model given by the stationary version of the Kirchhoff equation

$$\rho\frac{\partial^2 u}{\partial t^2} - \left(\frac{P_0}{h} + \frac{E}{2L}\int_0^L \left|\frac{\partial u}{\partial x}\right|^2 dx\right)\frac{\partial^2 u}{\partial x^2} = 0, \tag{4}$$

where ρ is the mass density, P_0 is the initial tension, h represents the area of the cross-section, E is the Young modulus of the material and L is the length of the string, see [20]. Problem (4) is often called a nonlocal problem because it contains an integral over Ω. This causes some mathematical difficulties which make the study of such a problem particularly interesting. The nonlocal problem models several physical and biological systems, where u describes a process which depends on the average of itself, such as the population density, see [21]. It is clear that nonlocal problems are more complicated because of the lack of compactness caused the presence of critical exponents. Together with Kirchhoff type problems involving critical exponents, there is a large number of papers concerning subcritical growth, the interested readers are referred to [22, 1, 2, 23, 24, 25, 26, 27, 28, 29, 30, 31, 32, 33].

From the aforementioned papers, we are interested in the existence of solutions for Kirchhoff type problem (1) with critical exponent and singular potentials involving Caffarelli-Kohn-Nirenberg inequalities. We are motivated by the recent results on these kinds of problems in the papers [34, 35, 36, 37, 38, 39, 40]. In [34, 35, 36, 37], the authors studied the existence and multiplicity of solutions for problem (1) with subcritical growth by using variational methods. In [40], Shen considered (1) in the special case $M(t) = a + bt^{\theta-1}$, $\theta > 1$ and $f(x, u)$ is sublinear at infinity. There, the author proved the existence of two positive solutions for (1) by using Nehari manifold and fibering map. In [38, 39], the authors considered the following Kirchhoff type problem

$$\begin{cases} -M\left(\int_\Omega |x|^{-ap}|\nabla u|^p\,dx\right)\operatorname{div}\left(|x|^{-ap}|\nabla u|^{p-2}\nabla u\right) = |x|^{-cp^*_{a,c}}|u|^{p^*_{a,c}-2}u \\ \qquad\qquad\qquad\qquad\qquad\qquad +\lambda|x|^{-\delta}f(x,u), \quad x \in \Omega, \\ u = 0, \quad x \in \partial\Omega, \end{cases} \tag{5}$$

where the nonlinear term $f(x, u)$ satisfies the Ambrosetti-Rabinowitz type condition and

($\mathcal{F}_1'$) $C_1'|t|^{r-1} \leq f(x,t) \leq C_2'|t|^{r-1}$ for all $(x,t) \in \Omega \times \mathbb{R}_0^+$, $C_1', C_2' > 0$, $p \leq r < p^*_{a,c}$ and $\delta \leq (a+1)r + N\left(1 - \frac{r}{p}\right)$.

With the help of the condition ($\mathcal{F}_1'$) and truncated arguments, the authors obtained some existence and multiplicity results for problem (1) without the condition ($\mathcal{M}_2$), see further [15] and the references therein for the non-singular case. In the next section, we will complement and generalize the above results to the case when the nonlinear term f is assumed to be almost critical at infinity and does not satisfy the condition ($\mathcal{F}_1'$) as in [38, 39]. We also notice that the parameter λ here is involved in the critical term $|x|^{-cp^*_{a,c}}|u|^{p^*_{a,c}-2}u$, which is different from those presented in [13, 14, 38, 39, 16, 17]. For this reason, our main tools come from the genus theory, introduced by Krasnoselskii combined with the Caffarelli-Kohn-Nirenberg inequalities and a variant of the mountain pass theorem in critical point theory.

In what follows, let us recall some useful results on weighted Sobolev spaces involving Caffarelli-Kohn-Nirenberg inequalities stated in [41, 42, 6]. We have known that for all $u \in C_0^\infty(\mathbb{R}^N)$, there exists a constant $S_{a,c} > 0$ such that

$$\left(\int_{\mathbb{R}^N} |x|^{-cp^*_{a,c}}|u|^{p^*_{a,c}}\,dx\right)^{\frac{p}{p^*_{a,c}}} \leq S_{a,c}\int_{\mathbb{R}^N} |x|^{-ap}|\nabla u|^p\,dx, \tag{6}$$

where $-\infty < a < \frac{N-p}{p}$, $a \le c \le a+1$, $p^*_{a,c} = \frac{Np}{N-(1+a-c)p}$. Let $W_0^{1,p}(\Omega, |x|^{-ap})$ be the completion of $C_0^\infty(\Omega)$ with respect to the norm

$$\|u\|_{a,p} = \left(\int_\Omega |x|^{-ap}|\nabla u|^p\, dx\right)^{\frac{1}{p}}.$$

Then $W_0^{1,p}(\Omega, |x|^{-ap})$ is a reflexive and separable Banach space. From the boundedness of Ω and the standard approximation argument, it is easy to see that (6) holds for any $u \in W_0^{1,p}(\Omega, |x|^{-ap})$ in the sense that

$$\left(\int_{\mathbb{R}^N} |x|^{-\alpha}|u|^l\, dx\right)^{\frac{p}{l}} \le S_{a,c}\int_{\mathbb{R}^N} |x|^{-ap}|\nabla u|^p\, dx, \tag{7}$$

for $1 \le l \le p^* = \frac{Np}{N-p}$, $\alpha \le (1+a)l + N\left(1 - \frac{l}{p}\right)$, that is, the embedding $W_0^{1,p}(\Omega, |x|^{-ap}) \hookrightarrow L^l(\Omega, |x|^{-\alpha})$ is continuous, where $L^l(\Omega, |x|^{-\alpha})$ is the weighted $L^l(\Omega)$ space with the norm

$$|u|_{l,\alpha} := |u|_{L^l(\Omega,|x|^{-\alpha})} = \left(\int_\Omega |x|^{-\alpha}|u|^l\, dx\right)^{\frac{1}{l}}.$$

In fact, we have the following compact embedding result which is an extension of the classical Rellich-Kondrachov compactness theorem.

Lemma 1.1 (see [6, Compactness embedding theorem]). *Suppose that $\Omega \subset \mathbb{R}^N$ is an open bounded domain with C^1 boundary and that $0 \in \Omega$, where $1 < p < N$, $-\infty < a < \frac{N-p}{p}$, $1 \le l < \frac{Np}{N-p}$ and $\alpha < (1+a)l + N\left(1 - \frac{l}{p}\right)$. Then the embedding $W_0^{1,p}(\Omega, |x|^{-ap}) \hookrightarrow L^l(\Omega, |x|^{-\alpha})$ is compact.*

In the next part of this section, we recall the weighted version of the Concentration Compactness Principle due to Xuan [7], the readers can see the original papers by Lions [8, 9] for the non-singular case. This is a key point which helps us to overcome the lack of compactness caused by critical exponents.

Proposition 1.2 (see [7]). *Let $1 < p < N$, $-\infty < a < \frac{N-p}{p}$, $a \le c \le a+1$, $p^*_{a,c} = \frac{Np}{N-(1+a-c)p}$, and let $\mathcal{M}^+(\mathbb{R})$ be the space of positive bounded measures on $\mathbb{R}^N$. Suppose that $\{u_n\} \subset W_0^{1,p}(\Omega, |x|^{-ap})$ is a sequence such that*

$$u_n \rightharpoonup u \text{ in } W_0^{1,p}(\Omega, |x|^{-ap}),$$

$$||x|^{-a}|\nabla u_n||^p \rightharpoonup \mu \text{ in } \mathcal{M}^+(\mathbb{R}^N),$$
$$||x|^{-c}|u_n||^{p^*_{a,c}} \rightharpoonup \nu \text{ in } \mathcal{M}^+(\mathbb{R}^N),$$
$$u_n(x) \to u(x) \text{ a.e. on } \mathbb{R}^N.$$

Then we have following statements:

(i) *There exists some at most countable set* Λ, *a family* $\{x_j : j \in \Lambda\}$ *of distinct points in* $\mathbb{R}^N$ *and a family* $\{\nu_j : j \in \Lambda\}$ *of positive numbers such that*

$$\nu = ||x|^{-c}|u||^{p^*_{a,c}} + \sum_{j\in\Lambda} \nu_j \delta_{x_j},$$

where δ_{x_j} *is the Dirac unitary mass concentrated at* $x \in \mathbb{R}^N$.

(ii) *The following inequality holds*

$$\mu \geq ||x|^{-a}|\nabla u||^p + \sum_{j\in\Lambda} \mu_j \delta_{x_j}$$

for some family $\{\mu_j : j \in \Lambda\}$ *of positive numbers satisfying* $S_{a,c}\nu_j^{\frac{p}{p^*_{a,c}}} \leq \mu_j$ *for all* $j \in \Lambda$. *In particular,* $\sum_{j\in\Lambda} \nu_j^{\frac{p}{p^*_{a,c}}} < +\infty$.

Let X be a real Banach space and $\sum$ the family of set $E \subset X\backslash\{0\}$ such that E is closed in X and symmetric with respect to 0, that is,

$$\sum = \{E \subset X\backslash\{0\} : \ E \text{ is closed in } X \text{ and } E = -E\}.$$

For each $E \in \sum$, we say genus of E is a number k denoted by $\gamma(E) = k$ if there is an odd map $h \in C(E, \mathbb{R}^n\backslash\{0\})$ and k is the smallest integer with this property.

Now, let us recall a version of the mountain pass theorem for even functionals, the readers can consult its proof in [43].

Proposition 1.3. *Let* X *be an infinite dimensional Banach space with* $X = V \oplus Y$, *where* V *is finite dimensional and let* $J \in C^1(X, \mathbb{R})$ *be an even functional with* $J(0) = 0$ *such that the following conditions hold:*

(I_1) *There exist positive constants* $\beta, \rho > 0$ *such that* $J(u) \geq \beta$ *for all* $u \in \partial B_\rho(0) \cap Y$.

(I_2) *There exists* $c^* > 0$ *such that* J *satisfies the* $(PS)_c$ *condition for* $0 < c < c^*$.

(I_3) *For each finite dimensional subspace* $\widehat{X} \subset X$, *there exists* $R = R(\widehat{X})$ *such that* $J(u) \leq 0$ *for all* $u \in \widehat{X} \backslash B_R(0)$.

Suppose that V *is* k *dimensional and* $V = span\{e_1, e_2, ..., e_k\}$. *For* $n \geq k$, *inductively choose* $e_{n+1} \notin X_n := span\{e_1, e_2, ..., e_n\}$. *Let* $R_n = R(X_n)$ *and* $D_n = B_{R_n}(0) \cap X_n$. *Define*

$$G_n := \{h \in C(D_n, X) : \ h \text{ is odd and } h(u) = u, \ \forall \partial B_{R_n}(0) \cap X_n\}$$

and

$$\Gamma_j := \left\{ h\left(\overline{D_n \backslash E}\right) : \ h \in G_n, \ n \geq j, \ E \in \sum \text{ and } \gamma(E) \leq n - j \right\}.$$

For each $j \in \mathbb{N}$, *let*

$$c_j := \inf_{K \in \Gamma_j} \max_{u \in K} J(u).$$

Then, $0 < \beta \leq c_j \leq c_{j+1}$ *for* $j > k$, *and if* $j > k$ *and* $c_j < c^*$, *then we conclude that* c_j *is the critical value of* J. *Moreover, if* $c_j = c_{j+1} = ... = c_{j+l} = c < c^*$ *for* $j > k$, *then* $\gamma(K_c) \geq l + 1$, *where*

$$K_c := \left\{ u \in E : \ J(u) = c \text{ and } J'(u) = 0 \right\}.$$

2. EXISTENCE OF SOLUTIONS

In this section, we will state and prove the main results. We denote by C_i general positive number whose value may change from line to line. Firstly, we make the definition of weak solutions as follows.

Definition 2.1. *We say that* $u \in W_0^{1,p}(\Omega, |x|^{-ap})$ *is a weak solution of problem* (1) *if*

$$M\left(\int_\Omega |x|^{-ap}|\nabla u|^p \, dx\right) \int_\Omega |x|^{-ap}|\nabla u|^{p-2}\nabla u \cdot \nabla v \, dx - \lambda \int_\Omega |x|^{-cp^*_{a,c}}|u|^{p^*_{a,c}-2}uv \, dx$$
$$- \int_\Omega |x|^{-bp} f(x, u) v \, dx = 0$$

for all $v \in W_0^{1,p}(\Omega, |x|^{-ap})$.

Theorem 2.2. *Let* $M : \mathbb{R}_0^+ \to \mathbb{R}$ *be a continuous and increasing function satisfying the conditions* $(\mathcal{M}_1)$-$(\mathcal{M}_2)$. *Moreover, we assume that the function* $f : \overline{\Omega} \times \mathbb{R} \to \mathbb{R}$ *is continuous and satisfies*

$(\mathcal{F}_1)$ $f(x,t)$ *is odd with respect to* $t \in \mathbb{R}$ *and*

$$f(x,t) = o(|t|^{p^*_{a,c}-1}), \quad \text{as } |t| \to +\infty \text{ uniformly in } x \in \Omega,$$

$$f(x,t) = o(|t|^{p-1}), \quad \text{as } |t| \to 0 \text{ uniformly in } x \in \Omega.$$

$(\mathcal{F}_2)$ $F(x,t) \leq \frac{\sigma}{p} f(x,t)t$ *for all* $x \in \Omega$ *and* $t \in \mathbb{R}$, *where* $F(x,t) = \int_0^t f(x,\tau)\,d\tau$ *and* σ *is given by* $(\mathcal{M}_2)$.

Then, there exists a decreasing sequence of real numbers $\{\lambda_k\} \subset (0,+\infty)$ *with* $\lambda_{k+1} < \lambda_k$ *such that, for each* $\lambda \in (\lambda_{k+1}, \lambda_k)$, *problem* (1) *has at least k pairs of nontrivial weak solutions.*

For each $\lambda > 0$, let us consider the functional $J_\lambda : X = W_0^{1,p}(\Omega, |x|^{-ap}) \to \mathbb{R}$ associated with problem (1) by the formula

$$J_\lambda(u) = \frac{1}{p}\widehat{M}\left(\int_\Omega |x|^{-ap}|\nabla u|^p\,dx\right) - \frac{\lambda}{p^*_{a,c}}\int_\Omega |x|^{-cp^*_{a,c}}|u|^{p^*_{a,c}}\,dx - \int_\Omega |x|^{-bp}F(x,u)\,dx,$$

where $\widehat{M}(t) = \int_0^t M(\tau)\,d\tau$ and $F(x,t) = \int_0^t f(x,\tau)\,d\tau$. Note that $J_\lambda \in C^1(X,\mathbb{R})$ and its derivative is given by

$$J_\lambda'(u)(v) = M\left(\int_\Omega |x|^{-ap}|\nabla u|^p\,dx\right)\int_\Omega |x|^{-ap}|\nabla u|^{p-2}\nabla u \cdot \nabla v\,dx -$$
$$-\lambda\int_\Omega |x|^{-cp^*_{a,c}}|u|^{p^*_{a,c}-2}uv\,dx - \int_\Omega |x|^{-bp}f(x,u)v\,dx$$

for all $u, v \in X$. This means that we can seek weak solutions of problem (1) as the critical points of the functional J_λ. Our goal is to verify the hypotheses of Proposition 1.3 to obtain the existence of infinitely many solutions for problem (1).

Lemma 2.3. *Assume that the conditions* $(\mathcal{M}_1)$ *and* $(\mathcal{F}_1)$ *hold, then the functional* J_λ *satisfies the hypothesis* (I_1).

Proof. By the first part of the condition $(\mathcal{F}_1)$, for any $\epsilon > 0$, there exists a positive constant $\eta_\epsilon > 0$ depending on ϵ such that

$$|f(x,t)| \leq \epsilon|t|^{p^*_{a,c}-1}, \quad \forall x \in \Omega, \quad |t| > \eta_\epsilon.$$

By the continuity of f, there exists a positive constant A_ϵ depending on ϵ such that $|f(x,t)| \leq A_\epsilon$ for all $(x,t) \in \Omega \times [-\eta_\epsilon, \eta_\epsilon]$. Hence,

$$|f(x,t)| \leq \epsilon |t|^{p^*_{a,c}-1} + A_\epsilon, \quad \forall (x,t) \in \Omega \times \mathbb{R}. \tag{8}$$

On the other hand, by the second part of $(\mathcal{F}_1)$, for any $\epsilon > 0$, there exists a positive constant $\delta_\epsilon \in (0,1)$ small enough such that

$$|f(x,t)| \leq \epsilon |t|^{p-1}, \quad \forall x \in \Omega, \quad |t| < \delta_\epsilon. \tag{9}$$

For $|t| \geq \delta_\epsilon$, it follows from (8) that

$$\frac{|f(x,t)|}{|t|^{p^*_{a,c}-1}} \leq \epsilon + \frac{A_\epsilon}{|t|^{p^*_{a,c}-1}} \leq \epsilon + \frac{A_\epsilon}{\delta_\epsilon^{p^*_{a,c}-1}} = \overline{A}_\epsilon,$$

that is,

$$|f(x,t)| \leq \overline{A}_\epsilon |t|^{p^*_{a,c}-1}, \quad \forall x \in \Omega, \quad |t| \geq \delta_\epsilon. \tag{10}$$

Combining (9) and (10) we get

$$|f(x,t)| \leq \epsilon |t|^{p-1} + \overline{A}_\epsilon |t|^{p^*_{a,c}-1}, \quad \forall (x,t) \in \Omega \times \mathbb{R}. \tag{11}$$

By integrating (11) we deduce that

$$|F(x,t)| \leq \frac{\epsilon}{p} |t|^p + \frac{\overline{A}_\epsilon}{p^*_{a,c}} |t|^{p^*_{a,c}}, \quad \forall (x,t) \in \Omega \times \mathbb{R}. \tag{12}$$

By $(\mathcal{M}_1)$ and relation (12), we have

$$\begin{aligned} J_\lambda(u) &= \frac{1}{p} \widehat{M} \left(\int_\Omega |x|^{-ap} |\nabla u|^p \, dx \right) - \frac{\lambda}{p^*_{a,c}} \int_\Omega |x|^{-cp^*_{a,c}} |u|^{p^*_{a,c}} \, dx - \int_\Omega |x|^{-bp} F(x,u) \, dx \\ &\geq \frac{m_0}{p} \int_\Omega |x|^{-ap} |\nabla u|^p \, dx - \frac{\lambda}{p^*_{a,c}} \int_\Omega |x|^{-cp^*_{a,c}} |u|^{p^*_{a,c}} \, dx \\ &\quad - \int_\Omega |x|^{-bp} \left(\frac{\epsilon}{p} |u|^p + \frac{\overline{A}_\epsilon}{p^*_{a,c}} |u|^{p^*_{a,c}} \right) dx. \end{aligned} \tag{13}$$

Since $a \leq b \leq c < a+1$, it follows from (3) that the embeddings $X \hookrightarrow L^p(\Omega, |x|^{-pb})$, $X \hookrightarrow L^{p^*_{a,c}}(\Omega, |x|^{-cp^*_{a,c}})$ and $X \hookrightarrow L^{p^*_{a,c}}(\Omega, |x|^{-bp})$ are continuous, and thus there exists $C_1 > 0$ such that

$$S_{a,c}^{\frac{1}{p}} |u|_{p^*_{a,c}, cp^*_{a,c}} \leq \|u\|_{a,c}, \quad C_1 |u|_{p^*_{a,c}, bp} \leq \|u\|_{a,c}, \quad C_1 |u|_{p,bp} \leq \|u\|_{a,c}, \quad \forall u \in X, \tag{14}$$

where $S_{a,c}$ is given by (3). Taking $\epsilon = \frac{m_0}{2C_1^p}$, from relations (13) and (14), for all $u \in X$ with $\|u\|_{a,c} = \rho$, $\rho \in (0,1)$, it follows that

$$J_\lambda(u) \geq \frac{m_0}{p}\|u\|_{a,c}^p - \frac{\epsilon C_1^p}{p}\|u\|_{a,c}^p - \frac{\lambda S_{a,c}^{\frac{p_{a,c}^*}{p}} + C_1^{p_{a,c}^*}\overline{A}_\epsilon}{p_{a,c}^*}\|u\|_{a,c}^{p_{a,c}^*}$$
$$= \left(\frac{m_0}{2p} - \frac{\lambda S_{a,c}^{\frac{p_{a,c}^*}{p}} + C_1^{p_{a,c}^*}\overline{A}_\epsilon}{p_{a,c}^*}\rho^{p_{a,c}^*-p}\right)\rho^p.$$

Since $1 < p < p_{a,c}^* = \frac{Np}{N-(1+a-c)p}$, there exists $\beta > 0$ such that $J_\lambda(u) \geq \beta$ for all $u \in X$ with $\|u\|_{a,c} = \rho$, where ρ is chosen sufficiently small. □

Lemma 2.4. *Assume that the conditions $(\mathcal{M}_2)$ and $(\mathcal{F}_1)$ hold, the functional J_λ satisfies the hypothesis (I_3).*

Proof. Let $\widehat{X}$ be a finite dimensional subspace of X. By (8), we have

$$|f(x,t)| \leq \epsilon|t|^{p_{a,c}^*-1} + A_\epsilon, \quad \forall (x,t) \in \Omega \times \mathbb{R}.$$

Integrating this inequality, we get

$$\frac{|F(x,t)|}{|t|^{p_{a,c}^*}} \leq \frac{\epsilon}{p_{a,c}^*} + \frac{A_\epsilon}{|t|^{p_{a,c}^*-1}},$$

hence for $\epsilon > 0$, there exists $\delta_\epsilon > 0$ such that

$$|F(x,t)| \leq \frac{\epsilon}{p_{a,c}^*}|t|^{p_{a,c}^*}, \quad \forall x \in \Omega, \quad |t| \geq \delta_\epsilon. \tag{15}$$

By the continuity of f, there exists $B_\epsilon > 0$ such that

$$|F(x,t)| \leq B_\epsilon, \quad \forall x \in \Omega, \quad |t| \leq \delta_\epsilon. \tag{16}$$

Combining (15) and (16), we get

$$F(x,t) \geq -B_\epsilon - \frac{\epsilon}{p_{a,c}^*}|t|^{p_{a,c}^*}, \quad \forall (x,t) \in \Omega \times \mathbb{R}. \tag{17}$$

From $(\mathcal{M}_2)$, there exists $C_2 > 0$ such that

$$\widehat{M}(t) \leq \frac{\widehat{M}(t_0)}{t_0^{\frac{1}{\sigma}}}t^{\frac{1}{\sigma}} = C_2 t^{\frac{1}{\sigma}}, \quad \forall t \geq t_0 > 0. \tag{18}$$

From (17) and (18), for any $u \in \widehat{X}$ with $\|u\|_{a,c}$ large enough, we have

$$\begin{aligned} J_\lambda(u) &= \frac{1}{p}\widehat{M}\left(\int_\Omega |x|^{-ap}|\nabla u|^p\,dx\right) - \frac{\lambda}{p^*_{a,c}}\int_\Omega |x|^{-cp^*_{a,c}}|u|^{p^*_{a,c}}\,dx - \int_\Omega |x|^{-bp}F(x,u)\,dx \\ &\le \frac{C_2}{p}\left(\int_\Omega |x|^{-ap}|\nabla u|^p\,dx\right)^{\frac{1}{\sigma}} - \frac{\lambda}{p^*_{a,c}}\int_\Omega |x|^{-cp^*_{a,c}}|u|^{p^*_{a,c}}\,dx + \\ &\quad + \frac{\epsilon}{p^*_{a,c}}\int_\Omega |x|^{-bp}|u|^{p^*_{a,c}}\,dx + B_\epsilon\int_\Omega |x|^{-bp}\,dx \\ &\le \frac{C_2}{p}\|u\|_{a,c}^{\frac{p}{\sigma}} - \frac{\lambda}{p^*_{a,c}}\int_\Omega |x|^{-cp^*_{a,c}}|u|^{p^*_{a,c}}\,dx + \frac{\epsilon}{p^*_{a,c}}\int_\Omega |x|^{-bp}|u|^{p^*_{a,c}}\,dx + \\ &\quad + B_\epsilon\int_\Omega |x|^{-bp}\,dx. \end{aligned}$$

Note that since $dim(\widehat{X}) < +\infty$, the norms $\|.\|_{a,c}$, $|.|_{p^*_{a,c},bp}$ and $|.|_{p^*_{a,c},cp^*_{a,c}}$ are equivalent in $\widehat{X}$. This means that there exist $C_3, C_4 > 0$ such that

$$C_3|u|_{p^*_{a,c},cp^*_{a,c}} \le \|u\|_{a,c} \le C_4|u|_{p^*_{a,c},cp^*_{a,c}},$$

$$C_3|u|_{p^*_{a,c},bp} \le \|u\|_{a,c} \le C_4|u|_{p^*_{a,c},bp}, \quad \forall u \in \widehat{X}.$$

Choosing $\epsilon = \frac{\lambda}{2}\left(\frac{C_3}{C_4}\right)^{p^*_{a,c}} > 0$, we obtain from the above information that

$$\begin{aligned} J_\lambda(u) &\le \frac{C_2}{p}\|u\|_{a,c}^{\frac{p}{\sigma}} - \frac{\lambda}{p^*_{a,c}C_4^{p^*_{a,c}}}\|u\|_{a,c}^{p^*_{a,c}} + \frac{\epsilon}{p^*_{a,c}C_3^{p^*_{a,c}}}\|u\|_{a,c}^{p^*_{a,c}} + B_\epsilon\int_\Omega |x|^{-bp}\,dx \\ &= \frac{C_2}{p}\|u\|_{a,c}^{\frac{p}{\sigma}} - \frac{\lambda}{2p^*_{a,c}C_4^{p^*_{a,c}}}\|u\|_{a,c}^{p^*_{a,c}} + B_\epsilon\int_\Omega |x|^{-bp}\,dx, \quad \forall u \in \widehat{X}. \end{aligned}$$

Since $1 < \frac{p}{\sigma} < p^*_{a,c}$, we conclude that $J_\lambda(u) < 0$ for all $u \in \widehat{X}$ with $\|u\|_{a,c} \ge R$, where R is chosen large enough. □

Lemma 2.5. *Assume that the conditions $(\mathcal{M}_1)$-$(\mathcal{M}_2)$ and $(\mathcal{F}_1)$-$(\mathcal{F}_2)$ hold. Then, any $(PS)_c$ sequence of J_λ is bounded in X.*

Proof. Let $\{u_n\} \subset X$ be a $(PS)_c$ sequence for J_λ, that is

$$J_\lambda(u_n) \to c, \quad J'_\lambda(u_n) \to 0 \text{ in } X^*, \quad n \to \infty. \tag{19}$$

By (8), the conditions $(\mathcal{M}_2)$ and $(\mathcal{F}_2)$, for n large enough, we deduce that

$c + 1 + \|u_n\|_{a,c}$

$$
\begin{aligned}
&\geq J_\lambda(u_n) - \frac{\sigma}{p} J'_\lambda(u_n)(u_n) \\
&= \frac{1}{p}\widehat{M}\left(\int_\Omega |x|^{-ap}|\nabla u_n|^p\,dx\right) - \frac{\lambda}{p^*_{a,c}}\int_\Omega |x|^{-cp^*_{a,c}}|u_n|^{p^*_{a,c}}\,dx - \int_\Omega |x|^{-bp}F(x,u)\,dx \\
&\quad - \frac{\sigma}{p} M\left(\int_\Omega |x|^{-ap}|\nabla u|^p\,dx\right)\int_\Omega |x|^{-ap}|\nabla u|^p\,dx + \frac{\lambda\sigma}{p}\int_\Omega |x|^{-cp^*_{a,c}}|u_n|^{p^*_{a,c}}\,dx \\
&\quad + \frac{\sigma}{p}\int_\Omega |x|^{-bp} f(x,u_n)u_n\,dx \\
&= \frac{1}{p}\left[\widehat{M}\left(\int_\Omega |x|^{-ap}|\nabla u_n|^p\,dx\right) - \sigma M\left(\int_\Omega |x|^{-ap}|\nabla u|^p\,dx\right)\int_\Omega |x|^{-ap}|\nabla u|^p\,dx\right] \\
&\quad + \lambda\left(\frac{\sigma}{p} - \frac{1}{p^*_{a,c}}\right)\int_\Omega |x|^{-cp^*_{a,c}}|u_n|^{p^*_{a,c}}\,dx + \int_\Omega \left(\frac{\sigma}{p} f(x,u_n)u_n - F(x,u_n)\right)dx \\
&\geq \lambda\left(\frac{\sigma}{p} - \frac{1}{p^*_{a,c}}\right)\int_\Omega |x|^{-cp^*_{a,c}}|u_n|^{p^*_{a,c}}\,dx,
\end{aligned}
$$

which yields

$$
\lambda\left(\frac{\sigma}{p} - \frac{1}{p^*_{a,c}}\right)\int_\Omega |x|^{-cp^*_{a,c}}|u_n|^{p^*_{a,c}}\,dx \leq c + 1 + \|u_n\|_{a,c}. \tag{20}
$$

Since $a \leq b \leq c < a+1$, $bp \leq cp^*_{a,c}$ and Ω is a bounded domain, we have the embedding $L^{p^*_{a,c}}(\Omega, |x|^{-cp^*_{a,c}}) \hookrightarrow L^{p^*_{a,c}}(\Omega, |x|^{-pb})$ is continuous, which leads to the fact that, there exists $C_5 > 0$ such that

$$
\int_\Omega |x|^{-bp}|u_n|^{p^*_{a,c}}\,dx \leq C_5\left(1 + \|u_n\|_{a,c}\right).
$$

By (15) and (16), we also have

$$
|F(x,t)| \leq \frac{\epsilon}{p^*_{a,c}}|t|^{p^*_{a,c}} + B_\epsilon, \quad \forall (x,t) \in \Omega \times \mathbb{R}. \tag{21}
$$

It follows from the condition $(\mathcal{M}_1)$, relations (19)-(21), and the definition of the functional J_λ that

$$
\begin{aligned}
&\frac{m_0}{p}\int_\Omega |x|^{-ap}|\nabla u_n|^p\,dx \\
&\leq \frac{1}{p}\widehat{M}\left(\int_\Omega |x|^{-ap}|\nabla u_n|^p\,dx\right) \\
&= J_\lambda(u_n) + \frac{\lambda}{p^*_{a,c}}\int_\Omega |x|^{-cp^*_{a,c}}|u_n|^{p^*_{a,c}}\,dx + \int_\Omega |x|^{-bp}F(x,u_n)\,dx
\end{aligned}
$$

$$\leq c + o_n(1) + \frac{\lambda}{p^*_{a,c}} \int_\Omega |x|^{-cp^*_{a,c}} |u_n|^{p^*_{a,c}} \, dx +$$

$$+ \frac{\epsilon}{p^*_{a,c}} \int_\Omega |x|^{-bp} |u_n|^{p^*_{a,c}} \, dx + B_\epsilon \int_\Omega |x|^{-bp} \, dx$$

$$\leq C_6 + o_n(1) + \frac{C_7}{p^*_{a,c}} (\lambda + \epsilon) (1 + \|u_n\|_{a,c}),$$

where

$$0 < C_6 = c + B_\epsilon \int_\Omega |x|^{-bp} \, dx < +\infty.$$

Hence,

$$\|u_n\|^p_{a,c} \leq C_8 (1 + \|u_n\|_{a,c}),$$

so that the sequence $\{u_n\}$ is bounded in X since $1 < p < N$. □

From Lemma 2.5, if $\{u_n\}$ is a $(PS)_c$ sequence for J_λ at level $c \in \mathbb{R}$, then we can extract a subsequence of $\{u_n\} \subset X$, still denoted by $\{u_n\}$ and a function $u \in X$ such that $\{u_n\}$ converges weakly to u and $u_n(x) \to u(x)$ for a.e. $x \in \Omega$ as $n \to \infty$.

Using Proposition 1.2, there exist two bounded nonnegative measures μ and ν on $\mathbb{R}^N$ and some at least countable family of points $(x_j)_{j\in\Lambda} \subset \overline{\Omega}$ and of positive numbers $(\nu_j)_{j\in\Lambda}$ and $(\mu_j)_{j\in\Lambda}$ such that

$$||x|^{-c}|u_n||^{p^*_{a,c}} \rightharpoonup \nu = ||x|^{-c}|u||^{p^*_{a,c}} + \sum_{j\in\Lambda} \nu_j \delta_{x_j}, \quad \nu_j > 0, \tag{22}$$

$$||x|^{-a}|\nabla u_n||^p \rightharpoonup \mu \geq ||x|^{-a}|\nabla u||^p + \sum_{j\in\Lambda} \mu_j \delta_{x_j}, \quad \mu_j > 0, \tag{23}$$

$$\nu_j^{\frac{p}{p^*_{a,c}}} \leq \frac{\mu_j}{S_{a,c}}, \tag{24}$$

for all $j \in \Lambda$, where δ_{x_j} is the Dirac mass at $x_j \in \overline{\Omega}$, where $S_{a,c}$ is the best constant of the embedding $X \hookrightarrow L^{p^*_{a,c}}(\Omega, |x|^{-cp^*_{a,c}})$ given by (6).

Lemma 2.6. *For any $j \in \Lambda$ we have $\nu_j > \left(\frac{m_0 S_{a,c}}{\lambda}\right)^{\frac{N}{(1+a-c)p}}$ or $\nu_j = 0$, where m_0 is given by the condition $(\mathcal{M}_1)$.*

Proof. For each $j \in \Lambda$ and $\epsilon > 0$, let us consider $\phi_j \in C_0^\infty(\Omega, [0,1])$ such that $\phi_j \equiv 1$ on $B_1(0)$, $\phi_j \equiv 0$ on $\Omega \backslash B_2(0)$ and $|\nabla \phi_j| \leq 1$. Defining $\phi_{j,\epsilon} =$

$\phi_j\left(\frac{x-x_j}{\epsilon}\right)$, where $\epsilon > 0$ and $x_j \in \overline{\Omega}$ belongs to the support of ν, we have that $\{u_n\phi_{j,\epsilon}\}$ is bounded in X. It then follows from (19) that $J'_\lambda(u_n)(u_n\phi_{j,\epsilon}) \to 0$ as $n \to \infty$, that is,

$$J'_\lambda(u_n)(u_n\phi_{j,\epsilon}) = M\left(\int_\Omega |x|^{-ap}|\nabla u_n|^p\,dx\right)\int_\Omega |x|^{-ap}|\nabla u_n|^{p-2}\nabla u_n\nabla(u_n\phi_{j,\epsilon})\,dx$$
$$-\lambda\int_\Omega |x|^{-cp^*_{a,c}}|u_n|^{p^*_{a,c}-2}u_n(u_n\phi_{j,\epsilon})\,dx - \int_\Omega |x|^{-bp}f(x,u_n)(u_n\phi_{j,\epsilon})\,dx \to 0.$$

Hence,

$$M\left(\int_\Omega |x|^{-ap}u_n|\nabla u_n|^p\right)\int_\Omega |x|^{-ap}u_n|\nabla u_n|^{p-2}\nabla u_n\cdot\nabla\phi_{j,\epsilon}\,dx$$
$$= -M\left(\int_\Omega |x|^{-ap}u_n|\nabla u_n|^p\,dx\right)\int_\Omega |x|^{-ap}|\nabla u_n|^p\phi_{j,\epsilon}\,dx + \lambda\int_\Omega |x|^{-cp^*_{a,c}}|u_n|^{p^*_{a,c}}\phi_{j,\epsilon}\,dx$$
$$+\int_\Omega |x|^{-bp}f(x,u_n)u_n\phi_{j,\epsilon}\,dx + o_n(1). \tag{25}$$

Since the support of $\phi_{j,\epsilon}$ is $B_{2\epsilon}(x_j)$, using the Hölder inequality and the Dominated Convergence Theorem we obtain

$$\left|\int_\Omega |x|^{-ap}u_n|\nabla u_n|^{p-2}\nabla u_n\cdot\nabla\phi_{j,\epsilon}\,dx\right|$$
$$\le \int_{B_{2\epsilon}(x_j)} |x|^{-ap}\nabla u_n|^{p-1}|u_n\nabla\phi_{j,\epsilon}|\,dx$$
$$= \int_{B_{2\epsilon}(x_j)} (|x|^{-a(p-1)}\nabla u_n|^{p-1})(|x|^{-a}|u_n\nabla\phi_{j,\epsilon}|)\,dx$$
$$\le \left(\int_{B_{2\epsilon}(x_j)} |x|^{-ap}|\nabla u_n|^p\,dx\right)^{\frac{p-1}{p}}\left(\int_{B_{2\epsilon}(x_j)} |x|^{-ap}|u_n\nabla\phi_{j,\epsilon}|^p\,dx\right)^{\frac{1}{p}}$$
$$\le C_9\left(\int_{B_{2\epsilon}(x_j)} |x|^{-ap}|u_n|^p|\nabla\phi_{j,\epsilon}|^p\,dx\right)^{\frac{1}{p}}$$
$$\le C_9\left(\int_{B_{2\epsilon}(x_j)} |x|^{-cp^*_{a,c}}|u_n|^{p^*_{a,c}}\,dx\right)^{\frac{1}{p^*_{a,c}}}\left(\int_{B_{2\epsilon}(x_j)} |x|^{-\frac{N(a-c)}{1+a-c}}|\nabla\phi_{j,\epsilon}|^{\frac{N}{1+a-c}}\,dx\right)^{\frac{1+a-c}{N}}$$
$$\le C_{10}\left(\int_{B_{2\epsilon}(x_j)} |x|^{-cp^*_{a,c}}|u_n|^{p^*_{a,c}}\,dx\right)^{\frac{1}{p^*_{a,c}}}$$
$$\to 0 \text{ as } n \to 0 \text{ and } \epsilon \to 0. \tag{26}$$

Since $\{u_n\}$ is bounded in X, we may assume that $\int_\Omega |x|^{-ap}|\nabla u_n|^p\,dx \to t_0 \geq 0$ as $n \to \infty$. Observing that $M(t)$ is continuous, we then have

$$M\left(\int_\Omega |x|^{-ap}|\nabla u_n|^p\,dx\right) \to M(t_0) \geq m_0 > 0, \quad \text{as } n \to \infty.$$

Hence, by (26),

$$M\left(\int_\Omega |x|^{-ap}|\nabla u_n|^p\,dx\right)\int_\Omega |x|^{-ap}u_n|\nabla u_n|^{p-2}\nabla u_n\cdot\nabla\phi_{j,\epsilon}\,dx \to 0, \quad \text{as } n\to\infty, \quad \epsilon \to 0. \tag{27}$$

On the other hand, by the compactness lemma of Strauss (see [44]), the boundedness of $\{u_n\}$ in X and Sobolev embedding, it follows that

$$\int_\Omega |x|^{-bp}f(x,u_n)u_n\phi_{j,\epsilon}\,dx = 0, \quad \text{as } n \to \infty, \quad \epsilon \to 0. \tag{28}$$

By relations (25) and (27)-(28), letting $n \to \infty$ in (25), we deduce that

$$\lambda\int_\Omega \phi_{j,\epsilon}\,d\nu \geq M(t_0)\int_\Omega \phi_{j,\epsilon}\,d\mu + o_\epsilon(1).$$

Letting $\epsilon \to 0$ and using the standard theory of Radon measures, we conclude that $\lambda\nu_j \geq M(t_0)\mu_j \geq m_0\mu_j$. Using (24) we have

$$\frac{\lambda\nu_j}{m_0 S_{a,c}} \geq \frac{\mu_j}{S_{a,c}} \geq \nu_j^{\frac{N-(1+a-c)p}{N}},$$

which implies that $\nu_j = 0$ or $\nu_j \geq \left(\frac{m_0 S}{\lambda}\right)^{\frac{N}{(1+a-c)p}}$ for all $j \in \Lambda$. The proof of Lemma 2.6 is now complete. □

Lemma 2.7. *Assume that the conditions* $(\mathcal{M}_1)$-$(\mathcal{M}_2)$ *and* $(\mathcal{F}_1)$-$(\mathcal{F}_2)$ *hold. Then the functional* J_λ *satisfies the* $(PS)_c$ *condition for* $c \in \left(0, \lambda^{1-\frac{N}{(1+a-c)p}}\left(\frac{\sigma}{p} - \frac{1}{p^*_{a,c}}\right)(m_0 S_{a,c})^{\frac{N}{(1+a-c)p}}\right)$.

Proof. Let $\{u_n\} \subset X$ be a sequence such that

$$J_\lambda(u_n) = c + o_n(1), \quad J'_\lambda(u_n) = o_n(1). \tag{29}$$

From the conditions $(\mathcal{M}_2)$, $(\mathcal{F}_2)$ and relation (29), we get

$$o_n(1) + c = J_\lambda(u_n) - \frac{\sigma}{p}J'_\lambda(u_n)(u_n)$$

$$
\begin{aligned}
&= \frac{1}{p}\widehat{M}\left(\int_\Omega |x|^{-ap}|\nabla u_n|^p\,dx\right) - \frac{\lambda}{p^*_{a,c}}\int_\Omega |x|^{-cp^*_{a,c}}|u_n|^{p^*_{a,c}}\,dx - \int_\Omega |x|^{-bp}F(x,u_n)\,dx \\
&\quad - \frac{\sigma}{p}M\left(\int_\Omega |x|^{-ap}|\nabla u_n|^p\,dx\right)\int_\Omega |x|^{-ap}|\nabla u_n|^p\,dx + \frac{\lambda\sigma}{p}\int_\Omega |x|^{-cp^*_{a,c}}|u_n|^{p^*_{a,c}}\,dx \\
&\quad + \frac{\sigma}{p}\int_\Omega |x|^{-bp}f(x,u_n)u_n\,dx \\
&= \frac{1}{p}\left[\widehat{M}\left(\int_\Omega |x|^{-ap}|\nabla u_n|^p\,dx\right) - \sigma M\left(\int_\Omega |x|^{-ap}|\nabla u_n|^p\,dx\right)\int_\Omega |x|^{-ap}|\nabla u_n|^p\,dx\right] \\
&\quad + \lambda\left(\frac{\sigma}{p} - \frac{1}{p^*_{a,c}}\right)\int_\Omega |x|^{-cp^*_{a,c}}|u_n|^{p^*_{a,c}}\,dx \\
&\quad + \int_\Omega |x|^{-bp}\left(\frac{\sigma}{p}f(x,u_n)u_n - F(x,u_n)\right)dx \\
&\geq \lambda\left(\frac{\sigma}{p} - \frac{1}{p^*_{a,c}}\right)\int_\Omega |x|^{-cp^*_{a,c}}|u_n|^{p^*_{a,c}}\,dx. \qquad (30)
\end{aligned}
$$

Letting $n \to \infty$ in (30), we get

$$
c \geq \lambda\left(\frac{\sigma}{p} - \frac{1}{p^*_{a,c}}\right)\lim_{n\to\infty}\int_\Omega |x|^{-cp^*_{a,c}}|u_n|^{p^*_{a,c}}\,dx. \qquad (31)
$$

Using (22), it implies that

$$
\lim_{n\to\infty}\int_\Omega |x|^{-cp^*_{a,c}}|u_n|^{p^*_{a,c}}\,dx = \int_\Omega |x|^{-cp^*_{a,c}}|u|^{p^*_{a,c}}\,dx + \sum_{j\in\Lambda}\nu_j \geq \nu_j, \quad \forall j \in \Lambda,
$$

if $\nu_s > 0$ for some $s \in \Lambda$, we deduce from (8) that

$$
c \geq \lambda\left(\frac{\sigma}{p} - \frac{1}{p^*_{a,c}}\right)\left(\frac{m_0 S_{a,c}}{\lambda}\right)^{\frac{N}{(1+a-c)p}} = \lambda^{1-\frac{N}{(1+a-c)p}}\left(\frac{\sigma}{p} - \frac{1}{p^*_{a,c}}\right)(m_0 S_{a,c})^{\frac{N}{(1+a-c)p}},
$$

which is an absurd. This leads to the fact that $\nu_j = 0$ for any $j \in \Lambda$, thus

$$
\lim_{n\to\infty}\int_\Omega |x|^{-cp^*_{a,c}}|u_n|^{p^*_{a,c}}\,dx = \int_\Omega |x|^{-cp^*_{a,c}}|u|^{p^*_{a,c}}\,dx \qquad (32)
$$

and by the Brezis-Lieb lemma [45], the sequence $\{u_n\}$ converges strongly to u in $L^{p^*_{a,c}}(\Omega, |x|^{-cp^*_{a,c}})$. Similarly, $\{u_n\}$ converges strongly to u in $L^{p^*_{a,c}}(\Omega, |x|^{-bp})$. Hence, by the Hölder inequality we deduce that

$$
\begin{aligned}
&\left|\int_\Omega |x|^{-cp^*_{a,c}}|u_n|^{p^*_{a,c}-2}u_n(u_n - u)\,dx\right| \\
&\leq \int_\Omega |x|^{-cp^*_{a,c}}|u_n|^{p^*_{a,c}-1}|u_n - u|\,dx \\
&\leq \left(\int_\Omega |x|^{-cp^*_{a,c}}|u_n|^{p^*_{a,c}}\,dx\right)^{\frac{p^*_{a,c}-1}{p^*_{a,c}}}\left(\int_\Omega |x|^{-cp^*_{a,c}}|u_n - u|^{p^*_{a,c}}\,dx\right)^{\frac{1}{p^*_{a,c}}}
\end{aligned}
$$

$$\to 0 \text{ as } n \to \infty. \tag{33}$$

and

$$\begin{aligned}
&\left|\int_\Omega |x|^{-bp} f(x,u_n)(u_n-u)\,dx\right| \\
&\le \int_\Omega |x|^{-bp}|f(x,u_n)||u_n-u|\,dx \\
&\le C_{11}\int_\Omega |x|^{-bp}(1+|u_n|^{p^*_{a,c}-1})|u_n-u|\,dx \\
&\le C_{11}\left(\int_\Omega |x|^{-bp}\,dx\right)^{\frac{p^*_{a,c}-1}{p^*_{a,c}}}\left(\int_\Omega |x|^{-bp}|u_n-u|^{p^*_{a,c}}\right)^{\frac{1}{p^*_{a,c}}} \\
&\quad + C_{11}\left(\int_\Omega |x|^{-bp}|u_n|^{p^*_{a,c}}\,dx\right)^{\frac{p^*_{a,c}-1}{p^*_{a,c}}}\left(\int_\Omega |x|^{-bp}|u_n-u|^{p^*_{a,c}}\right)^{\frac{1}{p^*_{a,c}}} \\
&\to 0 \text{ as } n\to\infty.
\end{aligned} \tag{34}$$

Since the sequence $\{u_n\}$ converges weakly to u in X, the sequence $\{u_n - u\}$ is bounded in X and $J'_\lambda(u_n)(u_n-u) \to 0$ as $n \to \infty$, that is,

$$\begin{aligned}
J'_\lambda(u_n)(u_n-u) &= M\left(\int_\Omega |x|^{-ap}|\nabla u_n|^p\,dx\right)\int_\Omega |x|^{-ap}|\nabla u_n|^{p-2}\nabla u_n\cdot\nabla(u_n-u)\,dx \\
&-\lambda\int_\Omega |x|^{-cp^*_{a,c}}|u_n|^{p^*_{a,c}-2}u_n(u_n-u)\,dx - \int_\Omega |x|^{-bp}f(x,u_n)(u_n-u)\,dx \to 0.
\end{aligned} \tag{35}$$

From (33)-(35), we have

$$\lim_{n\to\infty} M\left(\int_\Omega |x|^{-ap}|\nabla u_n|^p\,dx\right)\int_\Omega |x|^{-ap}|\nabla u_n|^{p-2}\nabla u_n\cdot\nabla(u_n-u)\,dx = 0,$$

and by $(\mathcal{M}_1)$ it follows that

$$\lim_{n\to\infty}\int_\Omega |x|^{-ap}|\nabla u_n|^{p-2}\nabla u_n\cdot\nabla(u_n-u)\,dx = 0. \tag{36}$$

From (36) and some standard arguments, $\{u_n\}$ converges strongly to u in X and thus, J_λ satisfies the $(PS)_c$ condition for $c \in \left(0, \lambda^{1-\frac{N}{(1+a-c)p}}\left(\frac{\sigma}{p}-\frac{1}{p^*_{a,c}}\right)(m_0 S_{a,c})^{\frac{N}{(1+a-c)p}}\right)$. □

Lemma 2.8. *Assume that the conditions* $(\mathcal{M}_1)$-$(\mathcal{M}_2)$ *and* $(\mathcal{F}_1)$-$(\mathcal{F}_2)$ *hold. Then, there exists a sequence* $\{M_n\} \subset (0,+\infty)$ *independent of* λ*, with* $M_n \leq M_{n+1}$*, such that for any* $\lambda > 0$*,*

$$c_n^\lambda := \inf_{K\in\Gamma_n} \max_{u\in K} J_\lambda(u) < M_n.$$

Proof. From the definition of c_n^λ, we deduce that

$$\begin{aligned} c_n^\lambda &= \inf_{K\in\Gamma_n} \max_{u\in K} J_\lambda(u) \\ &= \inf_{K\in\Gamma_n} \max_{u\in K} \left\{ \frac{1}{p}\widehat{M}\left(\int_\Omega |x|^{-ap}|\nabla u|^p\,dx\right) - \frac{\lambda}{p^*_{a,c}}\int_\Omega |x|^{-cp^*_{a,c}}|u|^{p^*_{a,c}}\,dx \right. \\ &\qquad \left. - \int_\Omega |x|^{-bp}F(x,u)\,dx \right\} \\ &\leq \inf_{K\in\Gamma_n} \max_{u\in K} \left\{ \frac{1}{p}\widehat{M}\left(\int_\Omega |x|^{-ap}|\nabla u|^p\,dx\right) - \int_\Omega |x|^{-bp}F(x,u)\,dx \right\}. \end{aligned}$$

Set

$$M_n := \inf_{K\in\Gamma_n} \max_{u\in K} \left\{ \frac{1}{p}\widehat{M}\left(\int_\Omega |x|^{-ap}|\nabla u|^p\,dx\right) - \int_\Omega |x|^{-bp}F(x,u)\,dx + 1 \right\},$$

then we conclude that $M_n < +\infty$ and $M_n \leq M_{n+1}$ by the definition of Γ_n, see relations (18) and (21). □

Proof of Theorem 2.2 By choosing for each $k \geq 1$, λ_k is sufficiently small, we construct a sequence of positive real numbers $\{\lambda_k\}$, with $\lambda_k < \lambda_{k-1}$ such that

$$M_k < \lambda_k^{1-\frac{N}{(1+a-c)p}} \left(\frac{\sigma}{p} - \frac{1}{p^*_{a,c}}\right) (m_0 S_{a,c})^{\frac{N}{(1+a-c)p}}.$$

Therefore, for $\lambda \in (\lambda_k, \lambda_{k-1}]$, by Lemma 2.8 we get

$$0 < c_1^\lambda \leq c_2^\lambda \leq \ldots \leq c_k^\lambda < M_k < \lambda^{1-\frac{N}{(1+a-c)p}} \left(\frac{\sigma}{p} - \frac{1}{p^*_{a,c}}\right) (m_0 S_{a,c})^{\frac{N}{(1+a-c)p}}.$$

From Proposition 1.3, the levels $c_1^\lambda \leq c_2^\lambda \leq \ldots \leq c_k^\lambda$ are critical values of J_λ. So, if $c_1^\lambda < c_2^\lambda < \ldots < c_k^\lambda$, the functional J_λ has at least k critical points. Now, if $c_j^\lambda = c_{j+1}^\lambda$ for some $j = 1, 2, \ldots, k-1$, again Proposition 1.3 implies that $K_{c_j^\lambda}$ is an infinite set [43, Chapter 7] and hence in this case, problem (1) has infinitely many weak solutions. In conclusion, problem (1) has at least k pair of weak solutions. □

ACKNOWLEDGMENTS

The author would like to thank the referees for their suggestions and helpful comments which improved the presentation of the original manuscript.

REFERENCES

[1] Assuncao R.B., et al. 2007. "Subcritical perturbations of a singular quasilinear elliptic equation involving the critical Hardy-Sobolev exponent". *Nonlinear Anal.* 66: 1351-1364.

[2] Avci M., et al. 2011. "Existence and multiplicity of the solutions of the $p(x)$-Kirchhoff type equation via genus theory". *Math. Methods Appl. Sci.* 34(14): 1751-1759.

[3] Bouchekif M., et al. 2011. "On singular elliptic equations involving a concave term and critical Caffarelli-Kohn-Nirenberg exponent". *Math. Nachr.* **284**(2-3): 177-185.

[4] Benmansour S., et al. 2016. "Multiple solutions for nonhomogeneous elliptic equations involving critical Caffarelli-Kohn-Nirenberg exponent". *Mediterr. J. Math.* 13(6): 4679-4691.

[5] Liang S., et al. 2012. "Multiplicity of solutions to the weighted critical quasilinear problems". *Proc. Edinburgh Math. Soc.* 55(1): 181-195.

[6] Xuan B.J. 2004. "The eigenvalue problem of a singular quasilinear elliptic equation". *Electron. J. Differ. Equ.* 16: 1-11.

[7] Xuan B.J. 2005. "The solvability of quasilinear Brezis-Nirenberg-type problems with singular weights". Nonlinear Anal. 62(4): 703-725.

[8] Lions P.L. 1985. "The concentration compactness principle in the calculus of variations, the limit case (I)". *Rev. Mat. Iberoamericana.* 1(1): 145-201.

[9] Lions P.L. 1985. "The concentration compactness principle in the calculus of variations, the limit case (II)". *Rev. Mat. Iberoamericana.* 1(2): 45-121.

[10] Kajikiya R. 2005. "A critical-point theorem related to the symmetric mountain pass lemma and its applications to elliptic equations". *J. Funct. Anal.* 225: 352-370.

[11] Miyagaki O.H., et al. 2007. "On positive solutions for a class of singular quasilinear elliptic systems". *J. Math. Anal. Appl.* 334: 818-833.

[12] Miyagaki O.H., et al. 2009. "On positive solution for a class of degenerate quasilinear elliptic positone/semipositone systems". *J. Math. Anal. Appl.* 70(1): 99-116.

[13] Alves C.O., et al. 2010. "On a class of nonlocal elliptic problems with critical growth". *Differ. Equ. Appl.* 2: 409-417.

[14] Chung N.T., et al. 2017. "Kirchhoff type problems involving p-biharmonic operators and critical exponents". *J. Appl. Anal. Comput.* 7(2): 659-669.

[15] Chung N.T. 2019. "Existence of positive solutions for a class of Kirchhoff type systems involving critical exponents". *Filomat.* **33**(1): 267-280.

[16] Hamydy A., et al. 2011. "Existence of solution for p-Kirchhoff type problems with critical exponents". *Electronic J. Differ. Equ.* 105: 1-8.

[17] Li H.Y. 2018. "Existence of positive ground state solutions for a critical Kirchhoff type problem with sign-changing potential". *Comput. Math. Appl.* 75(8): 2858-2873.

[18] Wang L., et al. 2018. "Existence and multiplicity of solutions for critical Kirchhoff-type p-Laplacian problems". *J. Math. Anal. Appl.* **458**: 361-378.

[19] Xie Q.L., et al. 2013. "Existence and multiplicity of solutions for Kirchhoff type problem with critical exponent". *Comm. Pure Appl. Anal.* 12(6): 2773-2786.

[20] Kirchhoff G. 1883. *Mechanik.* Germany: Teubner, Leipzig.

[21] Chipot M., et al. 1997. "Some remarks on nonlocal elliptic and parabolic problems". *Nonlinear Anal.* 30(7): 4619-4627.

[22] Afrouzi G.A., et al. 2014. "Existence and multiplicity of solutions for Kirchhoff type problems involving $p(x)$-biharmonic operators". *Z. Anal. Anwend.* 33: 289-303.

[23] Chung N.T. 2013. "Multiple solutions for a $p(x)$-Kirchhoff-type equation with sign-changing nonlinearities". *Complex Var. Elliptic Equ.* 58(12): 1637-1646.

[24] Chung N.T. 2018. "Existence of solutions for perturbed fourth order elliptic equations with variable exponents". *Electron. J. Qual. Theory Differ. Equ.* 96: 1-19.

[25] Chung N.T. 2013. "Three solutions for a class of nonlocal problems in Orlicz-Sobolev spaces". *J. Korean Math. Soc.* 50(6): 1257-1269.

[26] Chung N.T. 2015. "Existence of solutions for a class of Kirchhoff type problems in Orlicz-Sobolev spaces". *Ann. Polon. Math.* 113: 283-294.

[27] Chen C.Y., et al. 2011. "The Nehari manifold for a Kirchhoff type problem involving sign-changing weight functions". *J. Differ. Equ.* 250: 1876-1908.

[28] Colasuonno F., et al. 2011. "Multiplicity of solutions for $p(x)$-polyharmonic Kirchhoff equations". *Nonlinear Anal.* 74: 5962-5974.

[29] Corrêa F.J.S.A., et al. 2009. "On a p-Kirchhoff equation via Krasnoselskii's genus". *Appl. Math. Letters.* 22: 819-822.

[30] Dai G. 2013. "Three solutions for a nonlocal Dirichlet boundary value problem involving the $p(x)$-Laplacian". *Appl. Anal.* 92: 191-210.

[31] Ma T.F. 2005. "Remarks on an elliptic equation of Kirchhoff type". *Nonlinear Anal.* 63: 1967-1977.

[32] Nyamoradi N., et al. 2018. "Existence and multiplicity of solutions for fractional p-Laplacian Schrödinger-Kirchhoff type equations". Complex Var. Elliptic Equ. 63(3): 346-359.

[33] Pucci P., et al. 2016. "Existence and multiplicity of entire solutions for fractional p-Kirchhoff equations". *Adv. Nonlinear Anal.* 5: 27-55.

[34] Caristi G., et al. 2018. "Multiple solutions for degenerate nonlocal problems", *Appl. Math. Letters.* 84: 26-33.

[35] Chung, N.T., et al. 2013. "Existence and multiplicity of solutions for a class of degenerate nonlocal problems". *Electron. J. Differ. Equ.* 148: 1-13.

[36] Chung N.T., et al. 2014. "On a class of degenerate nonlocal problems with sign-changing nonlinearities". *Bull. Malaysian Math. Sci. Soc.* 37(4): 1157-1167.

[37] Chung N.T., et al. 2014. "Multiple solutions for a class of degenerate nonlocal problems involving sublinear nonlinearities". *Le Matematiche.* LXIX(Fasc. II): 171-182.

[38] Figueiredo G.M., et al. 2016. "Solutions for a Kirchhoff equation with weight and nonlinearity with subcritical and critical Caffarelli-Kohn-Nirenberg growth". *Proc. Edinburgh Math. Soc.* **59**(4): 925-944.

[39] Guimaraes M.B., et al. 2015. "Existence of solutions for Kirchhoff equations involving p-linear and p-superlinear terms and with critical growth". *Electron. J. Differ. Equ.* 38(12): 2650-2662.

[40] Shen L. 2018. "Existence and non-existence results for quasilinear Kirchhoff problems with the Hardy-Sobolev exponent". *Comput. Math. Appl.* 76: 1923-1937.

[41] Caffarelli L., et al. 1984. "First order interpolation inequalities with weights". *Composito Math.* 53; 259-275.

[42] Catrina F., et al. 2001. "On the Caffarelli-Kohn-Nirenberg inequalities: sharp constants, existstence (and non existstence) and symmetry of extremal functions", *Comm. Pure Appl. Math.* 54: 229-258.

[43] Rabinowitz P. H. 1984. "Minimax methods in critical point theory with applications to differential equations". in *CBMS Reg. Conf. Series in Math*, Vol. 65. USA: American Mathematical Society, Providence, Rhode Island.

[44] Chabrowski J. 1999. *Weak convergence methods for semilinear elliptic equations*. Singapore: World Scientific Pub. Co. Inc.

[45] Brezis H., et al. 1983. "A relation between pointwise convergence of functions and convergence of functionals". *Proc. Amer. Math. Soc.* 88: 437-477.

About the Editor

Mustafa Avci, PhD
Professor
Department of Science, Grande Prairie Regional College,
Grande Prairie, Canada
Email: avcixmustafa@gmail.com

Mustafa Avci is a professor of mathematics. He obtained his PhD in Mathematics from Dicle University, in 2011. He worked as mathematics professor in Batman University and Dicle University, Turkey for ten years. He was visiting professor at Morgan State University, USA. He worked as lecturer at Edwards School of Business, University of Saskatchewan, Canada. He is professor at Science Department, Grande Prairie Regional College.

His research interests focus on the nonlinear PDEs involving variable exponent growth conditions. These type equations are extremely attractive because they can be used to model dynamical phenomena which arise from the study of electrorheological fluids or elastic mechanics. Problems with variable exponent growth conditions also appear in the modeling of stationary thermorheological viscous flows of non-Newtonian fluids and in the mathematical description of the processes filtration of an ideal barotropic gas through a porous medium, in magnetostatics and in image restoration. He applies variational methods and nonvariational methods, such as monotone operator theory and topological methods to obtain the existence and uniqueness theorems.

He has been doing research in the mentioned areas for over ten years. He has many papers published in various highly reputed mathematical journals. He has been conducting editorial and reviewing services for mathematics journals indexed in Web of Science, Scopus and ISI.

INDEX

E

F

G

H

I

K

L

M

U

V

W